Water Supply Wells:
Siting, Drilling, and Construction

Water Supply Wells:
Siting, Drilling, and Construction

Second Edition

Frederick Bloetscher, Ph.D., P.E.
Albert Muniz, P.E.
and
John Largey

American Water Works
Association

Disclaimer

The authors, contributors, editors, and publisher do not assume responsibility for the validity of the content or any consequences of their use. In no event will AWWA be liable for direct, indirect, special, incidental, or consequential damages arising out of the use of information presented in this book. In particular, AWWA will not be responsible for any costs, including, but not limited to, those incurred as a result of lost revenue. In no event shall AWWA's liability exceed the amount paid for the purchase of this book.

Senior Editorial Manager – Book Products: Melissa Valentine
Technical Editor: Dianne Beirne
Product Manager: Tony Petrites
Production Manager: Gillian Wink
Compositor: Innodata
Cover Design: Michael Labruyere

Library of Congress Cataloging-in-Publication Data

Names: Bloetscher, Frederick, author. | Muniz, Albert, author. | Largey, John, author. | American Water Works Association, issuing body.
Title: Well location, drilling and operation / by Frederick Bloetscher, Albert Muniz, John Largey.
Description: Second edition. | Denver, CO : American Water Works Association, [2022] | First edition title: Siting, drilling, and construction of water supply wells (2007) | Includes bibliographical references and index. | Summary: "This book is intended for utilities and engineers who periodically design and construct water supply wells, as opposed to books designed for well drilling, voluminous well books that cover vast territory and details or highly theoretical books for course taught at the college level. This book is intended to help the reader understand what to expect from well drilling, including the equipment, terminology, and design concepts"— Provided by publisher.
Identifiers: LCCN 2022010480 | ISBN 9781647170707 (paperback)
Subjects: LCSH: Wells—Design and construction. | Handbooks and manuals.
Classification: LCC TD405 .B57 2022 | DDC 628.1/14—dc23/eng/20220502
LC record available at https://lccn.loc.gov/2022010480

Printed in the United States of America

American Water Works Association
6666 West Quincy Avenue
Denver, CO 80235-3098
ISBN 978-1-64717-070-7
ISBN, electronic 978-1-61300-604-7

CONTENTS

PREFACE

Welcome to Water Supply Wells: Siting, Drilling, and Construction. The first edition of this book (titled Well Siting, Drilling, and Construction) was published by the American Water Works Association in 2008. Unlike some of the other books in the field, this book is intended for utilities and engineers who periodically design and construct water supply wells, as opposed to books designed for well drilling—voluminous well books that cover vast territory and details or highly theoretical books for courses taught at the college level. Instead of being for academics, this book is intended to help the reader understand what to expect from well drilling, including the equipment, terminology, and design concepts. As one purchaser noted, "The fact that this book is so full of illustrations and photographs really helped me in the field." We have added more and better figures to this edition.

The second edition builds on the first while adding appendices that can help utilities and engineers as needed. The book has also been slightly rearranged. The first chapter begins with a short story about what can go wrong if considerations for siting a well are not performed. The rest of the chapter focuses on sources of groundwater, sustainability challenges, and consideration for siting wells.

Chapter 2 begins with identifying sources to research water supplies and then discusses how siting of a well should be performed. We have added information on ASTM 1527-21 Environmental Site Assessments as this process is useful for locating water supplies. Restrictions because of contamination and/or regulations are critical for long-term well sustainability just as water supply recharge is. Some discussion about site acquisition, monitoring wells, and modeling is included; however, since these are specific to location, they are not discussed in detail. Those interested in modeling are referred to other publications including AWWA's manual M73, *Groundwater Management.*

Chapter 3 covers geophysical logging and field testing. Little has changed from the first edition in this chapter because little has changed from a technology standpoint. Chapter 4 discusses drilling methods; these also have changed little aside from directional drilling, which is rarely used for water supplies. More information is provided on mud-rotary style wells, which are more common in many parts of North America.

Chapter 5 outlines basic mathematics used for the design of wells, well screens, and monitoring of wells. Concepts for design of wellfields (multiple wells), casings, water hammer, and specific capacity are outlined. Many of these items should be tracked to determine if there are problems occurring in the wells. This information was not consolidated in the first edition, but several readers asked for its inclusion.

Chapter 6 covers well construction and development. While little has changed in that area, portions of ANSI/AWWA Standard A100 were incorporated to demonstrate what wells will look like as they are completed. This includes information on casings, well completion, and development. The chapter includes several new illustrations. Chapter 7 was added because the operation of a well after installation is critical. The first edition did not include much on operations. This version expands that discussion so operators can track pertinent well operations data such as volumes, drawdowns, and specific capacity. Chapter 8 is the conclusion.

The new edition adds five appendices. Appendix A covers regulatory issues for drinking water quality, well operations, and the like. It includes information from the previous edition as well as new data. Appendix B discusses treatment options for groundwater. One issue that arises is that deeper water may require more treatment. So, a consideration in locating wells is the water quality, which may be a driver in making decisions about where to drill wells and how deep. That kind of data needs to be kept with the original well data. Appendix C includes a set of tables of collapse strength of well casings. Appendix D includes a method for calculations of screen lengths and entrance velocity. Appendix E includes useful information on geophysical logging.

Many new figures have been added, and some prior photographs have been updated to help the readers understand what they are looking at in the field. The authors believe that the second edition of the book will provide the readers with a valuable reference that they can use during the design, construction, and operational phases of their wells. Enjoy! Feel free to let us know if you find this helpful.

ACKNOWLEDGMENTS: The authors and AWWA wish to thank David Edson, Marjorie Craig, and Sandra Eberts for their time, expertise, and invaluable help in reviewing this manuscript.

CHAPTER 1

Groundwater Supply (Basic Concepts)

In this chapter, the goal is to outline the basic concepts with the use of groundwater, the hydrologic cycle, and the sustainability of groundwater. The largest growth in source waters in North America has been groundwater—mostly among small utilities. But the use of confined aquifers creates challenges to long-term sustainability of these utility systems.

GROUNDWATER UTILITIES

Most water systems in the United States and Canada are operated by local governments. Many of these systems were historically developed to accommodate the needs of industry and growth, with the goal to meet the water demands for quality and quantity expected by the customers of the areas served by the utility (Bloetscher 2019). Meeting this desire involves understanding the utility operating environment, performing planning activities, and making financial decisions that will provide appropriate funding to achieve the utility's needs, including the needs for future water supplies. The sustainability of these supplies is a critical component of a utility's infrastructure to ensure reliable, long-term operations and sustainability.

There are more than 52,000 community water systems in the United States, up from 16,000 in 1960 (Bloetscher 2019). Most of today's growth is occurring in small, rural utilities. The largest 500 water utilities serve half the nation's population, just as they did in 1960 (Bloetscher 2019). More than 41,000 utilities provide groundwater to their customers. The vast majority—40,000 or more—are small systems serving less than 1,000 customers. The goal of this book is to describe the location, drilling, and operation of the wells such utilities use. Many of the 41,000 groundwater utilities have similar issues with respect to siting, construction, and operating their water supply wells. Understanding the commonalities and challenges will give groundwater utilities an opportunity to improve long-term service and protect their valuable assets.

Small demands mean wells can supply adequate supplies, even if the aquifer is not very productive. But what happens when growth occurs? A short case study is worth reviewing; the case study is outlined in more detail in M73, AWWA's manual on groundwater management (AWWA 2021). The case study is in the eastern Carolinas, where groundwater was used for source water. The water required little to no treatment because the aquifer was 250–400 ft deep, confined, and very high quality (AWWA 2021). Many of the communities that relied on this water supply lacked central water and sewer service as late as the Great Depression. Most did not begin to grow until the 1940s. Examples include Jacksonville, N.C., which had a population of just over 1,000 people in 1940 but grew exponentially after Camp LeJeune was established. Richlands, N.C., which had 1,500 people in 1940, was located between Jacksonville and Kinston, N.C., which became a major lumber and textile center over the time period. They all used the water from this aquifer because it required only disinfection for treatment. The water quality was so good that Richlands, N.C., claimed to be the "Town of Perfect Water" as late as the 1990s.

As the eastern Carolinas began to grow in the 1960s, more industry and more people tied into the formation and water use began to increase rapidly (AWWA 2021). All these communities, from Greenville, N.C., to Mt. Pleasant, S.C., withdrew groundwater from the same aquifer (i.e., source). Throughout the 1970s and early 1980s, water levels from this historically productive aquifer began declining because of overpumping. Some utilities attempt to address overpumping by lowering pump settings. This was the case with the City of Jacksonville which, over a 14-year period, lowered its well pumps and deepened its new wells an average of 10 ft/yr (AWWA 2021).

The end of the tale of the eastern Carolinas is that this formation was no longer able to provide water in the late 1990s. A better understanding of wellfield operations, aquifer characteristics, pumping, and drilling might have helped prevent the ultimate loss of this water supply. Today, to meet the required reduction in withdrawals from the declining aquifers and continue to meet demand, groundwater utilities in North Carolina shifted to the Castle Hayne Aquifer which requires softening (AWWA 2021). The formation is lacking in South Carolina, so South Carolina communities migrated to surface waters in the 1990s since they were the only option (AWWA 2021).

The question at the time, though, was whether the wells or the water levels were the problem; that may be a similar challenge facing other groundwater utilities. To answer the question, a better understanding of the hydrologic cycle and water balance is needed. Without understanding the aquifer characteristics and adjusting to the real conditions, challenges will multiply. Monitoring is useful because withdrawals above safe sustainable yields will compromise an aquifer.

The case study highlights a key component of groundwater supply planning. Planning the utilization of water supplies requires an understanding of how the hydrologic cycle provides water to the groundwater service area, in what quantities, and with what reliability (i.e., natural variability). Reliability is a risk issue. Both water quantity and quality issues have significant fiscal impact on the potential users in the basin, and there are unrealized costs and benefits that are often overlooked in the current water management framework (AWWA 2021). The risks of a water supply can be estimated by the inputs and outputs of the hydrologic cycle.

HYDROLOGIC CYCLE

Figure 1-1 outlines the basic hydrologic cycle. The components of the hydrologic cycle are:

- evaporation and transpiration,
- precipitation,
- surface water/runoff, and
- groundwater.

There is no start to the cycle, but evaporation from water or transpiration from plants creates atmospheric moisture that returns to the earth as precipitation. The precipitation component has three options: travel across the land (runoff) to a surface water body where it can be reevaporated; uptake by plants (which will lead to transpiration); and seepage into groundwater. From a well perspective, it is the amount of water that seeps into the ground that is of interest as it is ultimately the source of water for the wells. Those formations that hold water are called aquifers.

AQUIFERS AND GROUNDWATER

Groundwater is water that may have recently entered the soil via rainfall or snowmelt, or it may be an ancient source found in geologic formations well below the surface. Groundwater generally flows vertically until it contacts a layer of impervious strata. After that, the groundwater will flow down gradient taking the path of least resistance unless impacted by other hydraulic conditions. Therefore, if

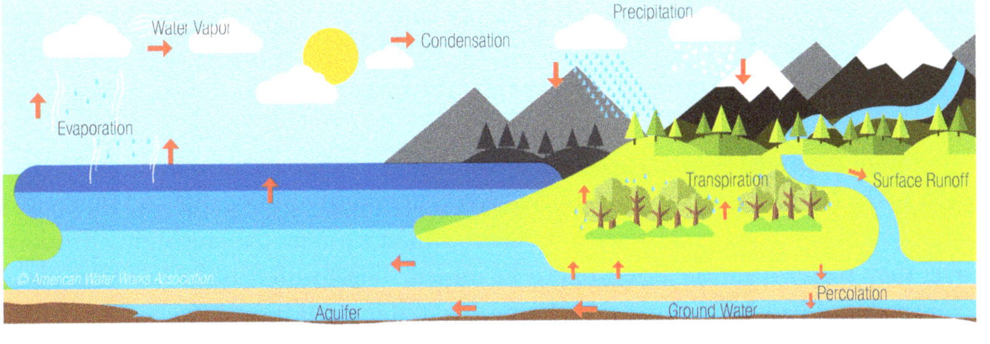

Source: DrinkTap.org by AWWA

Figure 1-1 Hydrologic cycle

high-permeability underground conduits or channels are present, the water will tend to flow along these pathways. These formations have the potential of yielding substantial quantities of water. If the grains of the aquifer formation are small, water may flow more slowly, and the aquifer will be less productive.

A body of rock that is sufficiently permeable to conduct groundwater to yield economically significant quantities of water to wells is called an *aquifer*. Water that is located near the land surface, exposed to atmospheric pressure, and has no overlying confinement is called the *water table* or *surficial aquifer*. The surface of a water table aquifer tends to follow the surface of the ground, although some conditions cause exceptions. Surficial aquifers will recharge creeks, lakes, or rivers whose bottoms are deeper than the top of the water table, and aquifers may be recharged by those same water bodies where the surface of the aquifer is beneath the creek, lake, or river bottom.

A *spring* forms when groundwater flows naturally from an aquifer to the surface, such as near a creek, lake, or river. Water flowing from a spring may have traveled hundreds of miles (kilometers) from where it seeped into the ground, or it could be from a surface water source only a few yards away. The only way to tell is dye testing or computer modeling.

Note that just because water seeps into the ground, it still may be subject to evaporation and transpiration if it is within 4 ft of the surface. At the same time, because there may be little resistance to migration of water into water table aquifers, they are also more susceptible to contamination than deeper or confined aquifers—a major consideration in locating wellfields.

Below the water table aquifer may be other aquifers. These aquifers will often be separated from direct surface recharge by a layer of dolomite, clay, or other material that prevents or limits the exchange of water between the aquifer layers. Such limiting formations are called *confining units* or *aquitards* (although the latter term is rarely used in modern reports). Aquifers that are located further beneath the surface and have low-permeability formations above them are called *confined aquifers*. These aquifers may be under pressure and are termed *artesian aquifers* if the pressure head of the water in the aquifer is located above the bottom of the overlying confining bed when exposed to atmospheric pressure. The term *flowing artesian aquifers* refers to those aquifers where the water surface rises above the overlying confining unit and may flow at land surface. An example of an artesian aquifer that also flows at land surface is the Floridan Aquifer in southeast Florida. The Floridan Aquifer will "flow" at nearly 30 ft above the ground surface yet is located 1,000 ft below the surface under a dense, 800-ft-thick clastic clay confining unit.

Some confined aquifers are buried river valleys or ancient lake beds. Examples exist in the Midwest, where such buried rivers are highly permeable and may yield large quantities of water. However, while confined aquifers may be highly productive, they suffer from the potential to be mined. The aquifer in the Carolinas was confined and had little vertical recharge. As water withdrawals increased, water levels and pressure in the aquifer dropped until no water could be economically withdrawn. Reilly et al. (2008) showed large areas of the United States have water-level declines over 40 ft in at least one confined aquifer since predevelopment or declines exceeding 25 ft in unconfined aquifers since predevelopment. The same report argued for a nationwide effort to organize available information on changes in groundwater storage—something that is only partly completed across the country. Since the withdrawal of this water appears to be a permanent loss of the resource in the long term, the implication is that water utilities and their customers will be acutely affected and that a major effort to understand and recommend solutions to the problem of groundwater shortages is required (AWWA 2021).

Evaluating Groundwater Supplies

When attempting to identify water supplies, one must evaluate the productivity and water quality of each aquifer because the quantity and quality of water in the aquifers may vary. The quantity and quality of groundwater depend on factors such as confinement, depth, aquifer thickness, rainfall, and geological formation. For example, thicker aquifers may hold more water, depending on the porosity of the formation, but secondary porosity and hydraulic conductivity/transmissivity indicate if water can flow through the formation. The more productive formations will have high porosity, a high specific yield (the amount of water released from the formation by pumping), and low specific retention (the amount of water that cannot be removed by pumping). Clay is an example of a formation with high porosity but limited yield because the tiny clay particles hold the water through electrostatic forces. Because their relatively high permeability provides significant productivity, sand, shell, and gravel aquifers are more suitable for public water system than clay, granite, or dolomite. Sandstone is porous and often yields water of good quality in sufficient quantity to supply public water systems. Limestone has moderate porosity but often contains cracks and cavities that can provide substantial quantities of water (AWWA 2014). Table 1-1 shows the porosity of rock formations.

The flow velocity and flow direction of groundwater depend on gradient (i.e., the elevation of the recharge source) and permeability of soil and rock layers.

Table 1-1 Porosity of rock formations

Strata	Porosity (%)	Specific Yield (%)	Specific Retention (%)
Soil	55	40	15
Clay	50	2	48
Sand	25	22	3
Gravel	20	19	1
Limestone	20	18	2
Sandstone	11	6	5
Granite	0.1	0.09	0.01
Basalt	11	8	3

Source: AWWA 2014

The movement of water through an aquifer is generally slow but can dissolve rock strata over time. Eventually, this process can result in cavities, which serve as conduits to interconnect and form underground rivers or caverns that can be tapped as a public water supply source. This connection between particles to permit ease of movement of the water is called *secondary porosity*.

Throughout North America, groundwater can generally be found from a few feet to hundreds of feet below the land surface. The point of withdrawal from deeper aquifers is likely to be more distant from the recharge region, so the water will be older and have more contact with the underground strata. Because deeper water is in contact with rock formations for longer periods, deeper waters tend to have poorer water quality as a result of minerals dissolving into the water over many years. A balance must be struck in such areas between the decreased water quality from deeper aquifers (which might require additional treatment) and the potential for more productivity. Deep wells also have higher construction costs but are less susceptible to contamination. Therefore, while some deep aquifers may be prolific, the quality of water obtained from a well will likely require more "treatment" to be usable for drinking water.

To assist utilities, the US Environmental Protection Agency (USEPA) has created designations for aquifers based on yield and water quality. While it is technically possible to treat almost all water to obtain acceptable quality, treatment may not be economical or practical for waters with greater than 10,000 mg/L of total dissolved solids. Such waters are not considered potential underground sources of drinking water. All other aquifers are regulated as they may be used for drinking water purposes at some point.

BENEFITS AND DISADVANTAGES OF GROUNDWATER USE

Groundwater supplies can have advantages over more traditional surface water uses, including:

- Water has less exposure to contamination (assuming it is not a water table aquifer).
- Water quality is stable.
- Water temperature is stable.
- Water quality changes are slow to occur.
- Evapotranspiration losses are insignificant.
- Less treatment is typically required.

There are also disadvantages with groundwater sources:

- difficult to clean up once contaminated;
- lack of early warning of contamination—unseen plumes of contaminants can migrate into a wellfield without warning unless sentinel wells are constructed;

- potential conflicts with land use with competing urban industrial, commercial, agricultural, irrigation, and ecosystem users in the same area;
- difficulty in determining sustainable yields;
- water levels are not obvious;
- recharge limitations;
- development patterns alter recharge;
- supplies are often limited in basins;
- control of aquifer recharge may be outside of users' jurisdiction;
- saltwater intrusion can occur in coastal areas; and
- aquifers may not be accessible in certain regions, similar to surficial sources.

SUSTAINABILITY OF WATER SUPPLY SOURCES

"Sustainable water supplies" is a buzzword made popular in the late 2000s. AWWA has committees and a division that focus on this issue that produced a compendium of issues in 2010 (Bloetscher 2010). The compendium found no consistent meaning, but the definition parties used tended to reflect their perspective, not an overarching view. The term *sustainable* has various connotations, but the context for this book is the ability to ensure long-term availability of the water supply. Unsustainable supplies lead to water supply deficits that cause a disruption of economic and social activity in a community (Reilly et al. 2008). The issues involved in determining sustainability include:

- ongoing, consistent recharge of the aquifer,
- water rights or permits,
- variability in water levels,
- water quality variability, and
- competing users.

Ongoing recharge of the aquifer can be identified/confirmed by hydrogeologic testing and evaluating historical records. Historical records enable comparison of water levels with rainfall to identify potential fluctuations and recharge relationships. For shallow sources, locations beside lakes and streams that flow consistently throughout the year may provide good sites if no adverse environmental effects are experienced. Step-drawdown testing also provides site-specific information to determine how significant the impact of the well may be to the aquifer locally and areally, although these may better determine yield in surficial aquifers than in confined aquifers over time.

The term *sustainable yield* occurs when the amount of water withdrawn the formation occurs at a rate that is less than the recharge. For sustainable yield, water quality and water levels do not drop over time. Where there are ongoing and increasing drawdowns, the groundwater supply is being overdrafted.

GROUNDWATER TERMS

The following terms (AWWA 2014) are used to explain the concepts found in this chapter and throughout the remainder of this book.

Anisotropic Aquifer: An aquifer in which the aquifer formation does not transmit water equally in all directions (i.e., the horizontal and vertical permeability are not equal, causing the water to move preferentially in one direction with respect to the other two).

Aquiclude: A low-permeability geologic unit that forms either the upper or lower boundary of a groundwater flow system. Clay is a typical aquiclude material. This term is rarely used in the field today but may exist in older hydrogeologic reference materials.

Aquifer: A geologic formation, or part of a group of formations, that is saturated and sufficiently permeable to transmit economic quantities of water to wells and springs.

Aquifer Storativity: The volume of water added to a unit horizontal area of the aquifer per unit rise in the water table elevation (i.e., the ability of the aquifer to increase its capacity to store water).

Aquitard: A low-permeability geologic unit that can store groundwater and also transmit it slowly from one aquifer to another. A leaky aquifer formation may be an aquitard. This term is rarely used in the field today, but it may exist in older hydrogeologic reference materials.

Effective Pore Space: An indication of how much of the void space within the rock or soil is capable of transmitting water. This is important because some rock formations may have considerable pore (or void) space but, because the pores are not interconnected, the rock or soil may have difficulty transmitting water.

Evaporation: The transformation of water from the liquid phase to a vapor phase in the atmosphere.

Evapotranspiration: The combination of evaporation and transpiration. Water less than 4 ft below the ground surface may be subject to evapotranspiration.

Groundwater: Water contained in interconnected pores located either below the water table in an unconfined aquifer or in a confined aquifer.

Hydraulic Conductivity: A measure of the ease with which fluid is transported through a porous matrix (also called coefficient of permeability).

Hydraulic Gradient: A measure of the vertical change of the aquifer over a given distance from a stationary base (i.e., the slope of the aquifer formation).

Hydrologic Cycle: The movement of water in the environment wherein water falls to the surface as rainfall, runs off over land to water bodies and/or infiltrates into the groundwater via percolation, and returns to the atmosphere via evaporation/transpiration, then returns as rainfall (Figure 1-1).

Isotropic Aquifer: An aquifer in which the formation transmits water equally in all directions within the aquifer formation.

Piezometric Surface: The level to which water will rise in an aquifer under natural conditions. In a water table aquifer, the piezometric surface is the water table level. In a confined aquifer, the piezometric section may be significantly above the top of the rock formation; the aquifer is under pressure. Artesian wells are confined aquifers with the piezometric level above the ground surface, meaning that water could flow to the surface unaided.

Potential Evapotranspiration: The amount of evapotranspiration that would occur, assuming the soil moisture was adequate at all times. This will not occur during drought conditions, and so actual evapotranspiration may be less. Because of the disparity between the rainfall throughout the year, the actual evapotranspiration in the winter months is less.

Specific Discharge Potential: The amount of water that an aquifer could discharge given its hydraulic conductivity and piezometric head.

Specific Retention: The volume fraction of water that is held back by adhesion and capillary forces when an aquifer is drained.

Transmissivity: The aquifer characteristic defined by the rate of flow per unit width through the entire thickness of an aquifer per unit gradient. This is valid only in two-dimensional flow.

Transpiration: The process in which plants absorb water from the soil and release it to the atmosphere through their leaves.

Viscosity: This combines with the fluid density to form kinematic viscosity, which is important in determining the hydraulic conductivity of the soil or rock matrix.

Void Space: The pore space within the soil or rock matrix that does not contain solid material.

REFERENCES

American Water Works Association (AWWA). 2014 (4th ed.). M21, *Groundwater*. Denver: AWWA.

AWWA. 2021. M73, *Groundwater Management*. Denver: AWWA.

Bloetscher, F. 2019. *Infrastructure Management.* Plantation, Fla.: J. Ross Publishing.

Bloetscher, F. (ed.). 2010. *Sustainability Compendium.* Denver: AWWA.

Reilly, T.E., K.F. Dennehy, W.M. Alley, and W.L. Cunningham, W.L. 2008. "Ground-Water Availability in the United States." Reston, Va.: US Geological Survey.

CHAPTER 2

Investigating Underground Conditions, Restrictions, and Siting the Well

The goal of this chapter is to outline how to locate potential well sites, identify useful information to make siting decisions, discuss the acquisition of preferred sites, and take actions needed to protect the well once located. This chapter will also cover wellhead protection. Wellhead protection is a result of modeling, which requires monitoring and some initial design concepts. The design concepts will be used and updated during the drilling, logging, construction, development, and design of the well, which will be covered in Chapters 3 to 6.

RESEARCH

Sources and Literature

There are several steps involved in identifying potential groundwater sources. The first is to locate the literature, prior investigations, local or regional water supply plans, and on-site activities of neighboring water purveyors regarding groundwater use. If groundwater is available, others will likely know about it. Oil and gas drilling logs often identify formations that may be potential sources. Agricultural interests are often the first to tap groundwater supplies.

The United States Geological Survey (USGS) has extensively studied underground formations for the presence of suitable water throughout the United States. For studies within the United States, the literature search should begin with the USGS's *Summary Appraisals of the Nation's Ground Water Resources (1978–1982)* or its most recent update. Individual reports covering the region of interest can also be obtained from USGS and state geological organizations. Much of this information had been moved online in recent years, making it more accessible to engineers and utility operations personnel. Other published reports on groundwater resources (available from federal and state agencies) provide a summary of quantity and quality of available groundwater in various geographic regions, including major water basins and state, county, and local regions.

Regional water authorities or neighboring water purveyors may study potential groundwater basins for their sources. Most of this information is available in the public record and provides an excellent start in determining the potential for water supply availability and for the design of a wellfield. Local engineers, geologists, or hydrogeologists will also likely have a significant amount of information on water quality parameters and drilling conditions. However, such reports should not be relied on exclusively as even fairly site-specific reports are often general in nature, and many local details may be omitted.

There are many sources of information for well conditions that vary by location and the amount of prior development or oil and gas exploration in the area. The following is a list of typically available database sites:

Geologic Data

- USGS National Geologic Map Database (http://ngmdb.usgs.gov/ngmdb/ngmdb_home.html)
- USGS Publications Warehouse (https://pubs.er.usgs.gov/)
- State and regional agencies and oil drillers

Groundwater Data

- National Ground-Water Monitoring Network for water elevations and trends
- USEPA STOrage and RETrieval (STORET) for water quality
- USEPA Water Quality eXchange (WQX) for water quality
- UEPA Safe Drinking Water Information System (SDWIS) for standards
- State and regional databases—typically from submittals for regulatory purposes/reporting
- Neighboring utilities
- Local consultants (engineers and hydrogeologists)

Surface Water Data

- USGS National Water Information System (http://waterdata.usgs.gov/nwis)
- State and regional agencies

Climatic/Weather Data

- NOAA, National Weather Service (www.weather.gov/)
- National Weather Service Climate Prediction Center (http://cpc.ncep.noaa.gov/products/predictions/long_range/seasonal.php?lead=5)

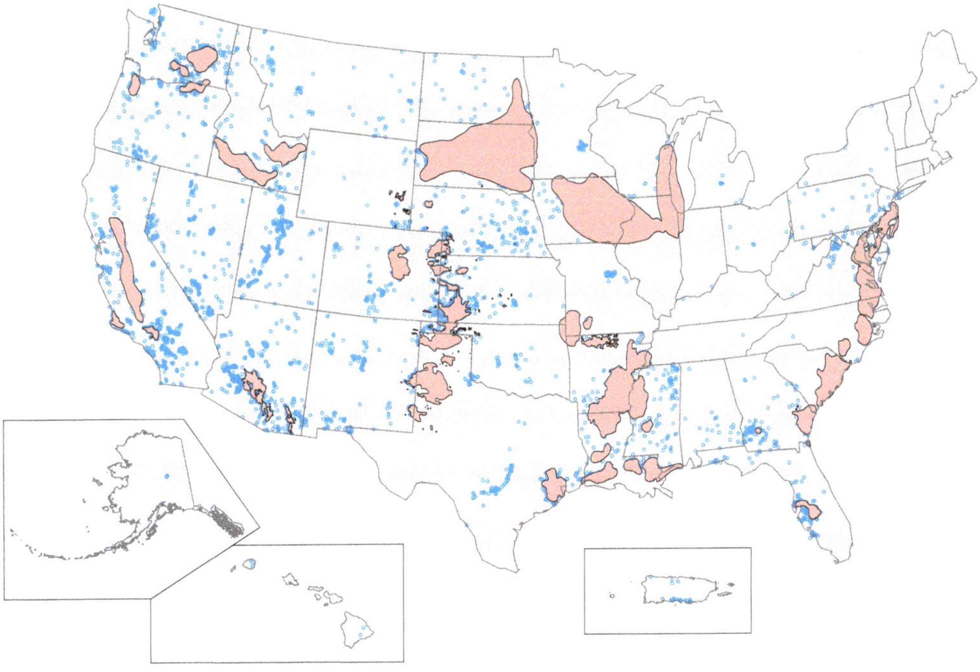

Source: Reilly et al. 2008

Figure 2-1 Water-level declines

Note: Red regions indicate areas in excess of 500 mi^2 that have water-level decline in excess of 40 ft in at least one confined aquifer since predevelopment or in excess of 25 ft of decline in unconfined aquifers since predevelopment. Blue dots are wells in the USGS National Water Information System database where the measured water-level difference over time is equal to or greater than 40 ft.

- Rainfall/weather data
- State and regional databases (Western Regional Climate Center, www.wrcc.dri.edu/)

Information from these reports can help determine if the water supply of the potential well site is sustainable as mentioned in Chapter 1. For example, Figure 2-1 shows areas where aquifer levels have declined from historical levels because of use. Field investigations (as outlined in Chapter 3) must be performed to confirm site-specific characteristics. However, as a part of the testing program, data can be gathered as a baseline to help the utility understand if the water supply is sustainable at the withdrawal rates proposed and, if not, the steps to mitigate the withdrawals. These tests will be discussed in Chapters 4 (testing) and 5 (calculations).

Water quality considerations are important when deciding whether a well site will provide water capable of potable use with reasonable treatment and delivery cost. Deterioration of water quality can have significant financial consequences because of higher treatment costs or the need to abandon the water supply wells. Therefore, some exploratory work should be done to provide needed details regarding water quality.

Environmental Site Assessments Before Purchase

Once a determination is made that a water supply might be available, the next step is identifying potential sites. Surficial activities can have impacts on aquifers. As land uses change, aquifer quality may deteriorate with development, so potential development and effects of that development should be considered, which require a good understanding of urban and industrial growth and zoning of the area associated with the groundwater supply. An example is instructive. In the 1930s, a small airstrip was built for a community of a few hundred people. That does not seem like a major issue. The first photograph of the site is from 1942 when the site had been converted to an active military base to support the war effort (Figure 2-2). Beyond the base, the site is basically virgin ground. The base was still active in 1943 (Figure 2-3). However, after the war, the base was abandoned and transferred to the state university system, where in 1964, it was converted to a university (Figure 2-4). Any consultant performing an environmental site assessment would flag the site as having the potential for Recognized

Source: Boca Raton Historical Society

Figure 2-2 Boca Raton Army Air Field, 1942

Environmental Conditions. ASTM E1527-21 defines *recognized environmental conditions* as "the presence or likely presence of any hazardous substances or petroleum products in, on, or at a property: (1) due to release to the environment; (2) under conditions indicative of a release to the environment; or (3) under conditions that pose a material threat of a future release to the environment. De minimis conditions are not recognized environmental conditions."

ASTM E1527-21 further defines "hazardous substance" as "a substance defined as a hazardous substance pursuant to CERCLA 42 U.S.C.§9601(14), as interpreted by EPA regulations and the courts: (A) any substance designated pursuant to section 1321(b)(2)(A) of Title 33, (B) any element, compound, mixture, solution, or substance designated pursuant to section 9602 of this title, (C) any hazardous waste having the characteristics identified under or listed pursuant to section 3001 of the Resource Conservation and Recovery Act of 1976 (RCRA), as amended, (42 U.S.C. §6921) (but not including any waste the regulation of which under RCRA (42 U.S.C.§§6901 et seq.) has been suspended by Act of Congress), (D) any toxic pollutant listed under section 1317(a) of Title 33, (E) any hazardous air pollutant listed under section 112 of the Clean Air Act (42 U.S.C. §7412), and (F) any imminently hazardous chemical substance or mixture with respect to which the Administrator (of EPA) has taken action pursuant to section 2606 of Title 15."

Source: Boca Raton Historical Society

Figure 2-3 Boca Raton Army Airfield, 1943

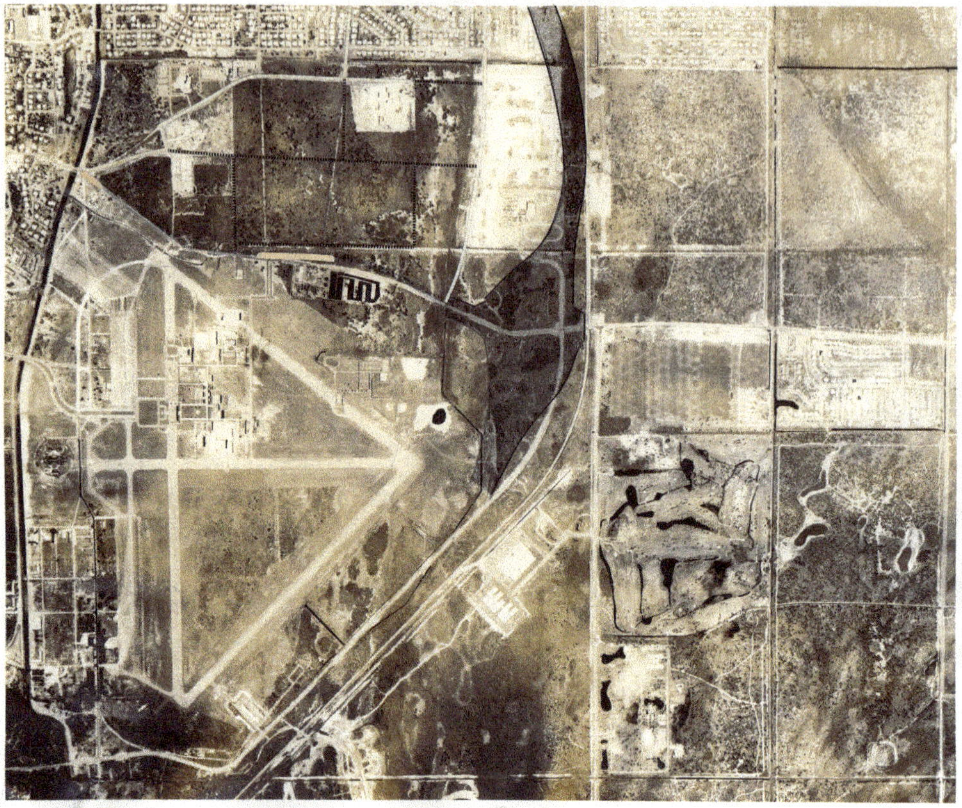

Source: Boca Raton Historical Society

Figure 2-4 Florida Atlantic University (FAU) emerges from former Boca Raton Army Air Field, 1970.

By 2012, the site was clearly a school, although the site is still recognizable as an airfield (Figure 2-5). That changed by 2019 when the site is a university that might raise little concern (Figure 2-6), yet the fact that the site was an airfield 70 years earlier does not obfuscate the fact that the groundwater could easily be contaminated to this day by materials buried on the site during World War II. There were few records, and old (and active) bases historically face this challenge. Military use aside, the question is whether a densely constructed school is a good place for a well. As Figure 2-7 shows, probably not.

In conjunction with the evaluation of the site, the following factors should be considered:

- the location of the well sites in areas where the demands are highest,
- the cost to develop the water supply for the region,
- the conveyance and treatment costs,
- the potential revenue generation as it relates to construction and operating costs,
- environmental concerns, such as for endangered species and critical habitats, and
- location of future growth.

Source: Peter Lorber and Courtenay Gilbert

Figure 2-5 FAU on the site of the former Boca Raton Army Air Field, 2012

Source: Market Media/Shutterstock.com

Figure 2-6 Florida Atlantic University today

Source: Photo by Fred Bloetscher

Figure 2-7 Street view of part of Florida Atlantic University campus

RESTRICTIONS

Considerations for Siting Wells

Considerations for siting production wells include the following issues: site availability, water supply, water quality, and wellhead protection limitations. Many small water systems have made cost the prime consideration in selecting sites for public water supply wells, but water quality may have a longer-term impact. Therefore, one must balance both water quantity and quality. The water supply system must balance factors including well depth, geology of the area, characteristics of the rock formations, and dissolved minerals in the aquifer.

In most places, locating productive wells may be accomplished using local, expert knowledge of the geology of the region; the experience with existing wells; and reviewing the site conditions as discussed in the prior two sections of this chapter. Having noted the benefits of groundwater supplies (Chapter 1), these principles can provide guidance in searching for well sites.

The next step is to review the land use. Figures 2-8 through 2-10 show three aerial photographs. Such photographic images are routinely available online. Figure 2-8 is a heavily industrialized area in the St. Louis, Mo., area. Land value in

Source: Photo by Fred Bloetscher

Figure 2-8 Commercial area photo taken in St. Louis, Mo.

such areas may be very expensive. The availability of land to locate a well may be severely restricted as a result of surface activities, such as stormwater management related to buildings and parking areas. In addition, such areas contain businesses that use a variety of chemicals and processes that may potentially contaminate the groundwaters. Reviews of industrial pretreatment records for wastewater plants, hazardous materials licenses, and chemical inventories provided to environmental agencies may provide information on the chemicals that may possibly be accidentally discharged to groundwater. Potential pollution sources of this type must be avoided if possible. As a result, industrial sites are rarely acceptable as potential well locations without incurring significant expense for monitoring wells, monitoring industrial practices, wastewater pretreatment processes, and ongoing dialogue with the businesses. Most utilities typically avoid such sites.

Figure 2-9 is a residential site. Wells may be located in residential sites, but many of the same problems of industrial sites exist with residential sites. Residents rarely want wells located in their yards or the well access that is required. Land costs may be expensive, and the utility cannot control surface activities. Residential areas that utilize septic tanks may pollute groundwater with household chemicals, microbiological contaminants, salts, and nitrates. Residents also use a variety of pesticides, fertilizers, solvents, and other chemicals that, while

Source: Photo by Fred Bloetscher

Figure 2-9 Residential aerial photo taken in Broward County (Fla.)

in much smaller quantities than industrial complexes, may also contaminate the aquifer. Monitoring programs rarely exist for residential development with regard to chemical contamination. As a result, care must be exercised in selecting well sites in residential areas; such sites should be pursued for deeper wells that are unlikely to be affected by surface activities.

Source: Photo by Fred Bloetscher

Figure 2-10 Undeveloped area aerial photo taken in Palm Beach, Fla.

Figure 2-10 shows a remote site that is well outside development. The land is unlikely to have been affected by industrial activity, but agricultural activity should be investigated. Agricultural use of land (Figure 2-11) can also affect groundwater quality because of pesticide, herbicide, and waste runoff. Sites that are well outside development areas should be less expensive than sites in urban areas, but this cost differential must be weighed against the cost of constructing transmission pipelines. Monitoring of remote sites should be included in any design as site visits will be less frequent. Ecosystem effects may be limiting factors if wetlands are in the projected cone of depression. In addition, the site may not remain outside of expanding development forever, as was Fort Lauderdale's Peele-Dixie wellfield, installed in 1926.

Aerial photographs are useful but may not provide all the information needed to assess whether the land is a potential well site. Figures 2-12 through 2-18 are examples of surface features in undeveloped areas, but what they indicate about groundwater availability is unclear. Figure 2-12 is a waterfall in north Georgia. It routinely has copious amounts of water, and the area is relatively undeveloped (north of Atlanta, Ga.). However, once on the site, it is clear the rock formations are granite, offering little capacity for groundwater sources. Also, the dammed lake in Figure 2-13 (also in north Georgia) appears to be a significant source of water to recharge the neighboring formation. However, a site visit demonstrates that the underlying rock is unlikely to have significant recharge potential unless direct accesses to the lake via fractures exist. In fact, when the lake was drained, it revealed that the underlying soils were clay—no recharge at all (Figure 2-14). Figure 2-15 is a familiar site in the desert of eastern Utah. There is little development within miles of this site, but the groundwater supply is far below the surface and provides little indication of its existence from the surface.

Figure 2-16 is the Grand Canyon—in the middle of the desert—but the north rim does have a supply of groundwater that is piped to the south-side visitor centers. Likewise, the Everglades is a swamp (Figure 2-17). It is also the recharge area for the Biscayne aquifer, so it would be considered a major groundwater source despite that water lies on the surface much of the year. Finally, the Rocky Mountain site is pretty (Figure 2-18), but the groundwater is limited. Small communities in the Rockies may use groundwater because it requires minimal treatment; but like the Carolinas, once the community grows, a reservoir (Figure 2-19) is needed.

The conclusion from each of these photographs is that aerial and on-site photographs are inappropriate for siting wells under most circumstances unless much more is known about the subsurface. That means that in addition to site visits to determine the potential for groundwater, the historical land use practices must be reviewed at the site.

Source: Photo by Fred Bloetscher

Figure 2-11 Agricultural property

Source: Photo by Fred Bloetscher

Figure 2-12 Helton Creek Falls, Blairsville, Ga.

Source: Photo by Fred Bloetscher

Figure 2-13 Chatuge Lake photographed from the North Carolina side

Source: Photo by Fred Bloetscher

Figure 2-14 Exposed soil beneath a dam in North Carolina

Source: Peresanz/Shutterstock.com

Figure 2-15 Delicate Arch, Arches National Park, Utah

Source: Photo by Fred Bloetscher

Figure 2-16 The Grand Canyon—looking at the North Rim

Source: Photo by Fred Bloetscher

Figure 2-17 The Everglades is a huge recharge area for a highly productive aquifer, but it is under water much of the year.

Source: Photo by Fred Bloetscher

Figure 2-18 Rocky Mountain National Park in Colorado has water, but there is little groundwater available.

Source: Photo by Fred Bloetscher

Figure 2-19 A small dam at Klamath Falls, Ore., could be used for water supply and downstream power generation.

Acquiring a Site

As soon as a property has been identified as attractive for detailed exploration, purchase or lease options should be obtained. The cost of land may be more reasonable before good water supplies are confirmed. Once the utility has decided on a general location for the wells, a site must be acquired. Most communities can readily determine ownership of property from the local county property appraiser's office (Figure 2-20).

If only the wellhead and perhaps some electrical panels are included, it may be possible to only acquire an easement. Easements prevent vertical construction over the site but otherwise interfere little with the use of the property. Many people will donate easements for the public benefit, but sometimes negotiations are required. Note that land used temporarily for drilling-site preparation and restoration can add costs to the groundwater development in the form of temporary easements, a cost that should not be overlooked. If a well is installed, the temporary easement may need to be converted to a permanent easement (Figures 2-21 and 2-22).

The neighbors may not want a wellhouse structure in their neighborhood; having only a well will be far less intrusive than a wellhouse (Figure 2-23).

PROPERTY APPRAISAL

Address: 86008 West Fig Tree Lane, Plantation, FL 33317

Owner: Joe Rose Trust

Mailing Address: 86008 West Fig Tree Lane, Plantation, FL 3331

Abbrev Legal Description: Plantation Gardens 3rd Sect 41-31 B Lot 8 Blk 1

PROPERTY VALUES

YEAR	LAND	BUILDING/ IMPROVEMENTS	MARKET VALUE	ASSESSED SOH VALUE	TAX
2022	$ 103,950	$ 323,500	$ 427,450	$ 333,480	$ 6,028
2021	$ 98,753	$ 307,325	$ 406,078	$ 316,806	$ 5,727
2020	$ 93,815	$ 291,959	$ 385,774	$ 300,966	$ 5,440
2019	$ 89,124	$ 277,361	$ 366,485	$ 285,917	$ 5,168

	COUNTY	SCHOOL BOARD	MUNICIPAL	INDEPENDENT
JUST VALUE	$ 427,450	$ 427,450	$ 427,450	$ 427,450
PORTABILITY	0	0	0	0
ASSESSED SOH	$ 333,480	$ 333,480	$ 333,480	$ 333,480
HOMESTEAD	$ 25,000	$ 25,000	$ 25,000	$ 25,000
ADD HOMESTEAD	$ 25,000	$ 25,000	$ 25,000	$ 25,000
WIDOW/VETR/DISAB EXEMPT	0	0	0	0
SENIOR EXEMPTION	0	0	0	0
EXEMPT TYPE	0	0	0	0
TAXABLE	$ 283,480	$ 283,480	$ 283,480	$ 283,480

Sales History

DATE	TYPE	PRICE	BOOK/PAGE
6/22/2013	DRR_T	100	12455421212
4/26/2013	WD-T	100	12425333214
7/23/2012	WD-Q	331000	45558/25881
5/25/1999	WD	158000	29587/25846
12/1/1993	WD	145000	21851/00046

PRICE	FACTOR	TYPE: SF
BLDG AREA	sf	2650
UNITS/BED/BA		1/3/3/

Figure 2-20 Example of property information from property appraiser's office

Fee-simple acquisition of the site may be necessary if a vertical structure is contemplated. A vertical structure will also require compliance with local setback requirements listed in zoning or land development ordinances. That may require a much larger piece of property if acquired. Normally, lots are purchased as fee simple—i.e., the utility owns the property. If the utility proceeds in this manner, a title search, property appraisal, and purchase documents must be created and recorded. Governing board approval will be required.

Source: Photo by Fred Bloetscher

Figure 2-21 Typical well installation

Source: Photo by Fred Bloetscher

Figure 2-22 Well on an easement

Source: Photo by Fred Bloetscher

Figure 2-23 Example of well building; it may require fee-simple purchase of the property.

Restrictions – Wellhead Protection

Once a well site is found, there is information that a water supply exists, and a decision to construct a well on the site has been made; efforts need to be undertaken to protect the water source. Wellhead or source water protection regulations were created to reduce the threat to water supplies from contaminants in runoff or as a result of surface activities. Watershed protection is a requirement of the Safe Drinking Water Act Amendments of 1986. Under Section 1428, each state must prepare a wellhead protection program and submit it to the USEPA for approval. The protection of public water supply wells from contamination through wellhead protection programs is considered an important component of comprehensive state groundwater protection programs. However, many states have promulgated minimal rules while delegating specific implementation of the program to regional water management districts and/or counties.

While the best practice is to first locate wellfields where contamination is unlikely, one must also protect the wellfield from surficial impacts. This is difficult to do without acquiring large amounts of property around the well. The issue becomes nearly impossible when the well needs to be located in a developed area.

As a result, states and provinces have implemented regulations to protect underground sources of water. These actions include:

- new requirements for installation and testing of underground storage tanks;
- increased regulation for handling, using, and transporting toxic chemicals to reduce the possibility of spills;
- greatly increased regulation of landfills and other waste disposal sites;
- tighter control of the use of pesticides and agricultural chemicals;
- sampling and monitoring of identified groundwater contamination locations; and
- action to remove contamination (Bloetscher et al. 2005).

All these actions will affect land use and local constituents. Delineation of a wellhead protection area is typically done using numerical computer groundwater modeling of travel time of the pollutant (solute) transport. These numerical models, which are very complex, indicate large areas where many land uses are restricted or prohibited—conflicting with private property rights objectives. Wellfield protection ordinances prohibit the use and/or storage of certain hazardous materials in zones around wells. In some states, the conflicts with source water protection exist within private property rights laws that indicate that if the property is damaged more than ten percent, the regulating agency must compensate the landowner according to condemnation or "taking" rules. This could be a significant impediment to the implementation of wellhead protection programs in developed areas.

Figure 2-24 is an example of the wellhead protection zones in Broward County, Fla. Broward County's Chapter 27 wellhead protection rule outlines three regulated areas around wells. Zone 1 is within the 10-day travel period of a contaminant within the cone of influence. All new nonresidential activities are prohibited in these zones, and many existing ones have been terminated. However, hazardous material wellfield licenses are granted for activities that have been grandfathered into the property rights and are subject to all requirements for Zone 2 licenses.

Zone 2 encompasses areas between the 10- and 30-day travel times within the cone of influence. Hazardous materials are allowed in Zone 2 but are subject to 100 percent containment; daily inventory records; records of storage and use; emergency plans for spills; acquisition of emergency control devices to continue the use; monitoring of groundwater; reporting of spills; and certain other paperwork requirements. Zone 3 permits any use but requires a hazardous materials license above a certain specified (or threshold) quantity. Zone 3 does not require notification of the utility whose wellfield might be affected.

The following is an example of why wellhead protection is so important for water suppliers. Figure 2-25 shows the Fort Lauderdale (Fla.) Peele-Dixie wellfield. At the time of construction (1926), the city had a population of approximately 8,000 and had little competition for water. The first production wells were located approximately 4 miles inland, away from any potential users or conflicts.

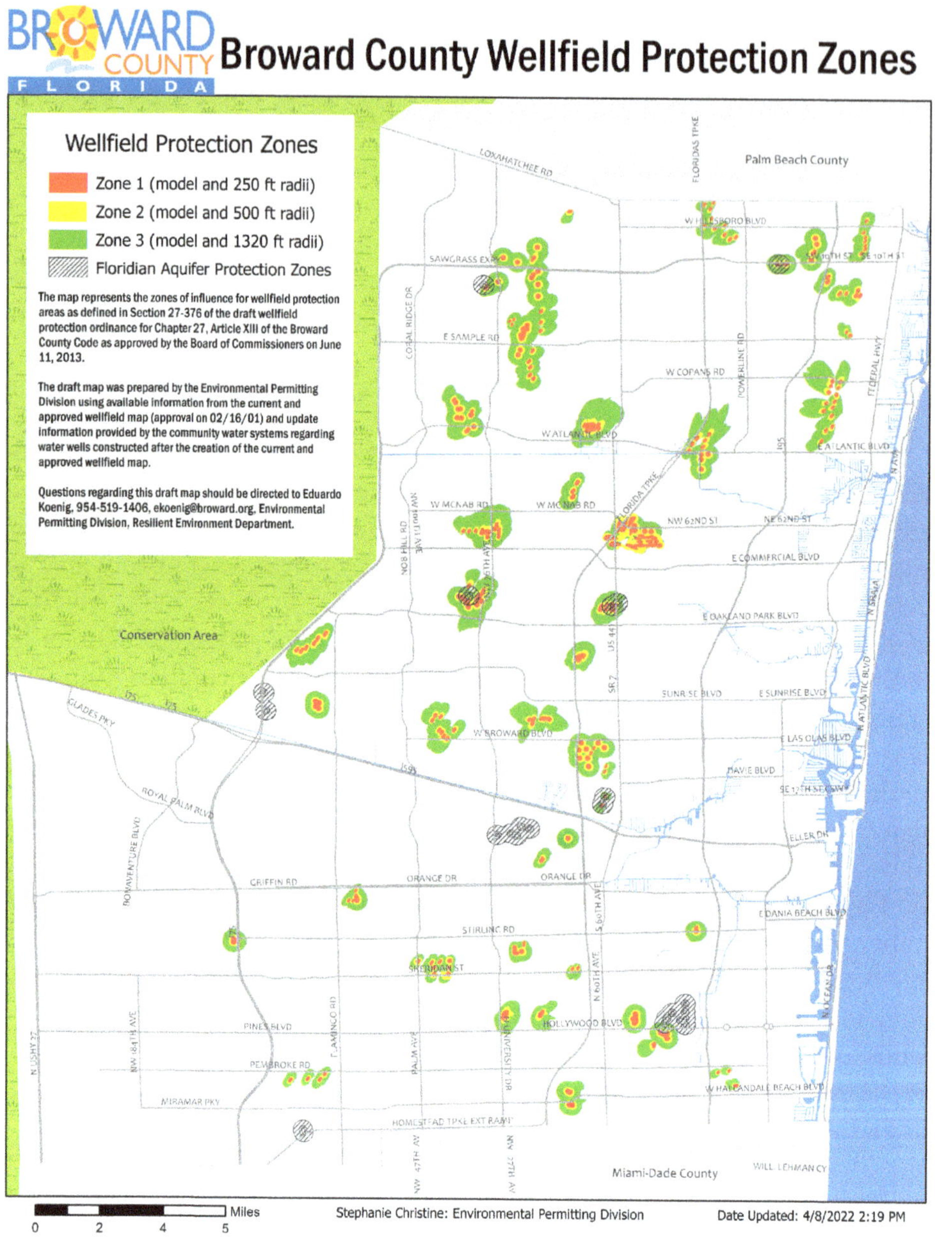

Source: Broward County, Chapter 27 Code of Ordinances

Figure 2-24 Broward County (Fla.) wellfield protection zones

This wellfield was located in an isolated area of the county and thought to be an ideal site for water supply. Figure 2-26 shows the area currently around this wellfield. The wellfield is in a neighborhood, and a golf course has been built around the wells.

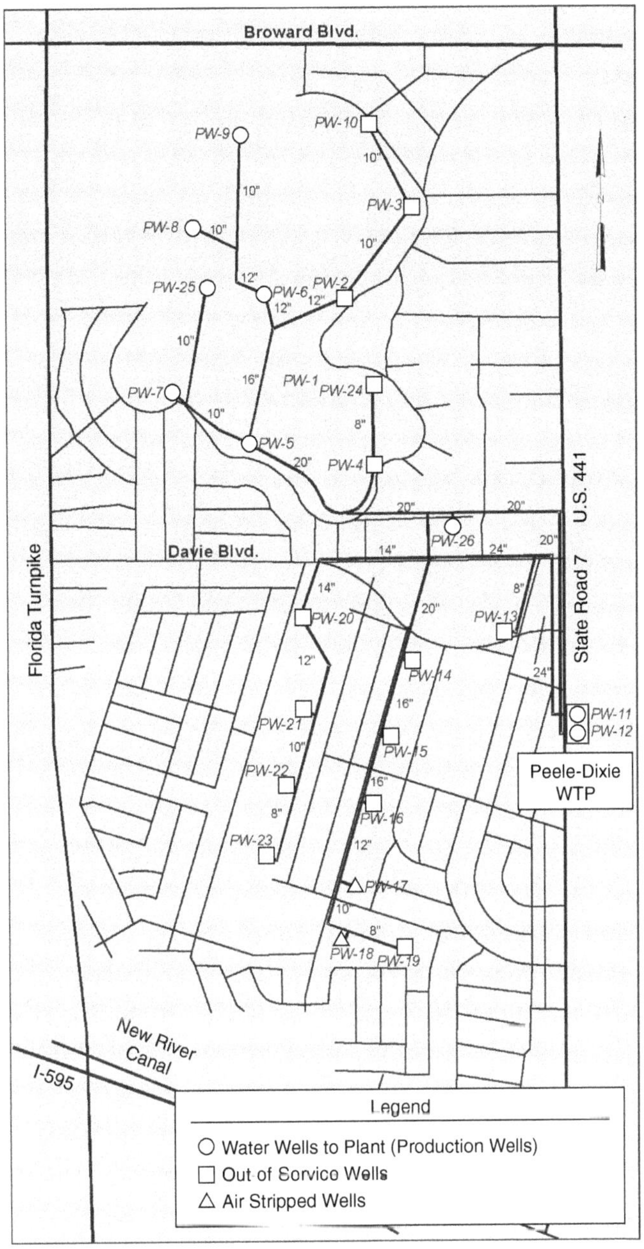

Source: Hazen and Sawyer, P.C., Boca Raton, Fla.

Figure 2-25 Diagram of well locations for Fort Lauderdale's Peele-Dixie Wellfield (pipe diameters are shown in inches).
Note: The northern half of the wells are the original wells installed in 1926 that are now on the golf course of Fort Lauderdale Country Club.

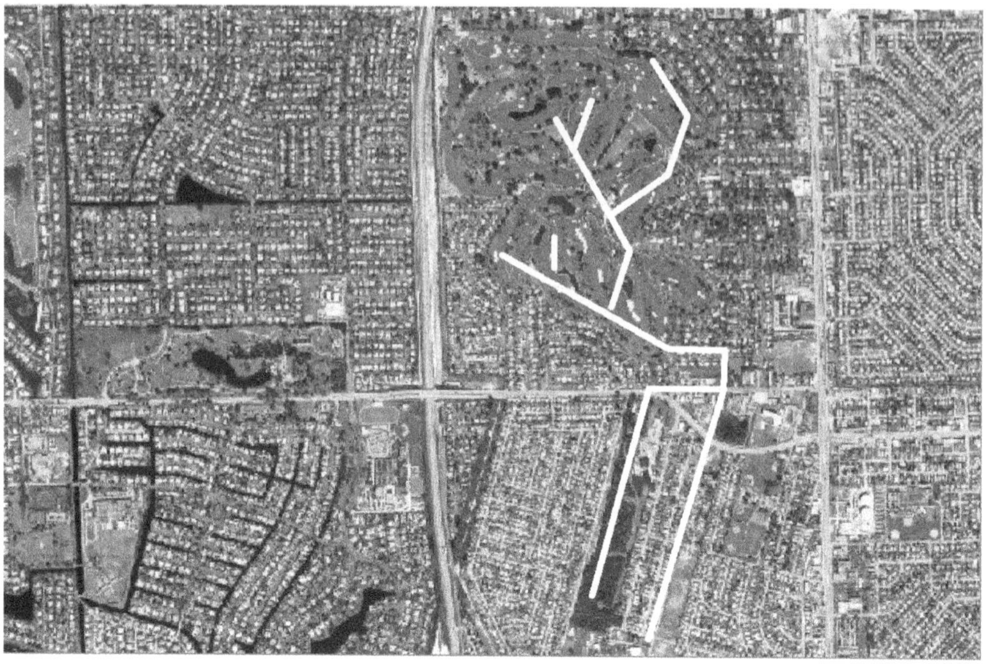

Source: Hazen and Sawyer, P.C., Boca Raton, Fla.

Figure 2-26 Aerial map of vicinity of Fort Lauderdale's Peele-Dixie Wellfield. Compare to Figure 2-27.

Data gathered from the well can be used to develop a computer model to represent the movement and withdrawal of water from the wells. (American Water Works Association's Manual M73 [AWWA 2021] provides further information on this topic.) The results are used to manage supplies and to track contaminants that might travel to the water supply wells. Figure 2-27 shows a model developed to indicate the drawdowns of the well. While the aquifer is very productive and the wells have drawdowns, the wells draw water from the south where there is a Superfund site that potentially contaminates the wellfield. If constructed today, the industrial site that may contaminate this wellfield would not be permitted because it is in the 30-day travel zone.

USEPA has also developed monitoring requirements to monitor and demonstrate compliance with the regulations. Water quality monitoring and recordkeeping are generally straightforward. However, water quality monitoring is directly related to water treatment needs. These monitoring requirements are permitted to change over time as improvements are made to treatment equipment, analytical techniques, and instruments. Changes in water quality may dictate changes in well locations and/or treatment needs. Adding treatment can be a significant issue for the 40,000 smaller utilities that may only apply chlorine to their water before distribution. Sentinel monitoring wells are used for this purpose.

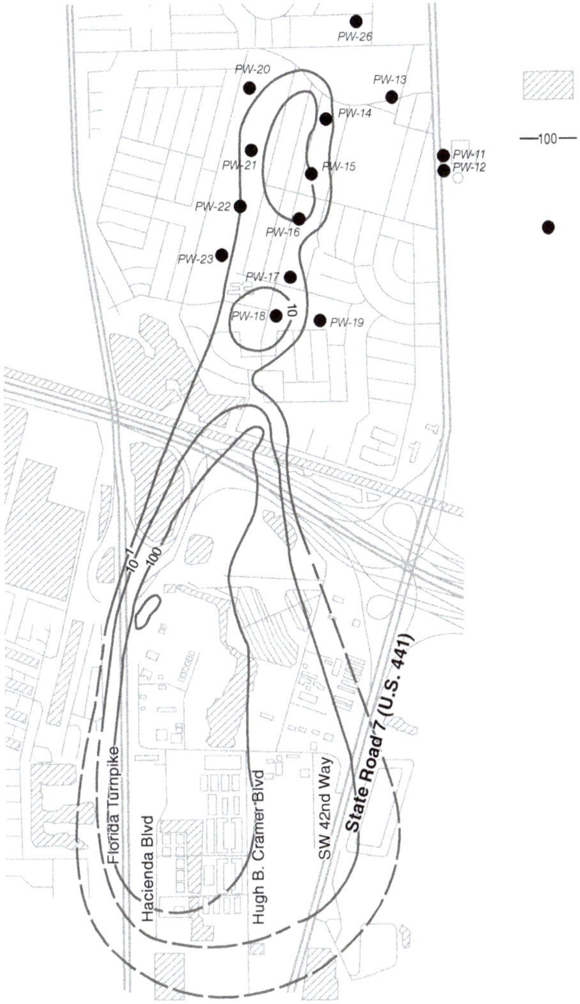

Source: Hazen and Sawyer, P.C., Boca Raton, Fla.

Figure 2-27 Drawdown map of Fort Lauderdale's Peele-Dixie wellfield site (potential contamination source is located south of interchange).

If a well becomes contaminated, the least expensive and most common strategy is blending the contaminated water with another source to meet drinking water standards. Options for blending include mixing with:

- a different aquifer zone,
- another aquifer, and
- a different supply.

However, this does not address the fact that the water is contaminated nor that it might become more contaminated with time. It also ignores the public perception issue that arises from the concept of "contaminated." Therefore, other options should be pursued. In the pump and treat strategy, groundwater is extracted from an aquifer and treated. Aeration may be an option for volatile organic compounds. Treatment is generally more expensive than blending and may not always be technologically feasible, so a new supply may be needed. Hence, the reason source water protection is so important. A discussion of treatment options is included in Appendix B.

Monitoring Wells

The objectives of and protocols for a groundwater monitoring program are typically documented in a groundwater monitoring plan. The USEPA website at www.epa.gov/osa/elements-quality-assurance-project-plan-qapp-collecting-identifying-and-evaluating-existing has references relating to the development and implementation of groundwater monitoring plans. The typical components of a groundwater monitoring plan include:

- description of the monitoring plan objectives,
- a history of groundwater monitoring in the study area,
- description of the physical, hydrologic, and hydrogeologic characteristics of the study area to be monitored,
- a map showing the well locations,
- a rationale for the location selection,
- the monitoring schedule,
- construction and development details for wells included in the monitoring network,
- sampling protocols,
- analysis and reporting (what is done with the data), and
- a means to revise the plan based on changes in the field.

No one can predict the number of monitoring wells needed for a project, but at least one should be installed in the monitoring zone. The point is to have enough wells to evaluate and characterize the local groundwater spatially across the basin and differentiated such that each separate water-bearing zone is monitored. The goals of the monitoring program are to identify and/or confirm groundwater trends, provide data for a computer model of the wellfield, track contaminants, and provide a means to optimizing well placement. To be of use, groundwater elevation and quality data should be combined with surface water data, meteorological data, and other data such as subsidence figures. The use of monitoring wells and computer models is particularly helpful in areas where contamination risks are high, drawdowns are increasing, or declining water levels may indicate mining.

Sentinel monitoring wells serve as an early-warning system to detect changes in water quality and water elevations before they affect the water supply wells. This also means that contamination previously considered downgradient from the well can become upgradient either after pumping begins or as influenced by nearby surface water. Therefore, monitoring wells must be installed upstream as well as downstream of the proposed water supply well. Note that if the utility is trying to assess a well's vulnerability to contamination, the utility staff would want to be familiar with the information in USGS Circular 1385 (https://pubs.usgs.gov/circ/1385/pdf/Cir1385.pdf), which can be used to devise improved monitoring programs to identify the most beneficial protection programs for a particular well (AWWA 2021).

Monitoring Well Installation

Clearly, wells and wellfields should be located upgradient or downgradient at an appropriate distance from potential threats to the water quality such as a military airfield. If the site cannot be avoided, more frequent testing for pollutants may be appropriate. Establishing early-warning monitoring (sentinel) wells at various depths may be required to monitor groundwater quality (AWWA 2014). Property appraisers' offices and local planning departments are good sources for evaluation of historical land uses.

Most states have well construction standards that should be followed in well design and construction; see M21, Bloetscher et al. (2008), ANSI/AWWA Standard A100-06, and Standard A100-97, and various ASTM standards such as *Standard practice for design and installation of groundwater monitoring wells in aquifers*, D5092-04 (2010). As a general rule, all monitoring wells should prevent the infiltration of surface water into a well, prevent cross-contamination within or between water-bearing zones, and be secured at the surface to prevent unauthorized access. In addition, per M73 (AWWA 2021):

- Monitoring wells should be designed with respect to the well's function within a well network.
- Wells and borings should be logged during drilling by someone experienced in hydrogeology.
- Well names and numbers should be permanently assigned and should never change.
- Replacement wells should have an "R" added to the well name to indicate that they are replacement wells.

M73 also outlines several types of monitoring wells (see next three chapters for installation methods):

- Traditional (vertical) wells with one single interval of well screen (Figure 2-28).
- Traditional (vertical) wells with multiple intervals of well screen: Same as above although with annular seals placed between filter pack sections adjacent to well screen intervals (Figure 2-29).
- Nested wells: Multiple vertical wells constructed in a single borehole. Nested wells are used to monitor multiple groundwater zones simultaneously. They allow for evaluation of vertical movement of groundwater. Another advantage is that they take up less space than multiple wells; however, they are tricky to install because of the need for annular space around wells (requires large-diameter borehole). There are commercially available multilevel monitoring systems. The cost of these systems is typically greater than the cost for conventional monitoring approaches, but they can provide better quality installation (Figure 2-30).
- Clustered wells: Multiple traditional wells with varying screen depths installed in the same general area (each well in its own borehole).

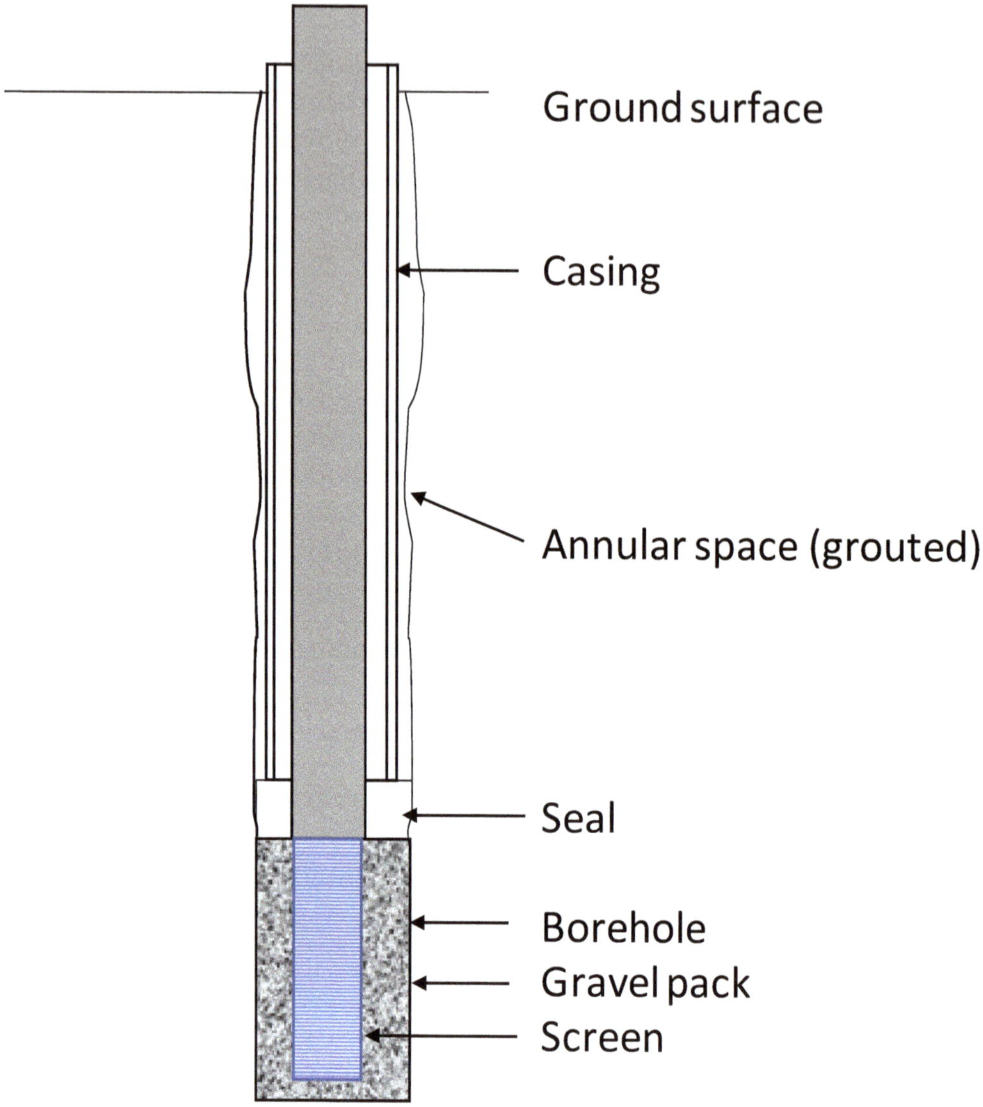

Figure 2-28 Single sentinel monitoring well

These wells obtain the same benefits of nested wells but without concerns about construction; however, they take up more space and can cost more because of multiple mobilizations of the drilling rig (Figure 2-31).

- Direct push (DP)/pre-packed wells: Similar in overall design and construction to typical groundwater monitoring wells except that the diameters of the borehole and well casing are generally narrower; pre-packed well screens may be used; and the well installation method is different. DP wells are small-diameter hollow steel rods or PVC pipes that are pushed, driven, or vibrated into the ground. They can be installed in a cluster or as individual wells and can be used as piezometers. DP wells installed in unconsolidated sediments can often be advanced to depths of around 100 ft. They are not suitable for coarse, consolidated, cemented, or lithified deposits or in bedrock. Care must be taken to ensure an adequate bentonite seal and annular seal during installation. Bentonite sleeves may reduce the risk of vertical cross-contamination. The biggest advantage is that they are fast and relatively inexpensive to install and have very little waste (cuttings).

- Angled (or inclined) wells: These use specialized drilling rigs to advance angled boreholes. Well construction in these boreholes is tricky because an even filter pack is needed in the annular space around the well.

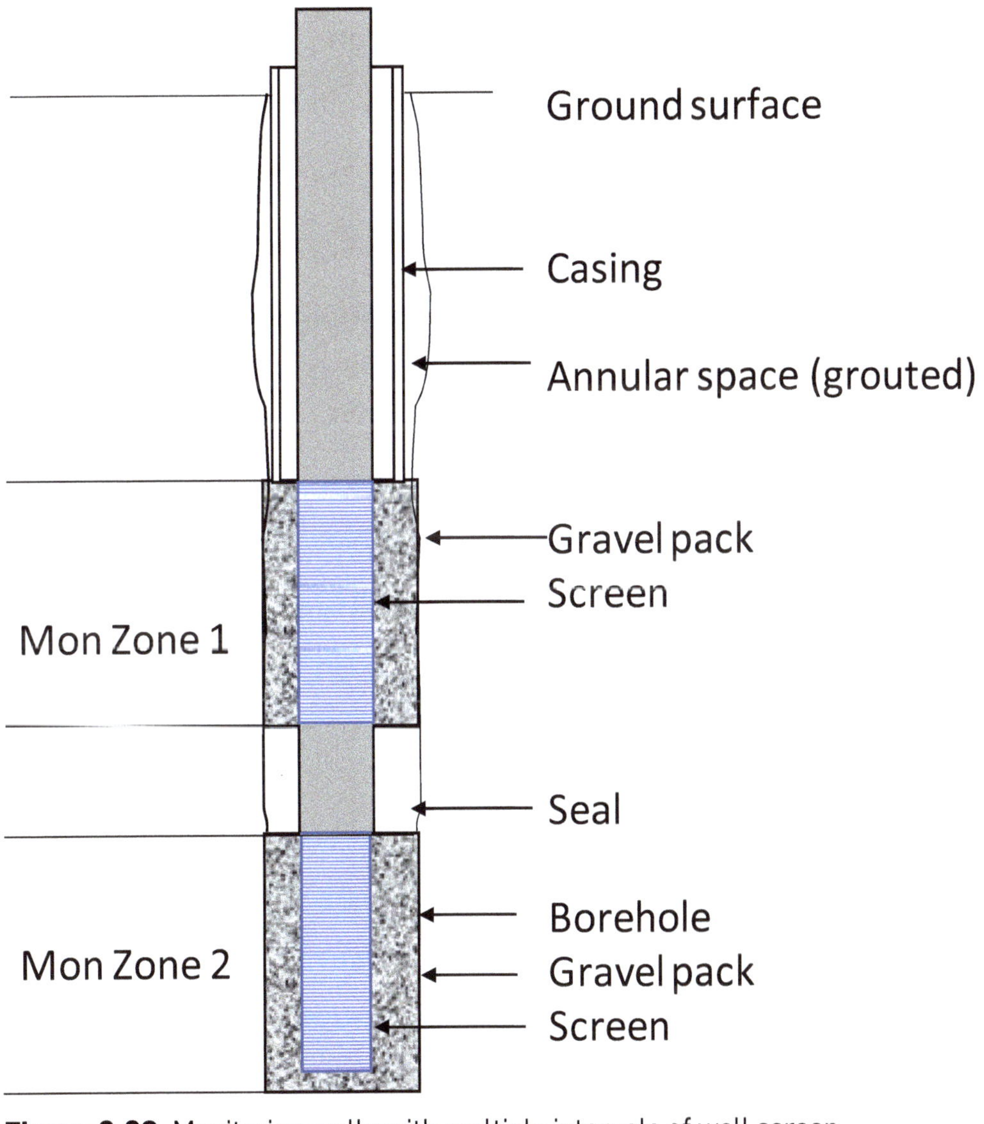

Figure 2-29 Monitoring wells with multiple intervals of well screen

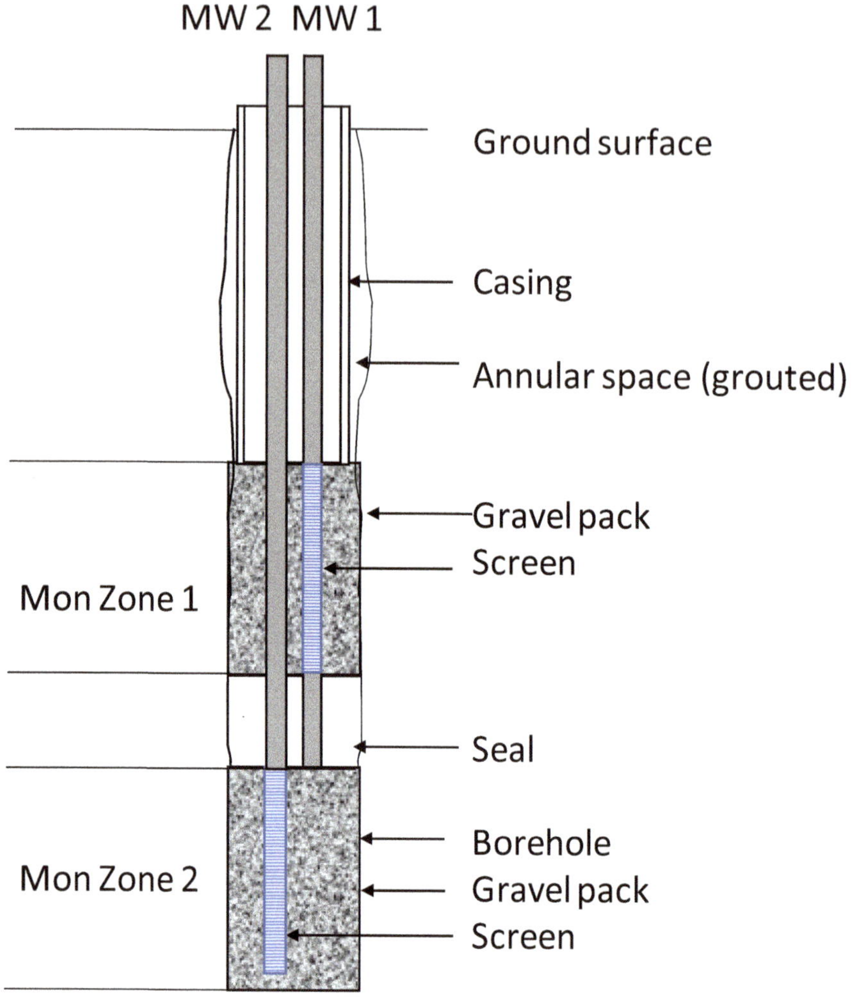

Figure 2-30 Nested monitoring wells

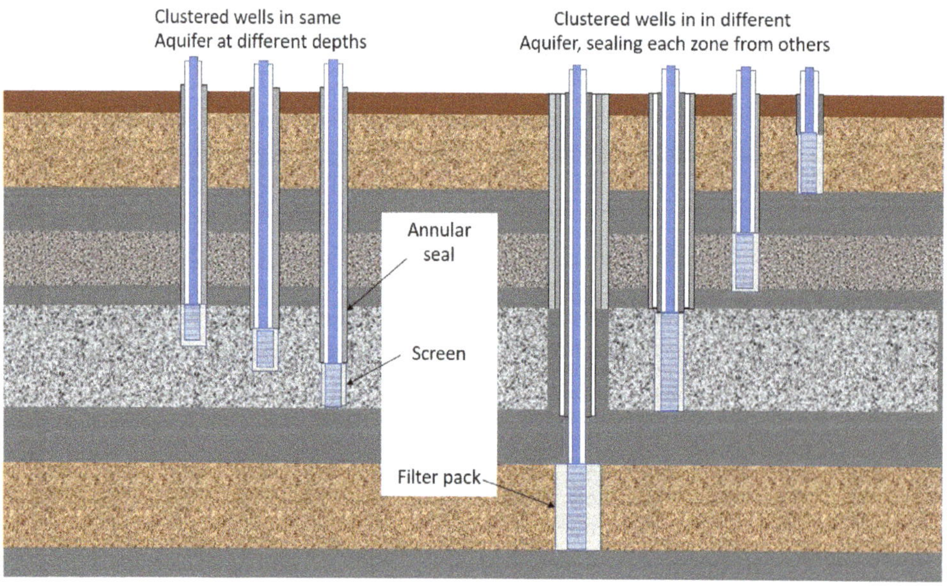

Figure 2-31 Clustered monitoring wells

Figure 2-32 shows a typical monitoring well installation.

Maintenance of Monitoring Network

Monitoring is a long-term proposition, so monitoring wells should be re-evaluated on a regular basis to ensure the appropriate data are being collected. New or updated monitoring or sampling techniques and methods should be considered and sampling and analysis plans updated to reflect current technology and industry standards. Equipment used in the program should be inspected and, if appropriate, recalibrated to ensure the accuracy of data.

Source: Photo by Fred Bloetscher

Figure 2-32 Typical monitoring well installation from surface

Professional Services

Up to this point, professionals should be involved with the process, but in some cases they are not. Professionals are needed from this point forward, however. Firms specializing in hydrogeology are valuable to the utility for their knowledge and experience in carrying out fieldwork and preparing the necessary reports. Some firms offer total services in groundwater development as well as highly specialized equipment or services. Budget constraints, the complexity of the project, and the adequacy of the utility's staff can determine the most appropriate team. Some states require registered engineers, certified geologists, and other professionals to verify the accuracy and completeness of fieldwork. Firms should be chosen based on qualifications, not price.

REFERENCES

American National Standards Institute (ANSI)/AWWA. 2006. A100-06. Water Wells. Denver: AWWA.

AWWA (American Water Works Association). 2021. M73, *Groundwater Management*. Denver: AWWA.

AWWA. 2014 (4th ed.). M21, *Groundwater*. Denver: AWWA.

ASTM. 2021. E1527-21. Standard Practice for Environmental Site Assessments: Phase I Environmental Site Assessment Process. West Conshohocken, Pa.: ASTM.

Bloetscher, F. 2008. *Water Basics for Decision Makers: What Local Officials Need to Know about Water and Wastewater Systems.* Denver: AWWA.

Bloetscher, F., A. Muniz, and J. Largey. 2008. *Siting, Drilling and Construction of Water Supply Wells.* Denver: AWWA.

Bloetscher, F., A. Muniz, and G.M. Witt. 2005. *Groundwater Injection: Modeling, Risks, and Regulations*. New York: McGraw-Hill.

Reilly, T.E., K.F. Dennehy, W.M. Alley, and W.L. Cunningham. 2008. "Ground-Water Availability in the United States." Reston, Va.: US Geological Survey.

CHAPTER 3

Geophysical Logging and Field Testing

Once the potential sites for the well location have been selected, initial subsurface investigations generally begin with construction of an exploratory or test well. Permits are required for this step. Once wells have been drilled, a series of tests should be run to better define the parameters of the subsurface environment, including the quantity and quality of available water. The goal of exploratory wells is to locate productive aquifers that yield sustainable high-quality water. The results of exploratory well drilling will provide the information needed for the proper location for casings and screens to maximize the efficiency of the well after construction. These initial exploratory test wells are used for lithologic logging, geophysical logging, aquifer testing, and collecting cores and cuttings of the formation.

Permits

Once land has been secured for a well, exploratory drilling permits must be secured and fees paid, often at the state or county level. Obtaining permits is usually the responsibility of the well driller or the engineer in charge. In addition to the driller, a qualified hydrogeologist often supervises the drilling and well-construction activities. The hydrogeologist's responsibilities will usually include identifying appropriate well construction materials, reviewing well logging data, conducting or overseeing geophysical logging, and interpreting logs. The engineer will design the well and certify the as-built drawings.

Part of the drilling permit will include proper abandonment of the well once the well has served its purpose. Regulatory agencies are concerned about well construction and post-use of a well because of the possibility of cross-contamination between shallow zones and deeper high-quality aquifers. The documentation of all field work can be valuable in later phases of groundwater development and operations, or for wellhead protection. Proper land survey location, global positioning system (GPS) locations, description of the wells, and complete as-built drawings of construction should be required.

Logging

Lithologic sampling is the protocol used to obtain core samples and define the formation type by depth. Geophysical logging consists of using a variety of electronic instruments to define the rock types, water quality, porousness, and other aquifer characteristics. Geophysical logs were developed by the oil and gas industries for subsurface exploration. Depending on the needs for water supplies, several formations may be capable of yielding sufficient quantities of water for a period of time. However, for water supply production wells, large, rechargeable aquifers are preferred. The following aquifer formations are more likely to provide sufficient water for public water supplies:

- shell, unconsolidated sands, and gravels of alluvial or glacial origin;
- sandstones and conglomerates;
- limestone; and
- porous or fractured volcanic rocks.

Exploratory wells are drilled to determine an aquifer's characteristics, including hydraulic conductivity, water quality, thickness, and areal extent. Newly drilled wells can be used for borehole geophysical logging and aquifer testing. The drilling method depends on the type of water samples desired and formation types that are expected to be encountered during drilling (see Chapter 4). As noted in Chapter 2, hydrogeologic reports already exist in many regions; between the reports, the results of the test wells, and geophysical logging, potential aquifers can be identified. However, the increase in the need for new water sources requires investigation in areas where little historical information is available. Exploratory drilling and geophysical logging efforts can fill in the details of the specific local hydrogeologic environment in which drilling is proposed. Figure 3-1 is an example of the results of a lithologic log that cataloged the type of formation encountered with depth.

CORES AND CUTTINGS

The information required to develop lithologic logs is gathered while drilling is in progress. Cores, cuttings, water samples, drill response, and other methods are used to identify the different formation types encountered while drilling. The result will be a lithologic log, which summarizes the formations, their water quality, and depth; these are relevant to determining the methods for drilling the well. For example, as the cuttings rise, the quality of the cuttings may be affected by many variables such as the drilling method, the time it takes for the samples to reach the surface (lag time), and contamination from formations already penetrated. Therefore, the driller must exercise care when collecting water and formation samples. Because cuttings come up from the drilling level, the actual

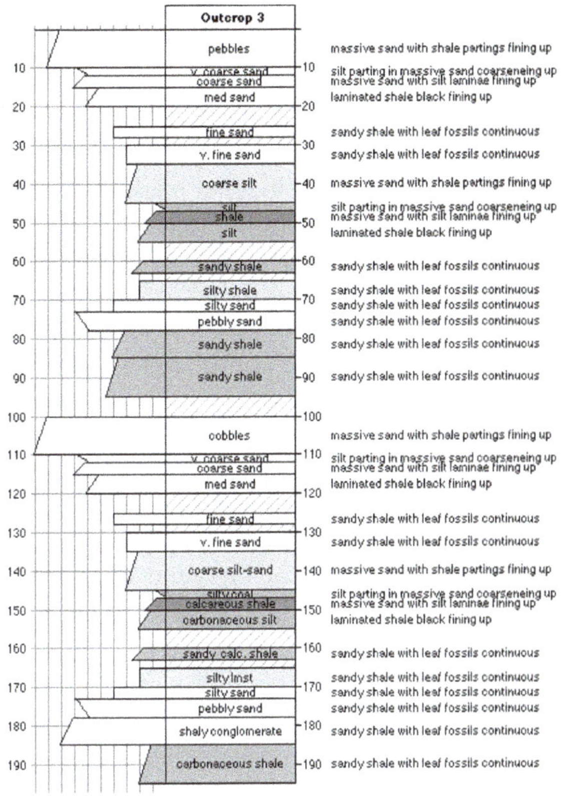

Source: Hazen and Sawyer, P.C., Boca Raton, Fla.

Figure 3-1 Example of lithologic log

changes in lithology may be only generally known. While useful for targeting further investigations, only limited information about the aquifer parameters can be gained from the cuttings and cores. Most hydrogeologists use lithologic logs to identify specific areas where usable water is likely to be encountered in large quantities and therefore require further investigation and geophysical logging.

BOREHOLE GEOPHYSICAL LOGGING

Geophysical logging is the science of applying the principles of physics to investigations related to the structure and properties of rock formations. Geophysical methods use electrical instruments whose readings represent different properties of the formations to identify physical properties of the rock, such as porosity, water content, and metallic content. The variation of the logging results may permit interpretation of the formation without seeing the actual formation (AWWA 2014, Keys 1989), provided experienced people are performing the interpretations. Compared with lithologic logging, geophysical logging is useful in identifying the appropriate depths to drill wells and set casings and can aid in accurately locating the depth of variations in hydrogeologic characteristics.

While there is no established order for application of exploration methods, a planned and balanced program of appropriate combinations will produce the most useful information. Knowledge gained through a geophysical investigation complements the information gained from test drilling and sampling. Advantages to geophysical logging include (AWWA 2014):

- rapid results,
- relatively low cost (for most logging methods),
- qualitative and quantitative results, and
- evaluation of the aquifer over a large depth.

Interpretation of the geophysics by an appropriate scientist is important because the results may be ambiguous or conflicting when first reviewed. Experienced people will be able to:

- interpret the logs based on assumptions of the materials, material homogeneity, and formation properties;
- validate calibration of modeling;
- understand the contrasts in the logs; and
- define the accuracy or precision of the logs.

Experience with logging formations in the same area or of the same type is especially useful to determine nuances in the logging results.

The purposes and uses of various geophysical logs are outlined in the following sections. Borehole logging is highly technical and quite involved, and no layperson should attempt to interpret logs based solely on the information provided herein. Log interpretation requires experts using specialized equipment. Accurate lithologic logging during drilling is crucial, so an experienced geologist, hydrogeologist, or groundwater engineer, and driller experienced in well logging should be employed. Often, it is wise to retain the expertise of a logging specialist to evaluate logs. The equipment that is used is extensive and expensive. Specialty companies are usually hired to perform the logging and interpretation at the well site (Figure 3-2). The discussion herein is designed to provide insight on the capabilities and limitations of the logging tools found to be most useful for groundwater exploration and to help water professionals understand why certain logs should be used.

Source: Photo by John Largey

Figure 3-2 Specialized equipment and companies perform borehole geophysical logging

Geophysical logging is extensively used in the oil industry, and most of the logging techniques have been developed for oil exploration. As a result, the focus of many logs and the geophysical logging industry in general is geared toward petroleum, not clean water, which is rarely encountered in conjunction with oil exploration. Most of the texts on borehole logging also present information tailored toward petroleum exploration. Given that water found in oil exploration tends to be highly saline, the presence of low or nonsaline water in a formation for water supply purposes requires the use of special analyses for interpretation. For a detailed discussion of this subject, please see *Borehole Geophysics Applied to Groundwater Investigations* (Keys 1989).

Single logs in a well are almost useless because there is limited basis on which to interpret the log variations. Therefore, the use of multiple logs (a suite of logs) in a single well will provide confidence in interpretations. Each type of log measures different physical properties, and a combined analysis may resolve ambiguities that might exist from a single log. The greater the number of wells logged in an area, the greater the statistical confidence in the data and interpretations as being representative of the subsurface environment. In addition, there are both surface and subsurface methods that may prove useful. Surface methods are used to guide subsurface logging.

SURFACE GEOPHYSICAL METHODS

Surface geophysical methods, principally electrical resistivity and seismic reflection and refraction logs, can be used to provide a picture of subsurface structure given some prior knowledge obtained from surface geology and lithologic logs. The lithologic log outlines a model of the subsurface geology as a basis for a proper interpretation of the surface geophysical data. The successful use of any surface geophysical method depends on the presence of sharp changes in the physical properties of the formation (such as clay to limestone). The detectable physical properties provide indirect estimates of the likelihood that the formation may yield water in sufficient quantities. The accuracy of geophysical estimates depends on how closely the physical properties can be separated from one another. Formations that often do not lend themselves to surface geophysical methods include areas consisting of large cobbles in alluvium, areas of severely distressed stratigraphy (areas where there is upheaval of the rock strata), or areas in which (in geologic time) high hydraulic energy was dissipated.

Electrical Resistivity

Electrical resistivity is probably the most commonly used surface geophysical method for groundwater investigations. To gather data, electrodes placed into the surface of the ground transmit current through the earth, and the voltage potential is measured between two points near the center of the generated field (Figure 3-3). With the most common electrode arrangements, such as the Schlumberger array and Wenner array, readings can be gathered using constant

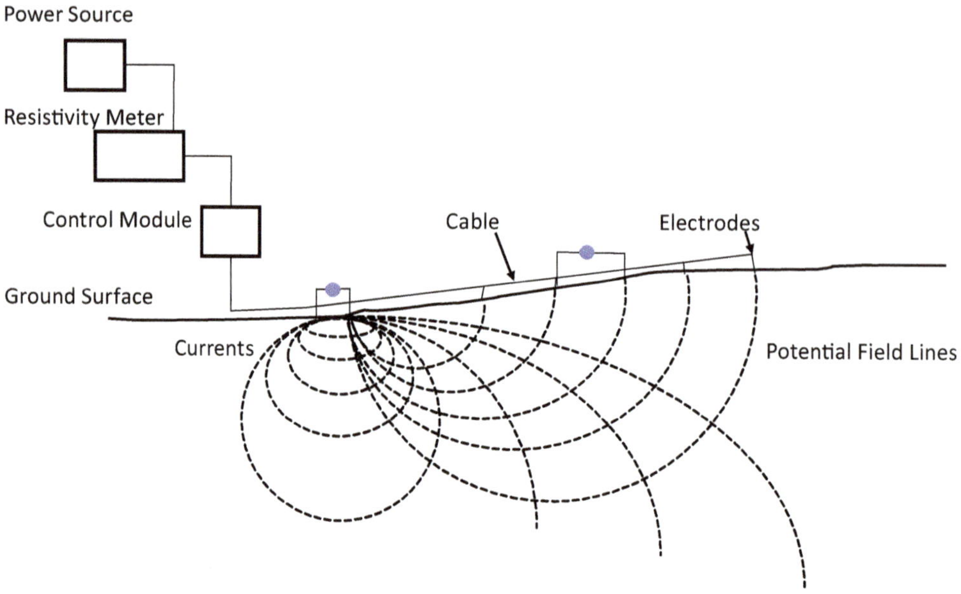

Figure 3-3 Surface resistivity method

electrode spacing (horizontal profiling), or the readings can be gathered at one location with expanding electrode spacing (electrical sounding). The first method will show apparent resistivities of materials at roughly the same depth along the transect (two-dimensional), while the second method produces a depth profile of resistivity (three-dimensional; AWWA 2003).

Electrical resistivity is strongly affected by water content. As a result, data collected involving the unsaturated zone can make interpretation of the saturated zone quite difficult. Resistivity is largely determined by the rock-matrix density and porosity or by the saturating-fluid salinity (electrical conductivity). Coarse sediments with low clay content will generally have higher resistivity than fine-grained sediments. Old streambeds are often laden with silt and clay; as a result, the variations in response allow for surface detection and mapping of buried stream channels or the depth profiling of shale–sandstone sequences.

Surface resistance methods are generally limited to use in:

- simple geologic environments, with two or three distinct layers;
- areas where depth of penetration is limited to about 1,500 ft (460 m);
- areas where the depth to groundwater is small, because of the complications of unsaturated materials; and
- nonurban or undeveloped areas as a result of the presence of buried metal pipes, wires, and similar obstructions, which dominate measurements with unwanted noise in developed areas.

Seismic Refraction and Reflection

The cost of seismic methods is high, but the information generated can be very useful to hydrogeologists. Seismic methods use the contrasts in the velocities of elastic wave propagation in different materials.

A variety of methods are used to generate seismic waves. One is the use of explosive *shots* in shallow borings. Truck-mounted hydraulic earth vibrators (*thumpers*) can also be used, although these are generally used for oil exploration and raise significant environmental concerns because of the damage they can inflict on the surface. Both are extensively used by energy companies trying to find oil and gas deposits.

To capture the reflected noise signal, geophones are distributed on the ground surface to detect and record the travel times for sound waves refracted or reflected from subsurface lithologic boundaries. The travel time records are analyzed to produce a model of the subsurface. As with all surface geophysical methods, the interpretation of seismic data requires an assumed model of subsurface structure; the more preexisting information from surface geologic data and borehole logs that is available, the more reliable the results will be from seismic surveying. Unconsolidated sands and gravels exhibit low propagation velocities, whereas crystalline rocks exhibit the highest propagation velocities. Propagation velocities are higher in saturated materials, providing detection of the water-bearing strata.

SUBSURFACE GEOPHYSICAL METHODS

Borehole logging techniques are one of many tools that should be used in evaluation and identification of underground environments. Geophysical logs that are commonly used in water exploration include:

- caliper,
- resistivity,
- spontaneous potential (SP),
- naturally occurring gamma radiation,
- neutron porosity,
- acoustic,
- fluid resistivity, and
- temperature.

Determining the types of logs to be used in an investigation is often difficult. Most groundwater investigations obtain adequate information using caliper, resistivity, SP, natural gamma, and lithologic logging. The cost of these techniques should be evaluated regarding the available time, accuracy needed, and the basic purpose of the survey. Resistivity, gamma, and caliper measurements are the most widely used.

All geophysical log measurements are obtained by lowering a probe down the borehole and recording continuous measurements with depth. Logging is normally performed during drilling operations and is often conducted in a small-diameter pilot hole. The results of analyses of the logs by qualified personnel provide the basis for decisions regarding well construction and completion, including depth of casing and screened intervals. Some types of logs can be conducted in existing, cased wells, which are useful if expansion of a wellfield is planned. Certain logs can only be performed in an uncased hole, which prevents the use of drilling methods requiring casing to be advanced during drilling if these logs are desired. It should be noted that the method of drilling may dictate the appropriate type of geophysical log employed. For example, temperature and fluid resistivity logs are not very useful in mudded boreholes.

Caliper Logs

Caliper logs are among the simplest logs and can reveal useful information for interpretation of other logs. A multiarmed probe is lowered into the well to determine the geometry of the borehole and variances in diameter. Caliper logs can provide indications of the presence of high permeability fractured or cavernous zones (by the arms extending outward indicating a larger and rougher borehole), as well as the occurrence of swelling clays (the arms will tighten and indicate that the borehole is smaller than the diameter of the bit used to drill as a result of the expanding clays) and locally well-lithified layers in friable or unconsolidated rock or sediment (the changes in the diameter of the borehole may be significant in a very narrow horizon; see Figure 3-4). Caliper logs are important because the interpretation of other geophysical logs varies with the borehole diameter. A good quality caliper log will identify where compensation must be made.

Caliper logging is conducted using a probe that usually has either three or four levered arms. As the probe is brought up through the hole, a record of the depth and degree of extension of the arms is made. The results provide information about the variations in the diameter of the borehole after drilling. Asymmetry of the borehole can be measured using the four-armed probe but not with the three-arm probe. The asymmetry may indicate cavities, fractures, or other parameters useful in water-bearing strata. They can also suggest areas that might experience a loss of drilling mud for rotary type drilling operations.

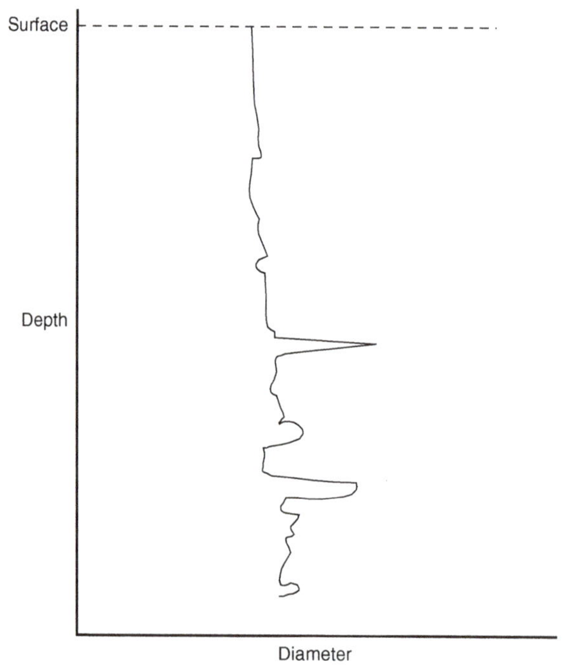

Figure 3-4 Example of caliper log

Therefore, caliper logs are not useful for estimating diameter inside a casing as the calipers only measure the diameter of the drilled hole. They can, however, be used to determine unknown well diameters and to determine the integrity of casings (i.e., collapsed casing). It is recommended, whenever possible, to run a caliper log before setting the casing as it will provide an indication of the annular space volume that must be filled with grout and provide an estimate of the cement requirements.

Electrical Resistivity Logs

Subsurface electrical resistivity logging is only slightly different than surface resistance logs. Resistivity logs are based on the principles of Ohm's law:

$$R = E/I \qquad \text{(Eq 3-1)}$$

Where:

R = resistance in ohms
E = potential in volts
I = current in amps

The simplest and least-expensive electric resistivity log is the single-point resistance log. The single-point resistivity log measures the potential drop between a surface electrode and a downhole electrode, which are also the current electrodes (Figure 3-5). The single-point resistivity log is used primarily for

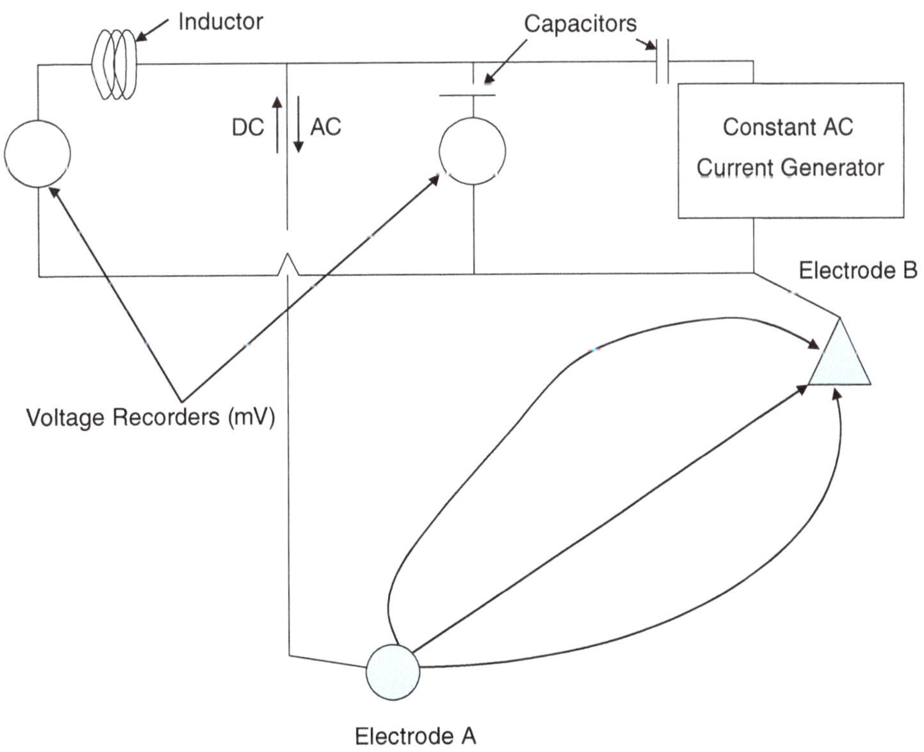

Figure 3-5 System used to make conventional single-point resistance and SP logs

geological correlation and the location of bed boundaries, changes in lithology (rock characteristics), and fracture zones. Single-point resistivity logs have a very good vertical resolution of lithologic changes but do not provide quantitative data on formation porosity or the salinity of formation water (AWWA 2014). High conductivity in the formation is proportional to the potential voltage differential created in the formation. Silt, clay, and shale tend to have the lowest resistivities; sands, granite, basalt, sandstones, and limestones with nonsaline pore waters have the highest resistivities. For single-point resistivity logs, interpretation is required to determine the meaning of the results. Single-point resistivity logs will change with variations in the borehole diameter requiring a caliper log to compensate for these differences. Table 3-1 outlines the resistivity of common formation materials.

Normal resistivity logs function on the following variant of Ohm's law:

$$R = r \times S/L \quad \text{(Eq. 3-2)}$$

Where:

R = resistivity
r = resistance
S = cross-sectional area normal to the flow
L = distance in the formation

Normal resistivity logs measure the apparent resistivity of a volume of the formation perpendicular to the borehole electrodes. The probes are commonly configured so that short normal (16-in. electrode spacing) and long normal (64-in. spacing) resistivities are measured simultaneously. Normal resistivity logs are commonly used in groundwater investigations as a source of qualitative information

Table 3-1 Electrical resistivity of selected aquifer materials

Material	Resistivity (ohms - m)
Granite	5,000–1,000,000
Basalt	1,000–1,000,000
Sandstone	100–4,000
Shale	20–2,000
Porous limestone	100–10,000
Dense limestone	1,000–1,000,000
Clay	1–20
Wet sand	20–200
Dry sand	500–100,000
Fresh water	10–100
Sea water	0.1

Source: Reprinted with permission of National Groundwater Association Press, copyright 1989.

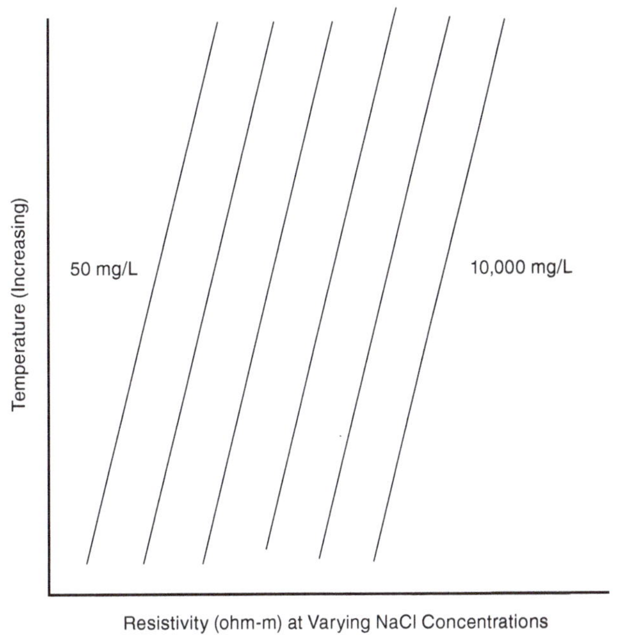

Figure 3-6 Example of temperature corrections required to electrical conductivity log as a result of increasing resistivity because of salt concentration in the aquifer water

on water quality. True formation resistivity and salinity can be calculated from the measured apparent resistivities, but the calculations require the application of several mathematical equations to reach the actual salinity and resistivity (AWWA 2003). Normal resistivity logs also require adjustments for temperature (Figure 3-6). Other resistivity log types that are less commonly used in groundwater investigations are discussed by Keys (1989).

Spontaneous Potential (SP) Logging

SP logs measure the natural electrical potential of the lithology in millivolts. In practice, SP logs will show increasing salinity by increasingly negative responses (Figure 3-7). The results are similar to a gamma log. Measurable differences in SPs happen where beds of different types of geological materials occur, such as between shale and sandstone beds. Negligible response may occur with certain localized sediments. SP logs provide information on bed-thickness determination and changes in lithology.

The SP logging apparatus consists of a surface and downhole electrode connected to a voltmeter. The SP logging equipment is usually incorporated into the electric resistivity log apparatus. Log definition depends on the contrast in fluid conductivity between the borehole and the geologic formation penetrated. While SP logs are the most common logs used, if the borehole contains water that is fresher than the formation, the SP logs will not yield useful results, so this log should be used in conjunction with other logs noted in this chapter.

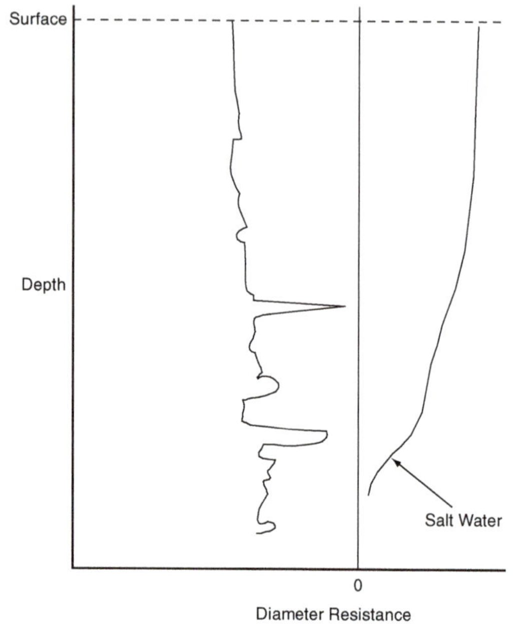

Figure 3-7 Example of caliper and electrical resistivity log for borehole

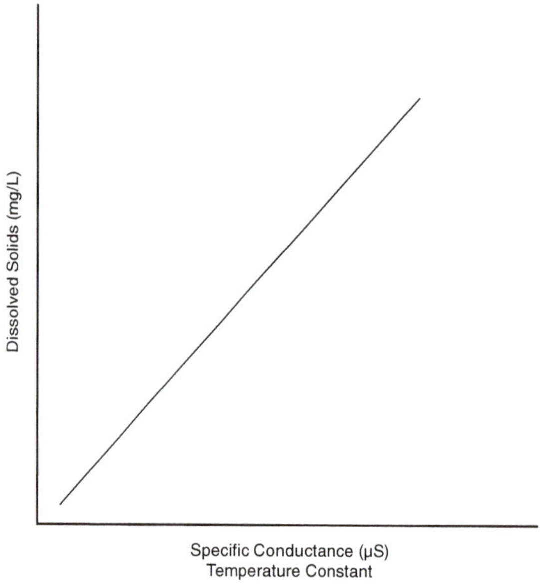

Figure 3-8 Conductivity relationship to dissolved solids for specific conductivity logs

Electrical Conductivity

Electrical conductivity (EC) logs are based on a relationship between the specific conductance and the concentration of dissolved solids in the water. In most cases, calibrated logs will show that there is a relatively linear relationship between the two—more dissolved solids means higher conductivity (Figure 3-8). The conductivity of specific ions is known, and a water quality analysis will permit investigators to determine if the logging results match the water quality and where there may be other constituents of concern.

Gamma Logs

Gamma logs measure the total gamma radiation that is naturally released from the formation. No radioactive materials are introduced into the well. Many naturally occurring elements have radioactive isotopes that are harmless to humans but measurable by gamma logging tools. The most significant natural source of gamma radiation is the decay of potassium-40 isotope and the daughter products of the uranium and thorium decay series. Rocks and sediments with relatively high concentrations of potassium, uranium, and thorium have high gamma response. Gamma logs measure the ratios of three basic isotopes to one another: uranium-238, thorium-232, and potassium-40. Table 3-2 shows the energy for each. Gamma results increase linearly with silt and clay content (Figure 3-9).

Table 3-2 Energy for gamma logs

Element	Energy of Major Gamma Peaks (million electronvolts)	Photons per Second per Gram	Average Percent in Shale	Percent of Total Gamma Intensity
Potassium-40	1.46	3.4	2%	19
Uranium-238	1.76	280,000	6 ppm	47
Thorium-232	2.62	10,000	12 ppm	34

Reprinted with permission of National Groundwater Association Press, copyright 1989.

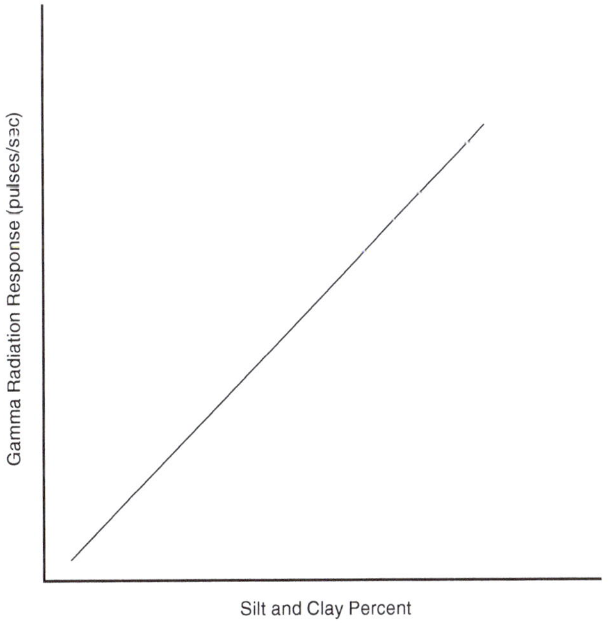

Figure 3-9 Gamma response to increasing clay and silt content

Clay-rich rocks, shales, and phosphatic rocks yield a high gamma response, whereas nonphosphatic limestones and dolomites, and quartz sandstones tend to yield low gamma responses (AWWA 2003). The gamma log is the most commonly used nuclear log. Gamma logs can be correlated with rock types based on the radioactivity of the formation, cuttings, and cores. The amplitude of gamma logs can be modified by changes in the density of the material through which the gamma rays pass. As a result, the gamma log measures porosity as proportional to bulk density. The following will inhibit the usefulness of gamma logs:

- large diameters wells,
- wells constructed with cable-tool methods, and
- wells already completed with casing and cement.

Gamma Radiation Response (pulses/sec)

Figure 3-10 shows an example of a gamma log in conjunction with a caliper log. Limestone, alluvium, coal, gypsum, and anhydrite are all materials where the gamma log shifts to the left, while granite and other hard rocks shift to the right. The latter are not rocks likely to have much water useful for production. Figure 3-11 shows that as the silt and clay content increases, like in shale, the gamma pulses also increase (shift right). SP logs used in conjunction with gamma logs will react similarly.

Gamma-gamma Logs

Gamma-gamma logs provide information on lithology and porosity. However, gamma-gamma logs should never be used for potable water supply wells because

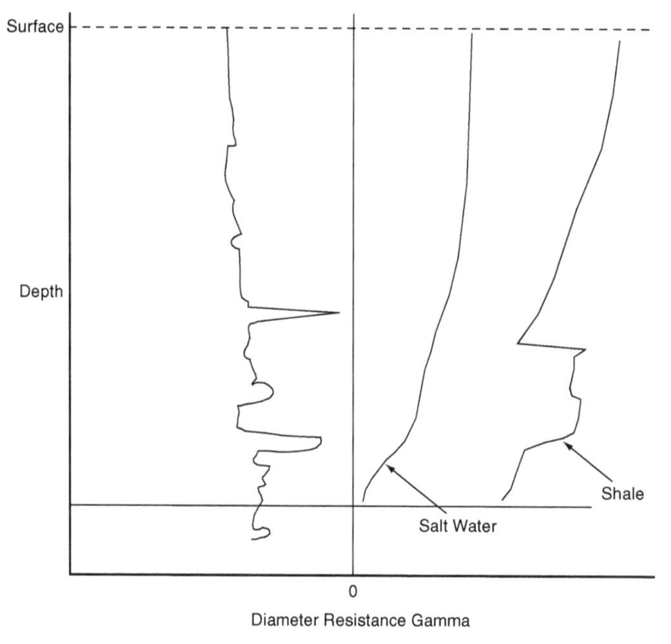

Figure 3-10 Example of comparison of gamma, resistivity, and caliper logs through shale

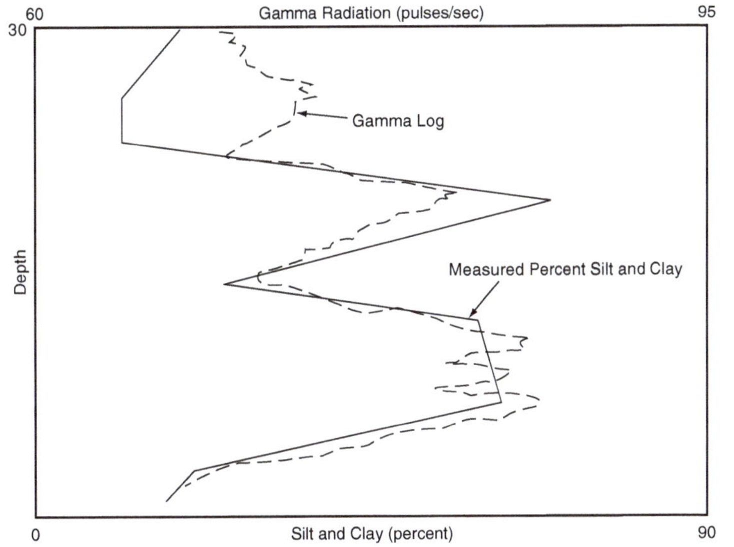

Source: Reprinted with permission of National Groundwater Association Press, copyright 1989

Figure 3-11 Relationship between gamma radiation and silt and sand

they are obtained by introducing a gamma radiation–emitting material in the borehole (usually cesium-137 or cobalt-60) and measuring the intensity of the backscattered radiation. The costs are high, and liabilities associated with the potential loss or rupturing of the radioactive source within the aquifer are very high.

Gamma Spectrometry Logs

Like gamma-gamma logs, gamma spectrometry logs should not be used for potable water supply well investigations because they also measure radioisotopes in the formation. Counting gamma pulses can be useful in identifying many additional, slowly degrading elements. By knowing their gamma energy characteristics, specific elements can be identified. By identifying the elements and the ratio of elements, clay and shale can be differentiated from limestones and other water-bearing formations. Figure 3-12 shows an example of how the calcium carbonate percent and the radium/potassium ratio can be used to identify specific formation types.

Neutron Logs

Neutron logs are used to find changes in porosity of the rock and sediments, which may be related to the amount of fluid in the formation. The concept is based on the emission of high energy neutrons from beryllium and an alpha-emitter (commonly americium). The emitted neutrons interact with hydrogen atoms and release gamma radiation, which is measured by a detector on the logging tool. The intensity of the measured gamma radiation is proportional to the hydrogen atom concentration and thus water content and the porosity of saturated rocks. Table 3-3 shows the neutron response for several common elements with an initial energy of 2 million electron volts. Calibration of neutron logs is critical to obtaining useful information.

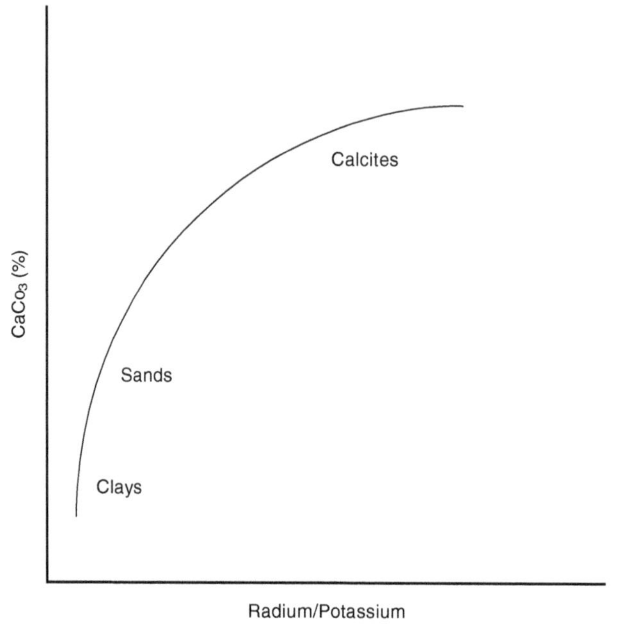

Figure 3-12 Comparison of radium/potassium gamma response to calcites

Table 3-3 Neutron collisions for selected materials

Element	Average Collisions per Neutron (million electron volts)	Maximum Energy Loss	Atomic Number	Atomic Weight
Calcium	371	8%	20	40.1
Hydrogen	18	100%	1	1.0
Oxygen	150	21%	8	16.0
Carbon	115	28%	6	12.0
Chloride	318	10%	17	35.3
Silicon	261	12%	14	28.1

Neutron logs work very well to indicate lithology by identifying suspected formations of sand, limestone, and sandstone that have pores and vugs that would indicate high porosity. These pores are usually filled with water, which contains hydrogen, a strong neutron absorber. However, coal, shale, and other formations with high hydrogen content will provide false readings despite yielding results that would indicate high porosity. The same problems that affect gamma-gamma logs also affect neutron logs. EC logs help to show where these formations provide false readings because the hydrogen effect is not a driving factor. As a result, the EC logs will deflect in the opposite direction from the neutron log. Shale or clay will be evident when a gamma log is used in conjunction with neutron logs because the bound water will reflect high water content, which the gamma log will not. Neutron logs are suggested to help determine how much of the conductivity is the result of salinity from salt (NaCl).

Table 3-4 Compression wave velocity transit time

Material	Velocity (feet per second)	Transit Time (μsec/feet)
Sandstone		
Slightly consolidated	15,000–17,000	58.8–66.7
Consolidated	19,000	52.6
Shale	6,000–16,000	62.5–167
Limestone	19,000–210,000+	47.6–52.6
Dolomite	21,000–24,000	42.6–47.6
Anhydrate	20,000	50
Granite	19,000–20,000	50–52.5
Gabbro	23,600	42.4
Fresh water	5,000	200
Sea water	5,300	189

Adapted from Keys, 1989

Acoustic Logging

Acoustic or sonic logging involves the recording of the transit time of acoustic pulses radiated from a tool in a borehole to one or more receivers also located on the tool. Acoustical techniques can be used to send sound through a formation to determine changes in lithology and types of formations. Table 3-4 shows an example from Keys (1989). Table 3-4 and Figure 3-13 show that acoustical logging cannot be used alone as granite and limestone have similar velocities for the sound waves (because sedimentary rocks, cement, and hardness have similar responses), while the water-bearing characteristics are vastly different. Transit times for the acoustic waves are related to formation mineralogy and the porosity of the rock. Most rock types have a limited range of acoustic travel times, which allows for acoustic logs to be used to determine lithology. The cement-bond log is a type of acoustic log that is used to determine how well a casing has been cemented to the formation.

Acoustical logging is useful for consolidated formations and provides useful information in uncased, fluid-filled boreholes. Figure 3-14 shows an example of an acoustic log in conjunction with a caliper log. Saltwater will shift the response slightly to the right. Limestone and granite provide similar responses so other logs must be used to confirm which is present. Limestone will generally yield water, while granite generally will not.

Fluid Logs

Fluid logs include temperature, fluid resistance, and flowmeter logs. The tool used for temperature logs usually contains a glass bead thermistor. Temperature logs can be used to identify the boundaries of aquifer zones in boreholes; as water flows through permeable zones, the normal geothermal gradient will vary. Temperature logs are developed using a downhole run. Temperature logs are useful in conjunction with EC logs as temperature changes often occur where there is

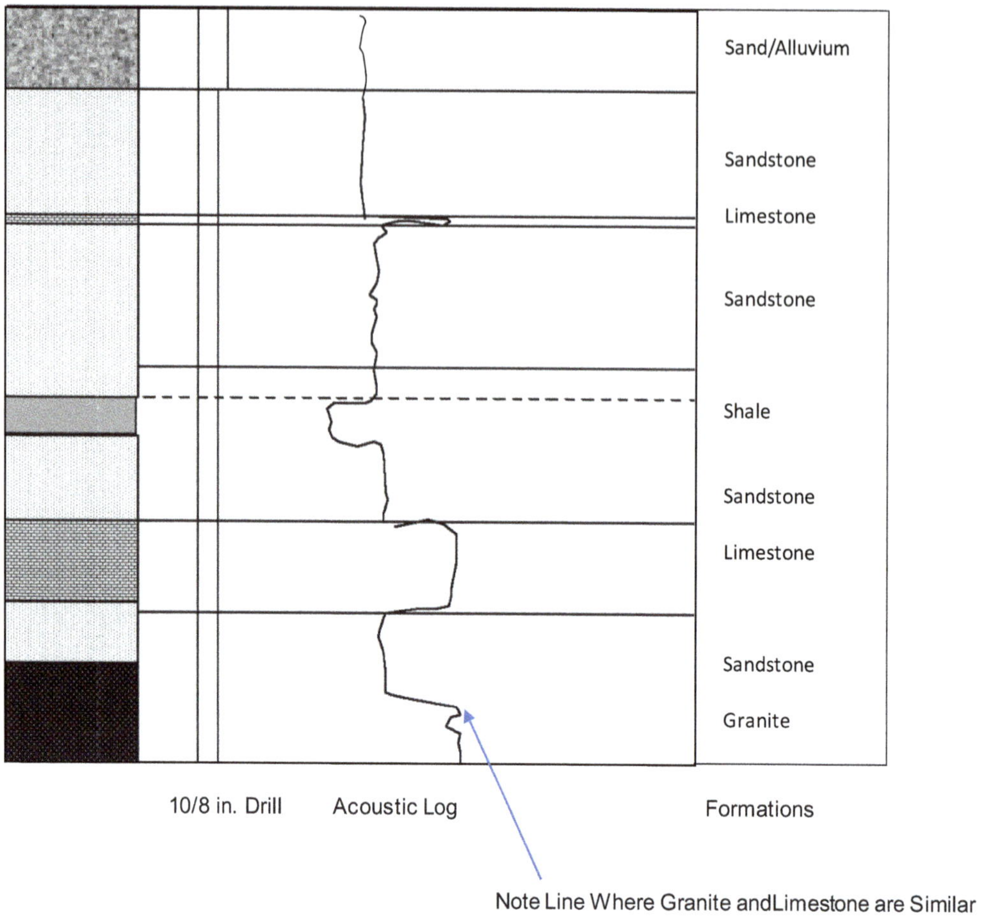

Note: The granite and limestone give similar results.

Figure 3-13 Response of sediments to acoustic waves

moving water. Changes in conductivity in conjunction with changes in temperature logs would tend to indicate flowing water across the borehole. Temperature logs can also be used to detect inter-aquifer flow. The presence of cement grout in the annular space of a well can be determined by running a temperature log within 24 hours of grouting because the heat of hydration of the cement raises the fluid temperature inside the casing in cemented areas.

Flowmeter logs are used to measure flow velocity within the wells. The most common flowmeter logging tool is the impeller-type, where the rate of rotation of the impeller is proportional to the relative flow velocity of the tool. The relative flow velocity includes the actual flow velocity of water in the well and the rate at which the probe is raised or lowered into the well. Caliper logs must be run in conjunction with the flowmeter log because the flow velocity is a function of the cross-sectional area of the borehole. The relative contribution of individual aquifer zones to the total flow from a well can be calculated using data from flowmeter and caliper logs (AWWA 2014).

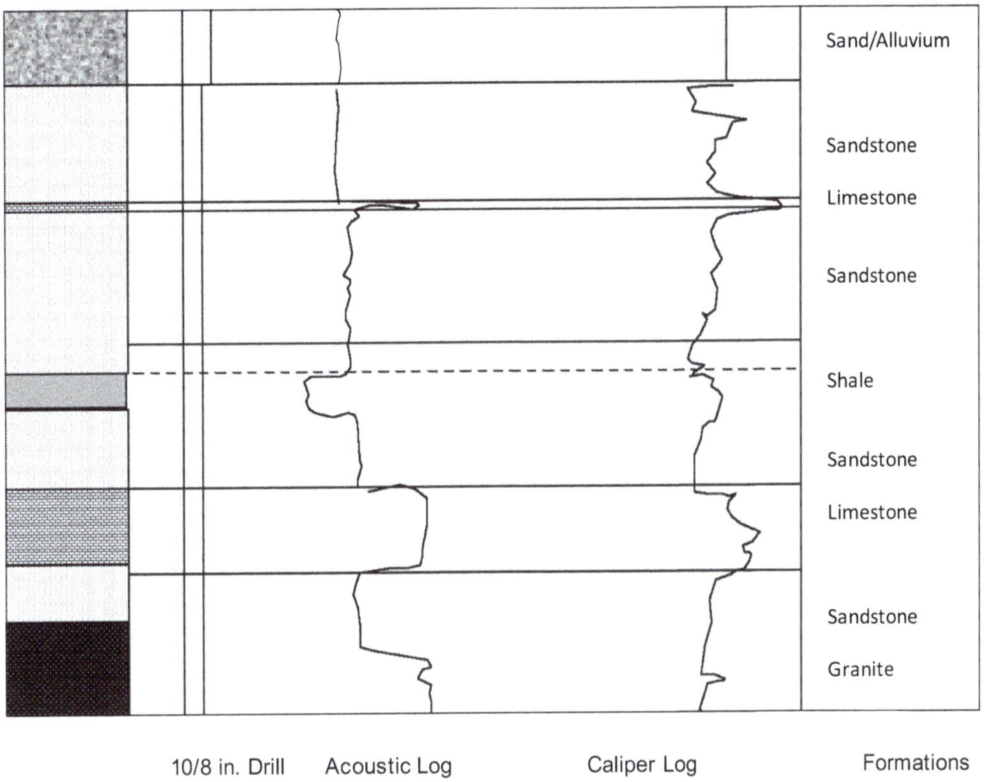

Figure 3-14 Acoustic velocity and caliper log example

Log Suites

It is necessary to use more than one log because one log cannot adequately describe the lithology and hydrogeology of the well. Individual geophysical logs do not provide unequivocal lithologic information. A high gamma response could be produced by a shale bed or phosphatic limestone layer. By running a suite of logs, more accurate qualitative and quantitative information of formation porosity, hydraulic characteristics, and fluid conductivity can be measured. Figure 3-15 shows a series of logs for a borehole (Bloetscher et al. 2005). In a log suite, occurrences such as shale layers are identifiable by a low resistivity and high gamma log response, where those logs singly would provide no specific information indicating that the layer was shale.

Further details of the logs are summarized in several reference books, most notably Keys (1989). In determining which logs and tests should be required, the following geologic logs should be considered for surface casing intended to protect underground sources of drinking water:

- resistivity (long and short, normal and single-point); spontaneous potential; gamma; and caliper logs before the casing is installed; and

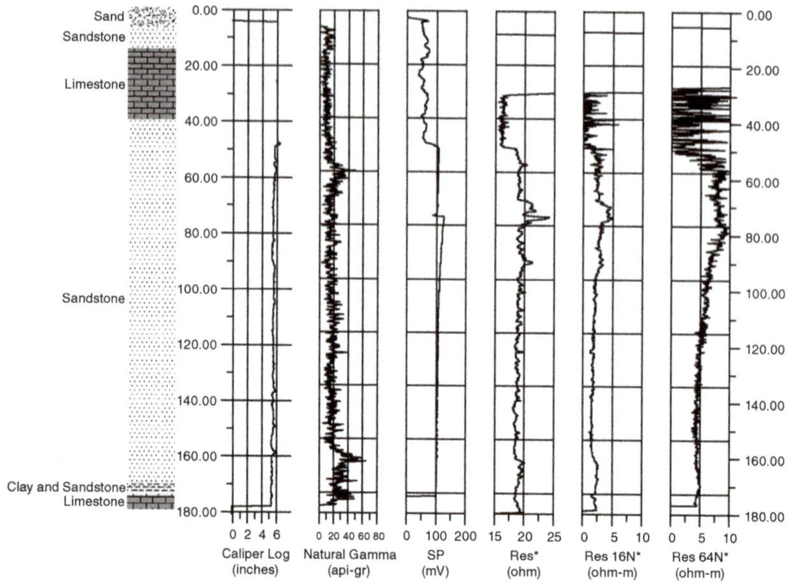

Source: Bloetscher et al. 2005

Figure 3-15 Typical suite of logs

- a cement bond, temperature, and density log after the casing is set and the annular space cemented, depending on casing material. If the hole is drilled without the use of drilling muds or other additives, the fluid resistivity flowmeter and temperature log should be performed under static and pumping conditions.

No radioactive source should be placed downhole in the underground source of drinking water or where water for potable use or human consumption is anticipated.

Table 3-5 outlines the application of types of geophysical methods that should be used based on the formation type and investigation desired. Table 3-6 summarizes the response of four logs to porosity, an important property when looking for water supply sources. Table 3-7 outlines a set of criteria suggested by Keys (1989) for selecting logs to be run on a given project. The same suite of logs may not be appropriate or cost effective in every instance. Instead, careful consideration should be given to selecting the logs that will provide the most useful information for making decisions based on the anticipated geological environment where the wells are proposed to be constructed. In all cases, a caliper log should be run. Other information that may be gathered includes a radial and side-view color television survey of the borehole under flowing and no-flow conditions to help identify flow zones and sanding problems (Figure 3-16). Other geophysical logs should be considered depending on the formation, availability of tools, cost of the project, and perceived hazards.

Table 3-5 Recommended geophysical logs

Material	Seismic Refraction	Electrical Resistivity	Penetrating Radar	Cross-hole Seismic	Electro-magnetics	Magnetics	Gravity
Subsurface geology	×						
Rock density	×						
Groundwater location		×	×		×		
Karst formation		×		×	×	×	×
Groundwater pollution		×			×		
Fracture zones		×					
Metallic objects buried			×		×	×	
Nonmetallic objects buried			×				
Seismic risk				×			

Source: Reprinted with permission of National Groundwater Association Press, copyright 1989

Table 3-6 Response of logs to porosity

Log	Property Measured	Response to Total Porosity	Response to Effective Porosity	Response to Secondary Porosity	Spurious Matrix and Fluid Responses
Resistivity	Resistivity and volume or fluid connected pores	No current flow in isolated pores	Response only to effective porosity	Detects secondary porosity, shape of pores effects results	Boundary effects
Gamma-gamma	Electron density	Best response with highly porous rocks	Does not distinguish	Does not distinguish from primary porosity	Matrix composition, high salinity = error
Neutron	Hydrogen content	Best response with minimal porosity	Does not distinguish	Does not distinguish from primary porosity	Bound water and other neutron absorbers. Pores must be saturated. High salinity.
Acoustic	Average compression wave speed	Relates only to total porosity when primary and intergranular	Does not distinguish	Does not distinguish from secondary porosity under most conditions	No signal in gas

Reprinted with permission of National Groundwater Association Press, copyright 1989.

Table 3-7 Criteria for selection of logs

Type of Log	Property Measured	Application	Borehole Condition	Limitations
Spontaneous potential	Electric potential caused by salinity	Lithology, water quality, and water content	Uncased borehole with water	Need salinity differences
Single-point resistance	Resistance of rock, water, and filled voids	High resolution lithology and fractures	Uncased borehole with water	Not quantitative, borehole diameter impacts results
Multi-electrode	Resistance of rock and fluids	Quantitative date on salinity and lithology	Uncased borehole with water	Normal logs do not measure thin bed thickness correctly
Gamma	Natural gamma radiation in rock	Lithology, use clay and silt content for permeability	Any borehole less than 24 in. diameter	None
Gamma-gamma	Electron density	Bulk density, porosity, moisture content, and lithology	Uncased borehole	Severe borehole diameter effects
Neutron	Hydrogen content	Saturated porosity, moisture content and lithology	Uncased borehole	Borehole diameter and chemical effects
Acoustic	Acoustic reflectivity of borehole wall	Orientation of fracture, dip of bed	3 to 16 in. fluid filled borehole	Does not detect secondary porosity
Caliper	Diameter of borehole	Lithology and voids	All conditions	None
Temperature	Temperature of water	Flow and temperature gradient	Fluid-filled borehole	Accuracy varies with probe
Conductivity	Fluid resistivity	Water quality, contamination location	Fluid-filled borehole	Accuracy varies with probe
Flow	Fluid movement	Flow in borehole	Fluid-filled borehole	Accuracy varies with probe

Reprinted with permission of National Groundwater Association Press, copyright 1989.

Source: Photo by Youngquist Bros. Inc.

Figure 3-16 Example of borehole photograph

FIELD TESTING

Aquifer testing is very useful during construction of test holes, and many effective methods are now available for performing such testing (AWWA 2014). In the simplest form, a record of the water flow rates produced at different depths while drilling using the reverse-air method is a form of aquifer testing, yielding valuable information. Packer tests are used to determine transmissivity and storage coefficients for isolating water-producing formations. They can be used on both tight and permeable formations for comparative purposes in identifying confining beds. The diameter of the cone of depression and the drawdown varies with the size of the well, pumping rate, and the flow rate of water through the aquifer. In porous sand and gravel, the cone of depression may be small. In tighter materials, the cone of depression may be very significant.

Wells are normally placed far enough apart so that their cones of depression do not overlap significantly. Wells should be sized to minimize drawdown so that the aquifer can rebound quickly. Permanent drops in water levels can occur when the aquifer is pumped too much, a phenomenon called *mining of the aquifer*. The Black Creek aquifer in eastern North Carolina discussed in Chapter 1 is such an example. When recharge to the aquifer is inadequate, a new water source will need to be located.

Ideally, wells should be pumped continuously without permanent drawdowns occurring. Where this is not possible, it is common practice to pump wells that have a significant drawdown for only a few hours each day to allow the aquifer to recover. An artesian well may be under enough pressure to cause the water to rise above the confining unit, which may eliminate the need for pumping initially. However, after an artesian well has been used for a period of time, the artesian pressure may reduce until the water no longer flows to the surface.

It is more important to determine the appropriate casing depth than to locate the pump depth. An improper casing depth may seal off highly productive zones. Determining the casing depth depends on the results of the design considerations, geophysical logs, and data collected during drilling. The data required to determine casing depth includes:

- lithology,
- core data,
- borehole geophysics,
- aquifer performance testing,
- water quality,
- drilling rate of penetration relative to weight on bit,
- cuttings, and
- lost circulation zones.

The first three were previously discussed. Determining transmissivity or the storage coefficient of an aquifer by any means other than actual performance tests in the field is expensive, time consuming, and of questionable accuracy. Field testing methods for determining these values have been developed and are well documented. All these methods apply a regulated stress (pumping) to the formation and measure the effects (changes in water level) produced. These data are analyzed, and the transmissivity and storage coefficient are calculated. They work best on surficial aquifers, which can react more quickly than confined aquifers; the latter theoretically must restabilize across the entire aquifer surface, which will take more time than the recovery period.

To obtain the required data, one or more monitor wells tapping the aquifer serve as observation points in the area of investigation. The location of all wells must be accurately plotted on the area map so that the lateral distance and direction from the pumping well and the relative position with respect to other wells can be included in the analysis. No set number of wells is required, but having more wells reduces the likelihood of making an error. For best results, the test well should fully penetrate the formation and be open only to the aquifer to be used as a source while the testing the well.

Chemical composition may be an issue—for instance, formations with high quantities of pyrite are often sources of leaching arsenic. Cuttings will indicate clay, voids, sand (that can leach into the well when pumped), and other useful information for clarifying geophysical results. The more data that can be gathered from different sources, the more efficient the investigation. Caliper logs will help identify zones where significant voids may exist. They often are flow zones as well.

Water Level Measurements

The first step in determining the transmissivity and specific capacity of the aquifer is to create a method to determine water levels before pumping. A surveyor's benchmark should be used to determine the elevations of the wells using a level and survey rod. By accurately measuring water levels with respect to surface elevations, groundwater gradients before commencement of pumping can be determined. Before a pump test is run, static water levels should be taken for several days to determine if there are any ongoing changes in the aquifer that should be considered. Once the static, prepumping water levels are determined, these water levels serve as a reference point that can be used for the collection of water level data during an aquifer performance test. During the aquifer test, water levels must be recorded for each well. Data sheets should be used to record the water level changes during the pump test that include the date, time, depth to water, pumping rate, and casing elevations. All data should be gathered and correlated to specific wells and formations.

Water levels can be measured using a hand-held tape with a weight attached to the end to hold it straight and taut. The tape should be metal and graduated in feet and in tenths and hundredths of a foot or a similar metric unit. Such graduations facilitate calculations by eliminating conversion of fractions to decimal equivalents. The old school solution is chalking the lower portion of the tape and lowering it into the water until an even foot graduation coincides exactly with the reference point, the precise distance to water from the reference is made by subtraction. The wetted chalk is easily identified, and direct readings to one hundredth of a foot can be made (AWWA 2014).

Other methods of collecting water-level data include an electric tape that has an insulated wire with an open-end weighted electrode on the end. When the electrode enters the water, it completes a circuit that actuates a light, buzzer, meter, or other signal device. The distance to water is then read directly from graduations on the wire line. However, the graduations are not usually fine enough to permit a very accurate reading without some supplementary device. Float-actuated recording devices provide a means of collecting data continuously, but the time drive is not fast enough for the early periods of a test program. Pressure transducers combined with a data logger can provide an excellent record of water levels. Air-line devices have little value for controlled tests, except where water-level fluctuations are very large.

Collection Schedule

While the aquifer test is run (pumped), water-level readings should be obtained using the following collection schedule:

1 reading at zero time	total elapsed time = 0 min; 1 reading each 1 min for 10 min
	total elapsed time = 10 min; 1 reading each 2 min for 10 min
	total elapsed time = 20 min; 1 reading each 5 min for 20 min
	total elapsed time = 40 min
1 reading each 10 min for 60 min	total elapsed time = 100 min; 1 reading each 20 min for 80 min
	total elapsed time = 180 min

Such a standard schedule can be easily followed and provides adequate data (AWWA 2014). All times are calculated from the precise instant that the pump is turned on or off, which is designated as zero. If the test extends beyond 24 hours, subsequent measurements can be made at 4-hours intervals. The timing of measurements at the onset of the test is critical. Each well should have at least one observer equipped with measuring devices and a synchronized stopwatch. After 180 min, measurements do not have to be made at a designated instant, but they must be accurate regarding the exact time that each measurement is maintained.

The data collected provide information on the aquifer performance, not well performance. Each method involves turning a pumped well on or off and observing what happens to the water level in nearby observation wells. All methods use the Theis nonequilibrium formula or modifications thereto, which considers the time that has elapsed since pumping began or ceased.

Ideally, all wells used in the analysis should fully penetrate the aquifer. Some departures from this requirement can be tolerated, but the construction details of the partially penetrating wells are required, and modifications to the equations are required. Any wells in the area that are not involved in the test should be stabilized before an aquifer test and maintained at the same pumping rate for the duration of the test. During the aquifer test, well pumping should be at a steady, unvarying rate and carefully measured. The pumping rate and water-level data should be carefully computed and plotted. Each method uses the Theis formula to analyze variations in drawdown with time or variations in drawdown with distance from the pumped well.

A family of curves has been developed to facilitate aquifer evaluation under a variety of conditions. The equations used for aquifer parameters are:

$$T = 114.6\,QW(u)/s \qquad \text{(Eq 3-3)}$$

Where:

T = the transmissivity of the aquifer, in gpd/ft
Q = the discharge rate of the well, in gpm
s = the drawdown at any point under study in the vicinity of the discharging well, in ft
u = for any given formation, is proportional to the ratio of r^2/T
$W(u)$ = the "well function of u," is determined from calculated tables for each value of u given:

$$u = 1.87r^2\,S/Tt \qquad \text{(Eq 3-4)}$$

Where:

r = the distance from the discharging well to the point where the drawdown is being observed, in ft
S = the aquifer storage coefficient
T = the transmissivity of the aquifer
t = the elapsed time since discharge began, in days

The equations assume an aquifer that:

- is a confined or an artesian aquifer;
- has confinement that occurs above and below by relatively impermeable materials; and
- is homogeneous and isotropic-uniform in structure, with the same physical and hydraulic properties in all directions (which is rarely the case).

In practical terms, the thickness and actual extent of the aquifer should be known to permit the best possible interpretation of the test data. The formula is modified for leaky artesian conditions based on the conditions for confined aquifers and the following assumptions:

- The aquifer is confined between an impermeable bed and a bed through which leakage can occur.
- Leakage is vertical into the aquifer and proportional to the drawdown.

- No water is stored in the confining bed.
- The hydraulic head in the deposits supplying leakage remains constant.

An unconfined, or water table, aquifer does not have water confined or under pressure beneath impermeable rocks. Water is derived from storage by gravity drainage of the interstices above the cone of depression, by compaction of the aquifer, and by expansion of water in the aquifer. Properties of an unconfined aquifer can be determined by the Theis method with some limiting conditions. One of the basic assumptions of the Theis solution is that water is released from storage instantaneously with a decline in head. In a water table aquifer, this is not always true, because water is derived partly from gravity drainage, and the effects of gravity drainage are not considered in the Theis formula. However, with long pumping periods, the effects of gravity drainage become negligible so that the Theis solution can be used.

Drawdown Method of Calculation

When the drawdown method of calculation is used to determine transmissivity, one well is pumped while the water levels are observed in two or more nearby wells (AWWA 2014). Figure 3-17 is a hydrograph—a plot of water level versus time—for observation well 1 (only the left half of Figure 3-17 should be considered at this point). Water-level measurements are taken in conformance with the schedule outlined previously. Baseline testing should occur for a minimum of 48 h before the start of the test to identify any preexisting trends that would need to be considered during the test. If no upward or downward trend of water levels is found in the wells, the measurements are plotted as a horizontal line.

The drawdown represents the difference between the water level observed in the well and the level at which the water would have stood had no pumping occurred. Aquifer transmissivity and storage coefficients can be determined by

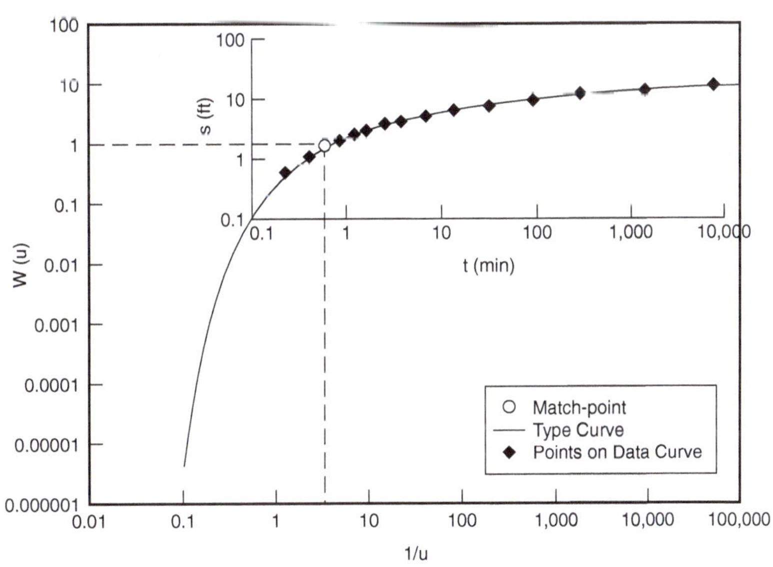

Figure 3-17 Typical curve for confined aquifer

comparing a logarithmic curve of time versus drawdown against one of a series of type curves developed from the Theis formula. The type curve is superposed over the field-data plot, keeping the respective graphical axes parallel. The curves are adjusted horizontally and vertically to obtain the best match of the two curves. An arbitrary match point is selected on the two graphs, and the field-curve and type-curve coordinates for substitution in the appropriate equation are selected. A different form of the type-curve solution is the distance–drawdown method. In this analysis, drawdowns in three or more observation wells at different distances from the pumped well are compared with one another to observe trends in the drawdown at each well in response to pumping.

An alternative solution is available for analyzing aquifer test data that is an approximate version of the type-curve solution. For this solution, well-test data is plotted on semilogarithmic paper, and variations of the basic formula are used to compute the aquifer transmissivity and storage coefficient. The drawdown data tend to follow a straight line when plotted on semi-log paper (Figure 3-18).

Recovery Method of Calculation

The recovery method is the same as the drawdown of calculation, only in reverse (AWWA 2014). The recovery method of analyzing aquifer test data involves shutting off a pumping well and observing the recovery of water levels in nearby observation wells. Recovery is the difference between the observed water level in the well at some time after pumping has stopped and the level at which the water would have been had pumping continued. The same curve options are available for this method of calculation as for the drawdown method, except that the concept has been inverted. The inverted curve indicates the rising levels in the observation wells. The recovery curve is compared with the inverted drawdown curve to determine the transmissivity and storage coefficient. The values should be similar

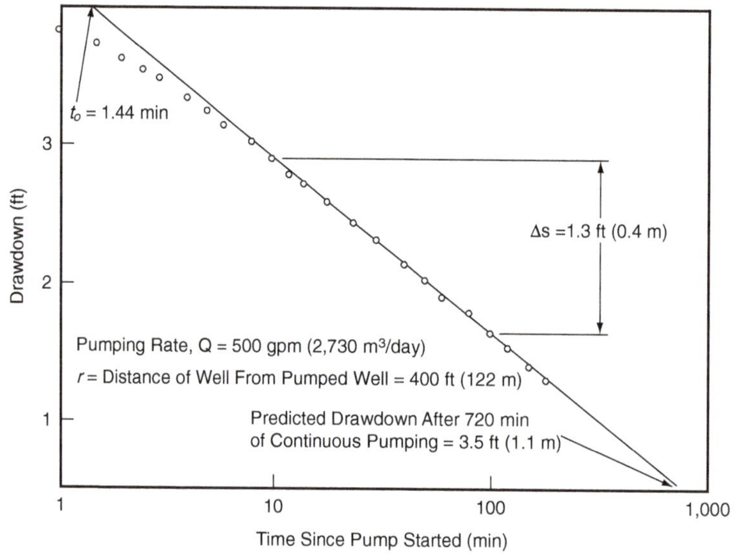

Source: Johnson Screens, Inc. 1986

Figure 3-18 Example of alternative solution for drawdown response

to those obtained using the drawdown calculation method. The same time periods for the recovering water levels should be observed as for the drawdown method.

A straight-line solution can also be used. As in the drawdown methods, the data curves are plotted on semi-log paper, showing a rising trend in the recovery period. With these modifications, the curves become straight lines. The same abbreviated equations are used to compute the transmissivity and storage coefficient. In practice, the recovery test is essentially the reverse of a drawdown test, therefore one drawdown curve and one straight-line plot will serve equally well for either kind of test data. Both kinds of data can be recorded on the same plot to check their agreement (AWWA 2014).

Use caution in analyzing recovery data. Water extracted from the aquifer may recover very slowly from drainage from the unsaturated zone above it. Therefore, when it is anticipated that the amount of time for full recovery is significantly greater than the duration of the pump test, a reduction in the pump test rate is recommended so that the test pumping rate is proportional to the ratio of pumping time to recovery time.

The reverse is also true. If a surface water source of infiltration or aquifer leakage is present, the recovery may occur more rapidly than the drawdown. In such cases, the pump test is unlikely to yield useful data. Further investigation is needed to determine the appropriate maximum water yield available.

Specific Capacity Method

An abbreviated well-performance evaluation can be performed using a relatively short test to determine the specific capacity of the well. The specific capacity method divides the total drawdown in a well into two components—drawdown in the aquifer and drawdown related to well loss. The drawdown in the aquifer depends on the aquifer's ability to transmit water. Drawdowns of this type do not usually change unless the aquifer is depleted. However, the drawdown caused by well loss may change considerably over time as a result of changes in flow regime (laminar to turbulent in the screen or column pipe) or changes in head loss as the water passes through the screen or well bore from corrosion, failure of the well screen or gravel pack, pump efficiency losses, water quality changes, and other factors. The potential for these changes to occur is the reason all wells should be tested annually for specific capacity to determine whether significant changes in well performance have occurred. Correction is recommended when capacity decreases by more than 50 percent. Tests for specific capacity should not be substituted for the more involved tests previously described when a more complete well and aquifer evaluation is necessary.

Identification of Aquifer Boundaries

No aquifer is infinite, although for the purposes of calculation aquifers that have extensive area may be assumed to be infinite. Most aquifers have, at some point, identifiable boundaries. Where the boundaries are close to the well, the drawdown test data will be plotted differently. While there are several potential boundary scenarios, there are two that are common: the impermeable barrier and the recharge barrier.

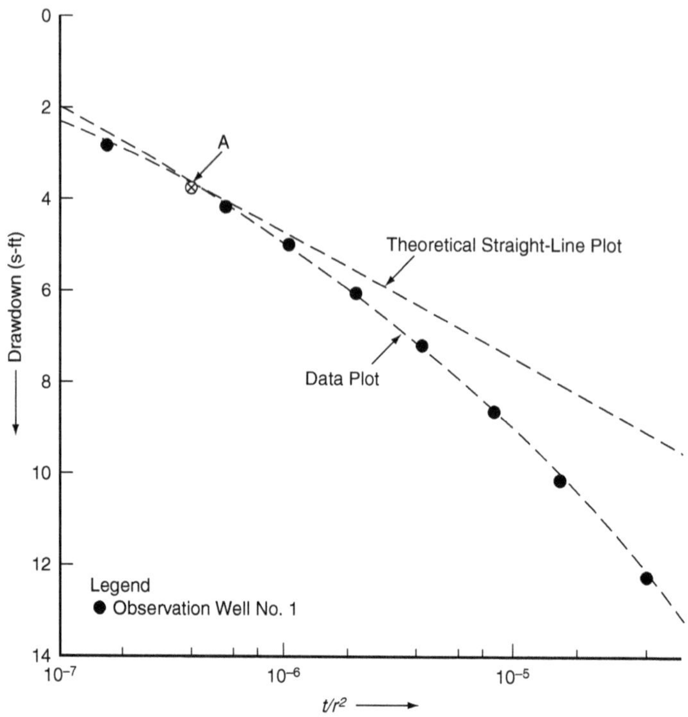

Source: AWWA 2014

Figure 3-19 Example of drawdown response at impermeable boundary

The effect of an impermeable barrier around an aquifer shows in the drawdown test when plotting in the manner previously outlined. The new graph, instead of staying on a straight line, now curves and eventually defines a new straight line having twice the slope of the original (Figure 3-19). This phenomenon is known as *reflection*. The aquifer is cut off by an impermeable barrier caused by the rising side of a buried valley or intruded, impermeable formation. This situation is quite common in the northern, once-glaciated parts of the United States. The effect is the same as having a second well located across the barrier at the same distance from the barrier as the actual well, pumping at the same rate. Data gathered and plotted in this manner will indicate the presence, location (distance), and type of aquifer boundary with respect to the pumped well.

A recharge zone creates the opposite phenomena of an impermeable boundary. Instead of being cut off by an aquifer barrier, the aquifer is cut off by a recharging stream—a situation that is often found in the field. The data plot begins as expected, with a curved portion leading into a straight line. However, instead of continuing on the straight line, as the data theoretically should for an infinite aquifer, the plotted data curves above it and eventually defines a horizontal line (Figure 3-20). The rate of drawdown caused by the withdrawal well lessens because of the water contributed to the aquifer by the stream and gradually approaches a fixed value. The effect is the same as if a well, identical to the pumping well, is introducing water to the aquifer instead of withdrawing it. The reflective well is located at an

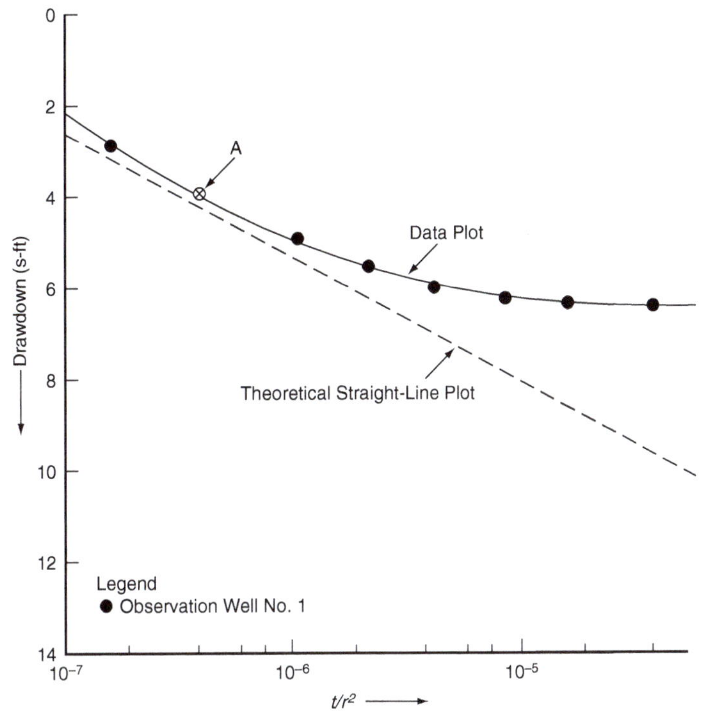

Source: AWWA 2014

Figure 3-20 Example of drawdown response at recharge boundary

equal distance from, and on the opposite side of, the recharge boundary from the withdrawal well.

Comparisons

To provide a baseline for testing the well design procedure, former AWWA groundwater Committee member and groundwater expert Carl Nuzman (1989) developed some rule-of-thumb ratios between transmissivity and well specific capacity (AWWA 2014):

$$\text{Confined Aquifer} \quad Q/s = T/2{,}200 \tag{Eq 3-5}$$

$$\text{Semiconfined Aquifer} \quad Q/s = T/1{,}700 \tag{Eq 3-6}$$

$$\text{Unconfined Aquifer} \quad Q/s = T/1{,}200 \tag{Eq 3-7}$$

These ratios were developed for a typical well radius of influence of 0.5 mi, and effective well diameter of 24 in., and assuming a storativity coefficient typical for the aquifer characteristics defined and the general assumptions of a theoretical aquifer (homogenous, isotropic, instant release from storage, infinite area extent, and no leakage or recharge). The equations are provided for purposes of comparison only and should not be used exclusive of good field testing. Once the geological data is secured, design information—size, materials, casing purpose—can be used to determine the appropriate depth of the casing and screening.

WATER QUALITY SAMPLING

Performance of groundwater monitoring and the development of periodic reports of such monitoring should be standard practice for all well systems. These analyses include the routine testing of the raw water quality of the proposed production zones (Bloetscher et al. 2005). Water quality sampling is an integral part of proper maintenance of wells. Water quality samples should be analyzed by a licensed, state-certified, or USEPA-certified laboratory. The sampling period should be determined by the stability of the water quality results in the well; how often, if ever, the parameters change and the amount of the change; and in accordance with federal, state, and local regulations. Changes in water quality are usually an indicator of contamination or water sources changes with wells and, if the changes are great, they may disrupt the water production process. In addition, certain changes in water quality can cause adverse environmental impacts. ASTM Standard D4195-08 requires water analyses to be performed on raw water designed for membranes processed for the parameters in Table 3-8. It is recommended that for all water production programs, analyses be conducted for the same parameters. Standard sampling methods as designated by USEPA of

Table 3-8 Summary of parameters to be analyzed in water quality testing

Parameters	
Aluminum (Al) (total and dissolved)	Oxygen (O_2)
Barium (Ba)	pH
Bicarbonate (HCO_3^-)	Phosphate (PO_4^-) (total)
Calcium (Ca)	Potassium (K)
Carbon dioxide (CO_2)	Silica as silica dioxide (SiO_2^-) (total and dissolved)
Carbonate (CO_3^-)	Silt density index (SDI)
Chloride (Cl^-)	Sodium (Na)
Fluoride (F^-)	Strontium (Sr)
Free chlorine (Cl_2)	Sulfate (SO_4^-)
Hydrogen sulfide (H_2S)	Temperature
Iron (Fe) (totaled, dissolved, and ferrous)	Total dissolved solids (TDS)
Magnesium (Mg)	Total organic carbon (TOC)
Manganese (Mn) (total and dissolved)	Turbidity (nephelometric method)
Nitrate (NO_3^-)	

Source: ASTM D-4195-03

state agencies should be used to collect and analyze the water samples. Temperature, pH, and silt density index (SDI) should be measured onsite at the time the sample is collected. Other chemical and physical parameters may require testing in certain regions based on specific groundwater quality of the area. Geochemical considerations may impact the long-term viability of the groundwater program in some aquifer systems. One concern is the potential for fresh and saltwater interaction. Also, fractures or other head or boundary conditions that lie outside the zone of the aquifer testing program may be issues. These interactions may include ion exchange, which can occur where significant clays may be present.

Redox processes may cause carbonate precipitation (or formation dissolution), suspended solids clogging, and biofouling. Redox processes are caused by oxygen consumption in the aquifer, typically an indication of either chemical oxidation or microbiological activity that is using the oxygen. Air binding, caused by either dissolved oxygen or entrained air, will lead to clogging problems. Excessive velocities may cause wearing and encrustation or excessive formation dissolution. Pressure differences between the formation and fluids may cause formation dissolution to occur or microfracturing. Temperature differences may also lead to dissolution problems.

SAND, SILT, AND COLLOIDS

Drilling methods and well development play a major role in identifying clogging potential. Proper selection of drilling equipment and performance of adequate pump tests, with optimal well development time, will indicate if sand, silt, or solids present an ongoing problem. The larger the pore size, the less potential there is for plugging, although filter theory indicates that particles 1/20 of the pore size may be effectively filtered out (i.e., may become part of the clogging matter). Inadequate well development is a frequent problem with new or refurbished wells (Bloetscher et al. 2005).

Sand, silt, and colloids are relevant to measuring potential impacts on the aquifer formation, including plugging and fouling problems. Plugging and fouling problems are caused by several hydrogeologic, geologic, engineering, and construction related factors, including:

- Hydrogeologic constraints that are not evaluated at the time of design and/or change over time, such as:
  - sand, clay, or rock layers that are unstable and collapse into the well boreholes;
  - naturally occurring and/or man-made fracturing/faulting;
  - long-term water quality changes caused by changes to the hydraulic regime, such as dams;
  - water hammer to the aquifer/formation;
  - man-induced influences (mining the aquifer, introduction of chemicals and/or microorganisms); and
  - naturally occurring phenomena (sinkholes, karst terrain features, and/or faulting);

- Poor well design and/or construction practices, including insufficient placement of grout, improper design of pumps, valves, and fittings; and excessive drawdown allowances;
- Poor operating and/or maintenance procedures;
- Mechanical failures, including failures of electrical motors and pumps, and failure of valves;
- Failure to develop the wells fully; or
- Interfingered sand or silt layers that have not or cannot be sealed off from the borehole or corrected in well design.

Most of these issues can be mitigated to some extent in the field. Determining the significance of resulting problems can be found by video survey of the wells; pump testing; and water quality testing for silt, sand, and colloidal material.

The purpose of sand testing is to determine the amount of sand being pumped from a well. This is important because sand, especially quartz sand, can adversely affect the longevity of pumps, motors, column pipes, and pipelines because of its ability to abrade steel. The abrasion can create points of potential corrosion by both electrolysis and bacteria. Sand testing should be performed on each production well on a yearly basis and should be performed in conjunction with step-drawdown testing. Two types of sand testing equipment can be used to perform sand testing. The first type is a Rossum sand tester, and the second type is a Lakos Laval sand separator.

Sand production is also an indicator that there may be structural concerns with the well and/or well screen. Continued sand production can cause catastrophic collapse of the formation around a well and is a serious concern. Under normal operating conditions, the concentration of sand produced by a water supply well should be less than the ANSI/AWWA A100-06 Standard for Water Wells of 5.0 mg/L during a 2-h pumping cycle when pumping at the design rate. Any recommendations for limiting sediment concentration must consider the water use, the method of treatment, the type of sediment, and the source of the sediment. Properly designed wells can meet 1.0 mg/L. The USEPA and the National Water Well Association have recommended the following limits:

- 1 mg/L—water to be used directly in contact with, or in the processing of, food and beverages (also suggested by the AWWA Groundwater Resources Committee for membrane applications).
- 5 mg/L—water for homes, institutions, municipalities, and industries.
- 10 mg/L—water for sprinkler irrigation systems, industrial evaporative cooling systems, and other uses where a moderate amount of sand is not especially harmful.
- 15 mg/L—water for flood-type irrigation and where the nature of the water-bearing formations and the overlying strata are such that pumping this amount of sand will not seriously shorten the useful life of the well.

The limits suggest reasonable goals that can be achieved if good well design, construction, and development practices are followed. In older wells or wells in problem aquifers, a well may pump unacceptable amounts of sediment. If the well cannot be redeveloped by conventional techniques, a special sand separator can be installed as a permanent part of the well system. Although sand separators are efficient, they may not remove all sediment and should not be used as a substitute for good well design and construction practices. In addition, if sufficient sand is removed, this removal could cause catastrophic collapse of the formation. The testing procedure for a Rossum sand separator is well documented in AWWA literature. While the Rossum sand tester is the method accepted by AWWA, the authors' experience has indicated that the use of a Lakos Laval sand separator provides a better method of quantifying sand produced from a well.

The amount of sand produced in milligrams per liter for each individual pumping rate is determined by the following equation (Witt and Andrews 1993):

$$S = \frac{1000 S_{wt}}{3.785\, Qt} \qquad \text{(Eq 3-8)}$$

Where:

S	=	sand content, mg/L
Swt	=	weight of sand, g
1,000	=	equation constant
3.785	=	equation constant
Q	=	rate through the sand separator, gpm
t	=	time, min

The well should be pumped at its design rate for 2 h, and sand samples should be collected at 5, 30, 60, and 120 min without stopping the pumping. Sand samples are removed from the sand separator and analyzed. The amount of sand pumped during normal operation is reflected in the fourth (120 min) sand sample. This sample is a realistic figure for the quantity of sand that will be produced during normal well operations. The desired maximum is 5.0 mg/L for drinking water applications (NGWA 1975). More sand, especially after proper development of the well, may indicate grouting problems or a sand vein in the production zone that may cause damage to the well and pumping equipment. Large discrepancies in the amount of sand collected at the 5-min sample as compared to the amount of the 120-min sample are of concern and may be an indication of water hammer to the formation. Water hammer is a phenomenon whereby the turn-on or turn-off of the well causes large pressure fluctuations in the aquifer. Such fluctuations will dislodge sand and silt and may cause damage to limestone and sandstone. Water hammer often indicates operational failure of the valves in a well and/or at the plant. It should be noted that the Lakos Laval sand separator only removes sand particles in the range of 74 µm with 98 percent efficiency. This means that particles less than 74 µm will pass through the sand separator.

Colloidal testing should be performed before well disinfection and after well disinfection. This procedure will allow evaluation of the effectiveness of the disinfection procedure. The filter pore spaces are 5.0 µm in size. This filter size allows for the capture of most clay- and silt-sized particles. Clay- and silt-sized particles can clog the gravel pack and well screens, causing increased drawdown, loss of production, and increased operating cost. Continued withdrawal of silts and clays can, like sand, cause sinkholes. The production of clays and silts may also be an indication of structural concerns with the well casing and the screen. A pressure gauge on each side (inflow and outflow) of the filter apparatus allows the measurement of the change in pressure across the filter cartridge. The following is an outline of the procedure to set up the cartridge filter:

1. Disinfect the filter holder with unscented Clorox® bleach and attach the colloidal test apparatus to the side of the discharge line.
2. Flush Clorox® out of the cartridge filter holder by opening the valve to the discharge line.
3. Insert the cartridge filters into the holder without touching the filter. (Open an end of the plastic bag and insert the filter, then remove plastic bag.)
4. Turn on water to run through the filter.
5. Check and note the discharge rate after 5 min; check and note the pressure in (P_{in}) and pressure out (P_{out}).
6. Check and note P_{in} and P_{out} at least once a day. Note any color changes in the filter and/or any growths on the filter. Once the pressure differential between P_{in} and P_{out} increases to more than 28 psi, remove the filter using plastic gloves.

The filter should be visually inspected and stored in a polyvinyl chloride (PVC) container and any odor noted. Culture swabs may be obtained and sent for microbiological analysis. The filters should be placed in PVC containers and sent for x-ray diffraction and microscopic analysis by a qualified professional.

SDI testing, as defined in ASTM Standard D4189-07, is an empirical measurement to test for the potential of silt, colloids, bacteria, colloidal silica, organic molecules, and/or corrosion products that foul well screens and gravel packs. The SDI test simply measures the decay in flow rate through a 47-mm diameter, 0.45-μm pore-size membrane. The 0.45-μm membrane is used because it is more susceptible to clogging from colloidal matter than from hard particles, such as sand and scale. Furthermore, the 0.45-μm size is smaller than the 5.0-μm size of the prefilter and therefore measures particles that would pass through the prefilter and clog the membrane. The measured decay in flow rate is converted to a number between 1 and 100.

The SDI number is a function of the rate at which the filter (membrane) clogs with colloidal material. The larger the SDI number, the greater the fouling tendency of the water. To perform the SDI test, a Millipore SDI or fouling index test kit or an equivalent is required. The SDI equipment includes the following: 47-mm filter holder, pressure regulator, pressure gauge, valves, fittings, tweezers, 0.45-μm membrane filter discs, a stopwatch, and a 500-mL graduated cylinder.

To calculate the SDI of a given water, the following formula is used:

$$SDI = \left(1 - \frac{T_1}{T_F}\right) \times 100 \div T_T \qquad \text{(Eq 3-9)}$$

Where:

SDI = silt density index (an empirical number between 1 and 100)
T_1 = the initial time to fill 500 mL, sec
T_F = the final time to fill 500 mL, sec
T_T = the total time test is performed, min

It is important to note the color of the filter because coloration is an indication of the clogging medium. If microorganisms are suspected, it is important to preserve the filter in a sterile container, such as a petri dish. The filter then should be sent for microbiological analysis.

One method of controlling the SDI is through the regulation of the uphole velocity of water in the well. Decreasing the velocity will decrease the SDI of the water. For water supply wells, an uphole velocity of less than 5.0 ft/sec is recommended. Suspended solids create difficulty in the aquifer, generally causing well plugging and a reduction in aquifer permeability, especially if the aquifer has any chemical reaction with suspended solids or the ions among the solids. In many cases, suspended solids can bypass filtration tests, thereby creating a buildup within the wells that is unknown on the surface. To solve suspended solids problems, it is best to redevelop or pump the wells to remove the suspended solids.

MICROBIOLOGICAL ISSUES

Understanding the microbiological activity can help with understanding the chemical changes of the raw water. In stark contrast to public perception that aquifers are "pristine" environments, bacteria naturally exist in most aquifer systems; most aquifers with an organic content will have some degree of bacteriological activity. The typical agents for microbiological fouling include iron, sulfur-reducing, and slime-producing organisms, although many others exist, some of which are opportunistic pathogens. As a result, biological contamination is an ongoing problem often overlooked by engineers and hydrogeologists.

Microorganisms occur naturally in most aquifer systems, especially surficial systems. Most microorganisms require nutrients, such as organic carbon, nitrate, and phosphate, to grow and flourish, and to meet cell-building needs and general food source demands, all of which are available in abundance from surface activities. Casing and column pipe materials should be carefully considered because iron and certain other metals can increase microbe populations. The organisms produce polysaccharide films that can cause microbiologically induced corrosion as they are generally acid-formers. Of special concern from an operational perspective is microbiological species that may impact human health or may cause biofilm growth that will clog the well screen or gravel pack. Some of these species will aggressively attach to the ferrous metals often used for casings and column pipes.

Iron bacteria, such as *Gallionella*, are common in aerobic environments where iron and oxygen are present in the groundwater, and where ferrous materials exist (such as steel or cast-iron wells). These bacteria attach themselves to the steel and create differentially charged points on the surface, which in turn create cathodic corrosion problems. The iron bacteria metabolize the iron that is solubilized in the process. Iron bacteria tend to be rust colored or cause rust-colored colonies on the pipe surfaces.

Sulfur-reducing bacteria are often responsible for the hydrogen sulfide smell released when raw water is aerated. These bacteria are common where sulfur naturally exists in the formation and will tend to form black colonies on pipe surfaces. While anaerobic, they will exist in environments where aerobic conditions can lead to symbiotic relationships with aerobic organisms.

The slime-producing bacteria are found in surface waters and in soil. Members of this genre are often used to protect farm crops from fungal growth and are to be expected in groundwater that has organics. However, the *Pseudomonas* genera are highly adaptive; research done several years ago indicated that *Pseudomonas* sp. would grow in any environment into which they were introduced. They are facultative anaerobes that can persist in oxygen-depleted environments by breaking down complex hydrocarbons for the oxygen and can permanently affix themselves to laser-polished 316L stainless steel in a matter of hours, so attaching to steel or lower grades of stainless steel is easily accomplished. Given that the *Pseudomonas* sp. are adhering bacteria, they can produce a polysaccharide matrix (biofilm) that can act as a barrier protecting the bacteria incorporated in the films from harmful substances such as disinfectants and, in some cases, oxygen. Biofilms also act to protect the bacteria from the shearing effect of turbulent flow and can provide an environment for other species.

The microbiological accumulations or biofilms pose several significant concerns. First, the accumulations on the metallic surfaces create anodes and, in conjunction with reactions caused by dissimilar metals, can lead to a steady cathodic deterioration over time (with or without iron bacteria). Because the *Pseudomonas* are acid-formers, ferrous materials are particularly vulnerable to deterioration, especially in the presence of iron bacteria.

Because of the potential for microbiological problems, routine microbiological sampling of production wells is important. Biological agents can cause corrosion of the well casing, pumps, column pipes, and valves. Bacteria can clog gravel packs, well screens, and the formation, causing excessive operational costs by requiring higher pressures (heads) to obtain and treat water. Microbiological agents can circumvent treatment systems and can be released into the distribution system, causing clogging of pipes and biofilms, creating public health concerns. It is important to sample raw and finished water for microbiological contaminants.

There are two types of microbiological sampling. One should be routinely performed; each well should be sampled for fecal and nonfecal coliform bacteria. The presence of coliform bacteria is an indication that a well is biologically contaminated and may pose a threat to human health. If coliform bacteria are detected, the well should be retested. Should a second test show this presence, the well should be disinfected and should not be used until it clears a subsequent coliform test. It is not uncommon to have sampling errors in bacteriological sampling. Therefore, sampling protocol is important. Sampling jars should be dedicated to each well, and each jar should be sterilized before sampling.

Likewise, HPC counts should be gathered regularly. HPC counts will indicate the presence of bacteria, although not necessarily fecal bacteria like coliform tests. Operators do not water any bacteria in the system, so if HPC counts are present, figuring out the species is important. BART® tests can be used for the presence of certain bacteria (iron and sulfur bacteria for example). If BART and HPC counts indicate bacteria, the next type of sampling should be pursued.

The next type of microbiological sampling is speciation of the bacteria. Such sampling analysis is necessary to determine the presence and/or absence of microorganisms that could adversely impact treatment systems and/or could potentially threaten human health if adequate treatment of the recovered water is not provided. Organisms to be analyzed include growths of certain types of slime-forming bacteria, fungi, and algae. This process of microbiological growth is commonly referred to as *biofouling*. In many cases, the main source of biofouling organisms is feed water. The presence of these organisms will adversely impact the treatment process. In addition, some organisms, if not removed in pretreatment and post-treatment, may adversely impact the water quality.

METHODS FOR MONITORING GROUNDWATER QUALITY

Before developing a groundwater supply, the water quality must be currently acceptable and expected to remain so in the foreseeable future. Otherwise, the investment is not warranted. After the initial water quality assessment is performed and groundwater development is assured, a system for monitoring water quality should be maintained, and a reassessment of upgradient contamination risks should be performed periodically. Where contamination risks are significant, or where the water supply is critical, sentinel monitoring wells should be installed (see Chapter 2 for discussion). Sentinel wells are located at various depths to define the initial groundwater assessment. In the future, sentinel wells can be used to detect changes in water quality and water elevations before they affect the water supply wells, serving as an early-warning system.

The most common size for monitoring wells is 4 in. in diameter, and they are usually constructed with PVC casings. However, the number of wells needed, and their locations, depths of completion, and construction details, must be specified as part of an integrated wellfield monitoring plan. Increases in water use may increase the contamination threat in the future, which may indicate a need for additional monitoring wells. This potential for development demonstrates the importance of continual evaluation of changes that might affect the groundwater supply. The wellfield monitoring plan should account for likely sources of contamination, local hydrogeology, and the hydraulic effects of the proposed groundwater development. For example, what was previously considered down gradient from the well can become up gradient either after pumping begins or as influenced by nearby surface water. These changes should be simulated with computer modeling to aid in designing a monitoring-well network. Samples taken from monitoring wells should be analyzed for suspected contaminants that may impact the end use of the water. Historically, it has been the mineral quality of water that limited possible water uses. For instance, water containing high concentrations of sodium or boron will be unsuitable for irrigation. Today, biological issues and endocrine disruptors have been added to the carcinogens and minerals of the 1970s and 1980s.

A list of the minimum required chemicals to be tested may be obtained from federal, state, and local regulatory officials. Indicator parameters, referred to as *priority pollutants*, often can be used to determine the likely presence or absence of chemicals that are a concern to groundwater development, but they are not guarantees of water quality (see Appendix A). Fortunately, groundwater quality does not generally change rapidly as a result of slow movement as compared with surface water quality. Therefore, once water quality has been established, the frequency of groundwater sampling normally need not exceed quarterly or even semiannual checks, except for potable water sources or areas of suspected contamination.

REFERENCES

American Water Works Association (AWWA). 2014. Manual M21, *Groundwater*. Denver: AWWA.

Bloetscher, F., A. Muniz, and G.M. Witt. 2005. *Groundwater Injection: Modeling, Risks, and Regulations*. New York: McGraw-Hill.

Keys, W.S. 1989. *Borehole Geophysics Applied to Groundwater Investigations*. Dublin, Ohio: National Water Well Association.

Nuzman, C.E. 1989. "Well Hydraulic Flow Concept." In *Recent Advances in Groundwater Hydrology*. Minneapolis, Minn.: American Institute of Hydrology.

National Groundwater Association (NGWA). 1975. *Manual of Water Well Construction Processes*. Westerville, Ohio: NGWA.

Witt, G.M., and L.J. Andrews. 1993. *Reverse Osmosis: Membrane Technology, Water Chemistry, and Industrial Applications*. (Z. Amjad, editor.) New York: Van Nostrand Reinhold.

CHAPTER 4
Drilling Methods

The purpose of this chapter is to identify suitable drilling methods that will permit the wells to serve their intended purpose. Wells can be viewed as large holes in the ground, deeper than they are wide. In addition to providing available water supply reliably, wells are typically in service for decades. Hence, the steps initially taken to design and drill the well play a large role in the longevity of a well. Selection of the proper well drilling method to meet the intent of the well along with sound well construction practices are important to accommodate existing underground conditions.

DRILLING WELLS

There are many available well drilling and construction methods that have been developed depending on the purpose for which the well is to be constructed. Some purposes for wells include:

- water production (the focus of this handbook),
- oil and gas wells,
- geothermal wells,
- injection/disposal wells,
- aquifer storage and recovery wells,
- environmental remediation wells, and
- monitoring wells.

Environmental remediation wells are generally shallow, with depths less than 50 ft, but oil and gas wells may be several thousand feet deep. Water supply and other wells fall somewhere in between. Just as there are many purposes for wells, there are even more options for their construction, including the following:

- hand dug
- California
- cable tool
- hollow stem auger
- large diameter auger
- screened stem auger
- solid flight auger
- hydraulic rotary
- reverse-air circulation
- casing hammer air rotary
- rotary down-the-hole hammer
- dual-tube rotary
- direct push/driven
- mud rotary
- reverse circulation (mud)
- bucket-type drilling
- direct-air circulation
- percussion hammer
- horizontal/directional/angle
- sonic
- jetting and driving

Many of these methods are used for construction of water supply wells. Factors such as the total depth of the well along with diameter and depth of casings being set may dictate the type of rig needed for construction. Availability of drilling rigs will also need to be considered. Selection of the proper drilling method will depend on many variables including:

- data needed during construction for final design of the well,
- diameter of well,
- casing depth and total depth of well,
- well completion (i.e., open hole or well screen),
- well diameter and depth, and
- site access (mobilization, height of drilling mast, etc.).

TYPES OF WELLS AND THEIR CONSTRUCTION

A variety of wells have been used for water supply purposes, although most are used only for small installations and under certain conditions. These include hand-dug, bored, driven, and drilled wells (AWWA 2014).

HAND-DUG WELLS

Hand digging wells is the oldest method of water well construction. Hand digging a water well is labor intensive, dangerous, and time consuming. A hand-dug well is frequently excavated using a pick and shovel, with a hoist with a bucket to remove materials encountered. They are large in diameter but relatively shallow (i.e., less than 30 ft in depth). Most hand-dug wells are circular because this shape adds strength and is usually easier to dig. Such wells may furnish relatively large supplies of water from shallow sources. Depending on the formation, the well may be able to stand without reinforcement; however, pilings, concrete, or bricks are commonly used to create support walls.

Once the top of the water table is encountered, it is difficult to advance the depth of the well into the aquifer. The limited production horizon severely limits the quantity of groundwater available for extraction. The narrow production zone also leaves the well vulnerable to changing water level elevations. During times of little precipitation or drought, the depth of the water table may increase because of lack of recharge. Eventually the top of the water table will be located below the bottom of the open hole, resulting in a dry well. The negative impacts of a well "gone dry" on a local community are obvious.

Hand-dug wells have little protection from contamination. Impacts to the well from surface runoff, airborne material, animals, insects, and objects falling into or finding entrance into the well are common. Concrete curbs are commonly constructed around the edges because dug wells are easily polluted by surface water. Although hand-dug wells are still constructed in many parts of the world, they are not commonly used as a source of groundwater for a modern water supply system in developed countries and are rarely used in the United States.

DRIVEN OR DIRECT PUSH WELLS

A driven or direct push well consists of a pointed steel screen (i.e., a drive or well point) and sections of pipe attached to the top of the well point (Figure 4-1). Well points are driven into the formation with a weight, derrick, and pulley system. The driver typically weighs 30–75 lbs, and well points may be mechanically driven as deep as 50 ft depending on underground conditions.

If the diameters are small, they could be installed by a hand-driven system. Figure 4-2 shows the installation of driven wells for dewatering purposes. When constructing a driven well, an outer casing is first installed by pounding the pointed screen into the formation. The steel tip breaks through pebbles and thin layers of material and opens a passageway for the screen. For small municipal water supplies, the driven well may be used in thin deposits of sand and gravel found at shallow depths. The outer casing protects the inner casing to which the pump is attached.

In sand and gravel formations, the outer casing should extend to just above the drive point. The outer casing can be driven with a sledgehammer; for larger

Source: Photo by Fred Bloetscher

Figure 4-1 Well points for driven/jetted wells

Source: Photo by Fred Bloetscher

Figure 4-2 Well point installation for construction dewatering

pipes, a tripod and pulley can be used. The tripod and pulley set-up raises and lowers a heavy block onto a drive cap placed on top of the outer well point casing. Extra-heavy pipe should be used to withstand the load. If the ground is clay, the outside casing should be set in a hole prepared by an auger before inserting the well point. After the outer casing is set, the annular space between the borehole and the outside of the casing should be sealed with cement grout.

Driven wells generally range in diameter from 1 to 4 in, and rarely are driven more than 30 ft below land surface. The screen is an integral part of the portion driven into the ground; therefore, only soft materials are easily penetrated. As a result, the production rate of driven wells is limited. To increase yield, a battery of well points—with the wells located a reasonable distance apart that are connected by a common header to the pump—is used. While this type of well may be used to supply raw water, they are more commonly used for dewatering purposes.

CALIFORNIA

The California, or stovepipe, method of well construction was developed in California for water wells drilled in unconsolidated alluvial materials. Wells ranging in diameter from 16 to 20 in. to depths up to 300 ft have been constructed using this drilling method.

The California drilling method uses the same general principles used for cable tool rigs. Short lengths of sheet metal, either riveted or welded together, are used for casing. After the casing is in place, it is perforated using a Mills knife or similar device that tears the metal. The openings should be managed with respect to size and area (Figure 4-3).

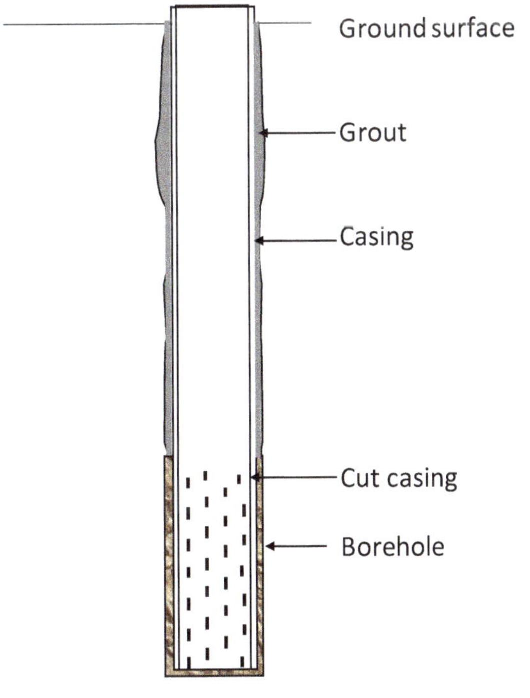

Figure 4-3 California well

JETTING

Jetting is often used to construct wells when water is found in sand at shallow depths, although the method is applicable for deeper wells. Jetting equipment consists of a drill pipe, or jetting pipe, equipped with a cutting bit on the bottom end. Water is pumped into the well through the drill pipe and out of the drill bit against the bottom of the drill hole. The casing usually is sunk as drilling proceeds. Jetted wells are usually small in diameter; therefore, deeper well construction becomes more difficult to manage and control. To solve this problem, several lengths of casing with different diameters may be telescoped one inside the other to reach the full depth (Figure 4-4).

Certain conditions make this method of well construction difficult. Rock and boulders are barriers that cannot be overcome by water pressure, regardless of how high the pressure might be. Formations of clay and hardpan become soft and sticky.

BUCKET-TYPE DRILLING

Bucket-type drilling uses a rotating cylindrical bucket. Cutting blades mounted on the bottom of the bucket cut the sediment, and the pieces of sediment are retained in the bucket. When the bucket is full, it is withdrawn from the borehole and swung to the side; the hinged bottom is opened, releasing the cuttings.

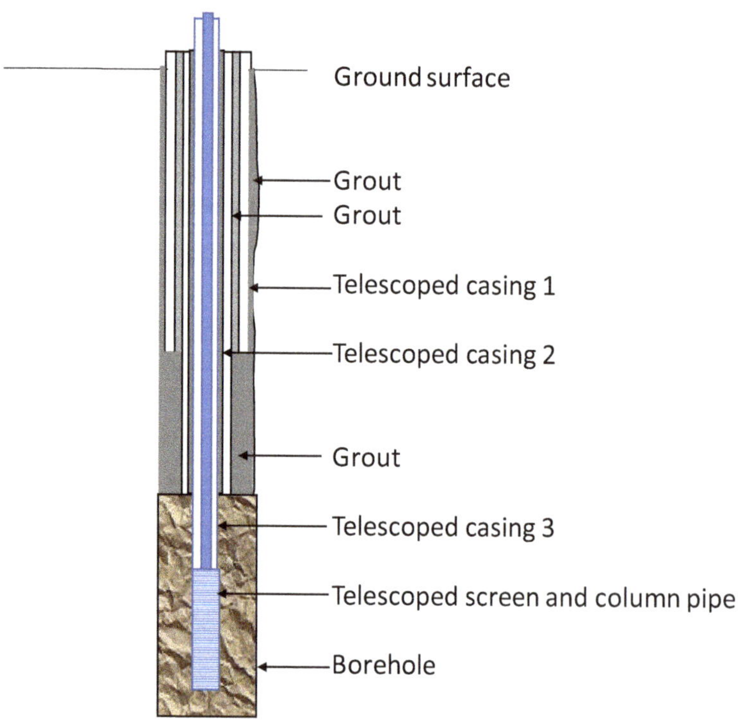

Figure 4-4 Telescoping of jetted wells

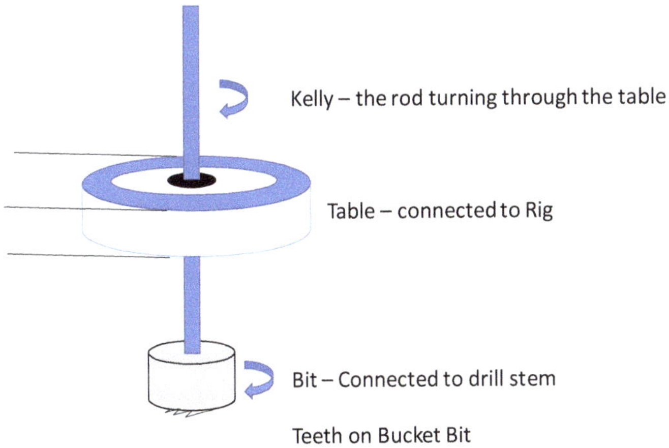

Source: Ingvar Tjostheim/Shutterstock.com

Figure 4-5 Bucket-type wells

Bucket-type drilling can be used to drill wells from 10 to 60 in. in diameter and to a maximum depth of about 100 ft.

Sediment samples recovered during borehole advancement are used to identify the penetrated formations. Bucket-type drilling requires little or no fluid to be added to the borehole during drilling, allowing for the recovery of water samples for analysis. Bucket-type rigs are relatively simple in design and are capable of drilling quickly. Figure 4-5 shows a bucket-type drilling rig. Often, casing is advanced into the borehole as part of the drilling activity. An advantage of this method is the reduced possibility of cross-contamination of fluids during drilling.

Bucket type drilling produces a large quantity of cuttings and, once the saturated zone has been penetrated, large volumes of fluids. Generally limited to poorly lithified cohesive sediments, bucket drilling may be continued into sand and sometimes gravel if the casing is advanced into the borehole during drilling or if the borehole is kept full of water. Once even moderately lithified sediments are encountered, they will usually prove to be too resistant for bucket-type rigs to penetrate.

The use of bucket-type drilling depends on the sediment types anticipated to be encountered. Good results can be realized on generally unlithified cohesive formations, such as glacial tills. Formations consisting of large boulders, competent shale, chalk, or sandstone are extremely difficult if not impossible to penetrate using this method, and it is impossible to penetrate well-lithified sedimentary, igneous, and metamorphic rocks.

CABLE-TOOL DRILLING

Cable-tool drilling methods are also called percussion, spudder, and solid tool methods. The cable-tool drilling method was used as early as 300 BC to construct salt wells in China to depths as great as 3,000 ft. Cable-tool drilling methods are commonly used today for shallow wells but can also be used for wells as deep as 1,000 ft.

Cable-tool methods are straightforward. The technical expertise required is less than for some other methods, although the details for construction and operation of the drilling machines vary widely. This drilling method is one of the oldest drilling methods and is well understood in the industry.

Cable-tool rigs create the borehole using the percussion and cutting action of a drill bit. Figure 4-6 shows a cable-tool drilling rig. Cable-tool drilling is accomplished by alternately raising and dropping a heavy drill bit suspended at the end of a cable, pulverizing the formation below (Figure 4-7). A casing is often used to keep the borehole open when drilling in loose material. The drill bit—a club-like, chisel-edged tool—breaks the formation into small fragments. The bit turns with

Source: USGS

Figure 4-6 Cable-tool drilling

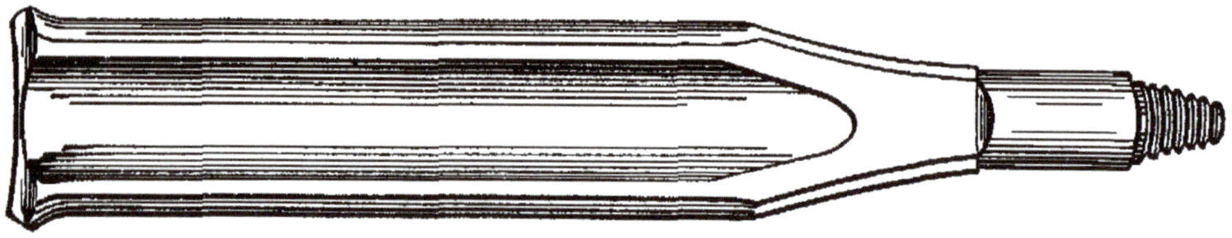

Source: USGS

Figure 4-7 Cable tool bit

each blow. The bit is then returned to the bottom of the borehole, and the process is repeated until the desired depth is achieved. Cable-tool wells can be 4–18 in. in diameter. The length of the drill cable is adjusted so that the bit will strike with the right amount of weight and stroke. The cable is monitored to determine how well the tools are operating. The length of stroke and rapidity of blows are continuously adjusted to maximize the rate of penetration.

After several feet of borehole are drilled, the bit and drill string are pulled from the hole and swung aside. A bailer is used to remove the slurry. The bailer consists of a 10- to 25-ft long section of tubing with a check valve in the bottom (Figure 4-8). The bailer is smaller in diameter than the drill hole so that it can move up and down freely (AWWA 2014). Once the borehole has been cleared of cuttings, the bit is returned to the bottom of the borehole, and the process is repeated until the desired depth is achieved.

The borehole must be straight and vertical (i.e., plumb) for ease of operation and for installation of casings, pumps, screens, and column pipes. In extreme cases, lowering a pump into a well or pulling it out may become impossible if the well is not plumb. The first indication that the borehole is not plumb is that the drilling tools begin to stick. When this happens, drilling should stop, and the borehole should be realigned and/or re-drilled.

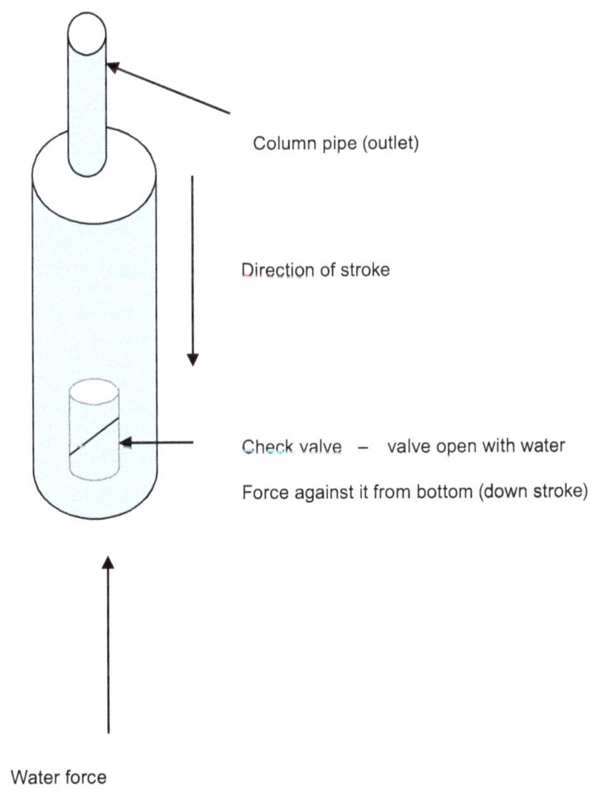

Figure 4-8 Cable tool bailer

Although used less frequently, cable-tool rigs have certain advantages. The ruggedness, weight, and simplicity of the drill rigs make them easy to move across rough terrain. Ease of repair also makes them ideal for isolated areas without access to repair parts or power. Because little or no outside drilling fluids are introduced to the well during drilling, representative water and formation samples can readily be recovered as drilling progresses. Water is usually not necessary as drilling mud is not used during well construction, thus minimizing plugging of the formation, which simplifies development of the well. The use of a bailer to remove the cuttings from the well aids in keeping the native formations clean of sediments produced during drilling. Each time the bailer is removed from the borehole, it is filled with cuttings and formation fluids. Removal of these solids and fluids allows the formation fluids to flow from the formation into the borehole, keeping the borehole wall clean. Most types of formations can be drilled unless they are either very hard or very soft (clays). Limited damage to the formation adjacent to the borehole occurs as the borehole force is downward.

A disadvantage of cable-tool drilling methods is efficiency. Cable-tool drilling is slow, providing poor productivity in hard formations, hard sand, or deep wells. The limitation of diameter may be a concern for high productivity needs. The casing must often be advanced with the drill bit to maintain the competency of the borehole and to prevent formation fluids and previously drilled sediments from migrating within the hole. The action of the bit can damage the borehole wall, resulting in a lower well efficiency than desired.

CAISSON WELL

Caissons are used in shallow, very loose, and permeable alluvial formations because of the potential to lose large quantities of drilling fluid or mud into the loose formation. A hole is made by sinking a large-diameter casing (typically concrete) to about 15 ft. The next casing is installed in a concentric manner, one size smaller in diameter than the previous, using the same bailing method (Figure 4-9) until it is extended another 15 ft below the first casing. This process is continued until the bottom of the desired formation or the desired depth is reached. The concept of ever-smaller casings inside and below the prior casing is called telescoping. The smallest diameter casing will be the innermost one. Drilling may only be involved in the later stages of caisson construction as well as radial wells (an extension to caisson construction).

HOLLOW STEM AUGER

Several types of augers are used in the drilling industry including large-diameter augers, solid flight augers, screened stem augers, and hollow stem augers. Of these various auger methods available, the hollow stem auger is the most used method for construction of water supply wells (Figure 4-10).

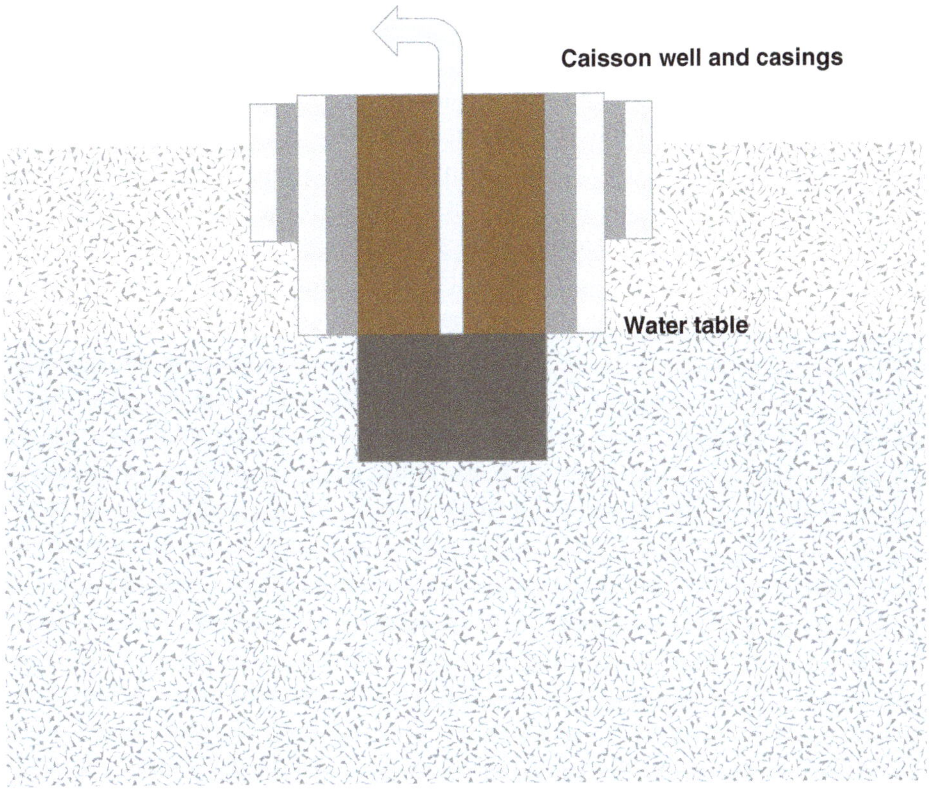

Figure 4-9 Caisson well and casings

Source: Photo by Fred Bloetscher

Figure 4-10 Hollow stem auger and flights

Hollow stem augers use continuous flight augers generally up to 14 in. in diameter that use mechanical methods to remove drilled materials from the borehole. Axial openings in the center of the auger, up to 10.25 in. in diameter, allow access to the bottom of the hole without removing the augers. The augers also act as a temporary casing during well construction allowing for recovery of soil and water samples as well as aiding in construction of the well. This well drilling method is generally used for small-diameter, shallow wells up to 150 ft in depth.

Hollow stem auger methods of drilling are commonly used for short-term wells, such as for construction dewatering or for recovery wells for cleaning up groundwater contamination. Each section, or flight, of augers is usually 5 ft in length. As shown in Figure 4-11, a cutting bit is attached to the first flight. Either a plug, or more often a pilot bit, is inserted through the hollow stem of the augers. The pilot bit, held in place by drill rods, helps cut the formation and prevents cuttings from entering the bottom of the auger.

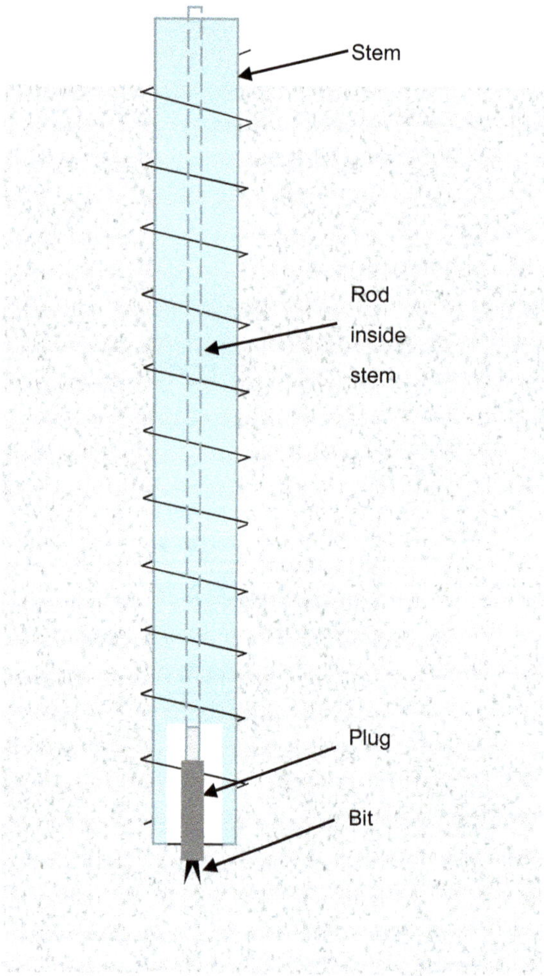

Source: Reprinted from *Groundwater Monitoring & Remediation* with permission of the National Ground Water Association. Copyright 1987.

Figure 4-11 Hollow stem auger

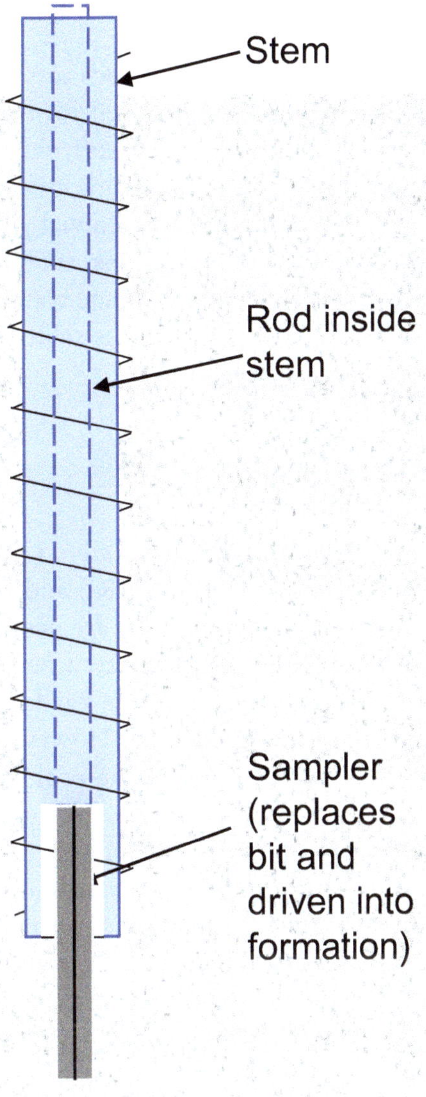

Figure 4-12 Split spoon sampling device

Once the desired depth is reached, the drill rods and pilot bit can be removed from inside the augers, allowing split spoons or thin-walled samplers to be used for sample recovery (Figure 4-12). Water samples may also be recovered from inside the augers using a sampling device such as a Hydro-punch. The pilot bit is placed back inside the auger, and drilling can resume. Once the desired total depth has been achieved and the pilot bit has been removed, the well screen and riser can be easily lowered down through the center of the augers. The well-screen gravel pack is placed as the augers are pulled, not rotated, out of the borehole. Advantages of the hollow stem auger drilling method include:

- Drilling fluids are not introduced.
- The borehole is drilled, and the casing can be placed simultaneously.
- Representative samples of aquifer fluids and formation material are easily collected in uncontaminated condition.

- The drilling technique is fast.

However, hollow stem drilling methods are limited to softer formations, such as loose sand, gravel, or unconsolidated formations. An auger can be used only where formations, though relatively soft, will permit an open hole to be bored to depths ranging from 25 to 60 ft without caving. The most suitable formations for bored wells are glacial till and alluvial valley deposits. The depth and diameter are limited as well. Wet formations, such as running sands, will make drilling nearly impossible and will contaminate water quality samples and formation samples. This type of drilling method is seldomly used for construction of water supply wells.

HYDRAULIC ROTARY DRILLING

Rotary drill rigs have a long history. They were used by the Egyptians for cutting stone for the pyramids (Lehr et al. 1988) and in the 19th century for well drilling. Hydraulic rotary drilling methods are used for larger production wells and where construction methods must proceed faster. Extensive use of hydraulic rotary drilling for oil and gas exploration has led to many advances, some of which are applicable to water well drilling. However, the method requires more technical equipment and expertise than cable-tool and hollow stem auger methods. Hydraulic rotary methods, as the name suggests, involves the introduction of fluids to improve drilling efficiency. Water and drilling mud are typically used, but in some cases, air or foam are used as the drilling fluid. When the fluid used is drilling mud, the method is usually referred to as mud rotary.

The drill string usually consists of a bit, drill collars, stabilizers, and drill pipe. Rotation is transmitted from the surface along the drill string to the bit. As the bit is rotated, drilling fluid is pumped down through the drill string and out of the bit (Figure 4-13 and Figure 4-14). The cuttings are returned to the surface by the drilling fluid as it rises in the annular space between the drill string and the drilled formation. The drilling fluid is cleaned and recirculated through the drill string.

Bit selection depends on the anticipated formations to be encountered. Figure 4-15 shows a cutaway view of a typical tricone mill-tooth rotary drill bit. The bit size is roughly the diameter of the borehole drilled. Different types of bits are used depending on rock hardness and composition. Figure 4-16 shows a chart for bit selection. Figures 4-17 to 4-19 show various bits and the type of rock for their appropriate use. Drag bits have no moving parts; the blades (Figure 4-17) are designed to cut into soft formations, such as clay and poorly consolidated sands. Mill tooth tricone bits (Figure 4-18) are used for moderately hard formations, while tricone bits with carbide button inserts are used for harder formations, such as well-cemented sandstone, micritic limestones, and dolomites. For very large diameter drilling applications, flat bottom bits are used (Figure 4-19). These are large drill bits with many cutting heads.

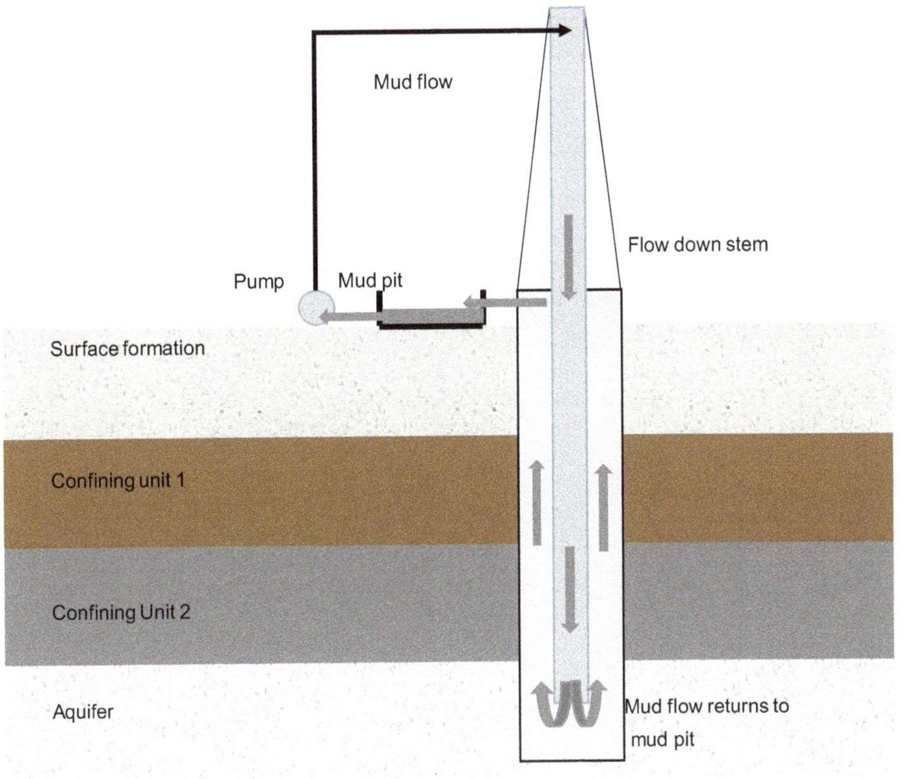

Figure 4-13 Mud rotary circulation

Figure 4-14 Mud rotary method with cuttings carried to surface

Source: Photo by John Largey

Figure 4-15 Typical tricone carbide button step drill bit

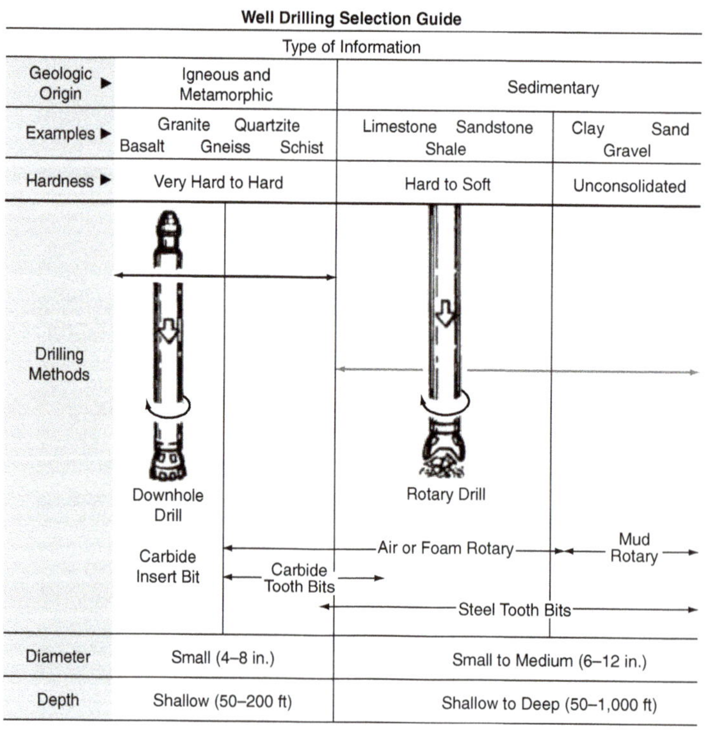

Source: www.globalsecurity.org

Figure 4-16 Chart for bit selection

Source: Photo by Fred Bloetscher

Figure 4-17 Large-diameter drag bit

Source: Photo by John Largey

Figure 4-18 Mill tooth tricone bit

Source: Photo by Fred Bloetscher

Figure 4-19 Flat-bottom bit

Source: Photo by John Largey

Figure 4-20 Drill pipe and drill collars

The drill pipe is typically 20–40 ft sections referred to as joints of pipe (Figure 4-20). It is usually made of manganese–carbon steel or a molybdenum alloy steel. Drill pipe used for water well drilling is generally 2.375–4.500 in. in diameter. Proper size is important to minimize friction loss and therefore reduce power required to rotate the drill string. The joints of drill pipes are threaded to facilitate connections. The threads should be cleaned and lubricated before each connection to prevent the threads from seizing. Care should be taken in picking up the drill pipe so as not to damage the threaded ends required for connection (Figure 4-21).

Drill collars are heavy walled pipe machined from solid bars intended to keep the borehole straight and plumb by placing weight directly on top of the bit. Drill collars allow weight to be placed directly downward to keep pressure on the bit and force it into the rock. The combination of directing weight downward and preventing bit wandering improves the efficiency of the bit action in crushing the rock. Stabilizers (Figure 4-22) are positioned at various locations within the lower part of the drill string. They are used to maintain borehole geometry by keeping the bit and drill collars straight and plumb.

As previously mentioned, the drill string is hollow to allow fluid to be pumped to the bit at the base of the borehole (Figure 4-23). The fluid migrates up the annulus carrying the cuttings to the surface. When drilling in clay soil or mud, fluid of sufficient viscosity is required to lift cuttings to the surface. For soft formations, less-dense fluids can be used; however, drilling mud is used in many cases. In addition to removing the cuttings from the borehole, drilling mud serves to cool and lubricate the bit and drill string; build a filter cake on the borehole; prevent fluid loss into upper formations; and prevent borehole collapse. Figure 4-24

Source: John Largey

Figure 4-21 Connecting drill pipes

Source: Hazen and Sawyer, P.C., Boca Raton, Fla.

Figure 4-22 A 42-in. drill bit with stabilizers

Source: Photo by Fred Bloetscher

Figure 4-23 Typical drill pipe is hollow to permit mud circulation.

Source: Photo by Fred Bloetscher

Figure 4-24 Mud circulation system—drill rod, overflow pipe, settling basin, and overflow mud pit. Note: The overflow pipe, settling basin, and mud pit generally sit on the surface (as shown) or are cut 2–3 ft into the surface.

shows a typical basic mud circulation system that has been used for quite some time. It must be noted that ditches, settling pits, and storage pits are simply dug into the ground and not lined. In many locations, this type of circulation system is prohibited to prevent discharges that may affect a surficial aquifer.

Drilling fluids perform four primary functions (Lehr et al. 1988):

- remove the cuttings below the rotating bit,
- transport the cuttings up the borehole to the surface,
- maintain borehole stability, and
- cool the bit.

In addition, additives such as drilling mud (Lehr et al. 1988) assist to:

- prevent fluid entry from the porous rocks,
- reduce drilling fluid losses into the formation,
- lubricate the mud pump, bit bearings, and drill string,
- reduce wear on the drilling equipment, and
- control formation fluid pressures.

One of the primary purposes of the drilling mud is to form a seal, known as a filter cake, on the borehole wall. The density of the mud is related to the ability to form a proper filter cake, thus increasing borehole stability and reducing fluid loss to the formations being penetrated. The density of the mud, measured in pounds per gallon, is generally designed to be pumped into the borehole at a slightly higher hydraulic head than the surrounding formation pore pressure, thereby forming the filter cake and controlling artesian pressure. Proper monitoring of the mud density (mud weight) is important as too low of a mud weight will prevent the proper formation of a filter cake and fail to bring cuttings to the surface, while too high a mud weight may result in high fluid loss and plugging of porous formations. When highly porous formations are encountered, the drilling mud may stop returning to the surface and flow directly into the surrounding sediments. This condition is known as lost circulation. Several additives, known as lost circulation material, are available to plug the formation pores and help regain circulation of the mud to the surface.

The drilling mud is usually prepared using a hopper attached to a suction pit located near the drill rig. The preferred drilling mud material is commercial colloidal material (bentonite clay base) that is purchased in powdered form and mixed with water. Controlling the mud during drilling is an important factor in borehole control and conditioning the borehole for logging. Too much mud caked on the borehole will make logging inaccurate, while too little will not properly seal the borehole to prevent the intrusion of mud into upper formations or loss of the mud into fissures above the bit. Key factors in preparing drilling mud include:

- mud weight,
- mud viscosity,
- gel strength, and
- sand content of the mud.

The lightest mud weight possible should be used considering potential formation pressures and potential for borehole collapse. Lighter-weight drilling mud will intrude into permeable zones less than heavier mud, making development of the well easier. Additionally, unwanted mud may be formed by normal drilling operations when native clays are encountered. In other formations, clay additives may be used initially to increase mud viscosity.

Viscosity is related to uphole velocity of the drilling mud and the ability of the drilling mud to carry cuttings to the surface. The intent is to have a mud that is viscous enough to carry the cuttings to the surface given a certain uphole velocity but also drops the cuttings quickly once it is returned to the mud system. Drilling fluid viscosity is measured with a Marsh funnel. The drilling fluid is poured into the funnel, and the viscosity is determined by the amount of time (in seconds) it takes to drain one quart out of the funnel. Lehr et al. (1988) indicates that a good Marsh funnel viscosity ranges from 35 to 45 s. If the circulating mud begins to pick up sand, the weight will increase and viscosity will decrease. Under such conditions, additional water or other additives should be mixed into the mud.

Settling is needed to remove the cuttings and allow recovery of the cuttings for professionals to determine formation type and to make decisions about the formations most likely to yield water. Once the mud is returned to the surface, it is directed to the cleaning portion of the mud system. Shale shakers are used to first remove the large cuttings (Figure 4-25). Upon returning to the surface, the drilling mud is directed over vibrating screens. In sandy formations, de-sanders can also be used to remove excessive sand from the drilling mud (Figure 4-26).

Source: John Largey

Figure 4-25 Collection of cuttings at the shaker

Source: John Largey

Figure 4-26 De-sanding operations

Drilling mud that is too thick will be difficult to clean and to remove from the borehole wall once drilling is complete, extending the amount of time required for well development. Low viscosity mud often has 10–15 percent solids, composed mostly of heavy sand. If the viscosity is too low, cuttings will not be brought to the surface, and lubrication will be inadequate for the drill bit and collars. Such a situation increases the likelihood of the drill string getting stuck and the loss of drilling fluids into the formation. Observation of the recovered cuttings can provide insight into the efficiency of the mud-cleaning system. Because the drilling mud continuously circulates through the hollow drill pipe to the bit and back to the surface, sand grains that are recirculating will become smaller and more rounded.

In highly porous formations, the low viscosity mud can migrate into cavities in the formation prior to the mud building up on the borehole wall. This loss of drilling mud can be a significant problem for the drill crew and may create regulatory compliance concerns. At the same time, the sand content may decrease where clays or other fines are introduced into the mud and not settled in the settling basin. In such cases, the mud may need to be reconditioned.

Gel strength is a measure of the ability of the mud to suspend the cuttings (and related to viscosity). Proper gel-strength drilling mud will suspend the cuttings in the mud when circulation is stopped to add another section of drill pipe (make a connection), thereby preventing the cuttings from settling on the bit and potentially causing it to seize. However, if the gel strength is too high, large volumes of mud will be lost over the shale shakers, and the cuttings will not settle in the settling pit.

Source: Hazen and Sawyer, P.C., Boca Raton, Fla.

Figure 4-27 Top-head hydraulic drive system

Mud rotary drilling systems use two basic drive systems. Rotary table drives rely on a rotary table recessed into the floor of the drilling rig. A Kelly bushing fits into the rotary table. A multisided pipe, called the Kelly, is attached to the top of the drill string and slides through the Kelly bushing. The rotational forces are transferred from the rotary table to the Kelly bushing, to the Kelly, and through the drill string. Top-drive rigs (Figure 4-27) have a hydraulic motor suspended from the traveling block. The drill string is directly connected to the hydraulic drive motor. Top-head drive systems tend to operate with lower rotational speeds and higher torque, and therefore tend to advance the drilling operation more slowly than rotary table systems (but they generally drill a more vertical hole). Rotary table systems use higher rotation speeds with lower torque resulting in higher rates of penetration, but they also have a greater tendency for the borehole to deviate from a vertical track.

Mud rotary drilling systems have several advantages over cable-tool and auger systems. They are faster and can drill deeper holes with much larger diameters. Good quality lithologic samples can be gained if careful attention is paid to the amount of time it takes for the drilling fluid to reach the surface (lag time). Because a casing is not installed during the drilling operation, high-quality geophysical logs can be conducted on open-hole sections.

However, mud rotary drilling also has disadvantages over cable-tool and auger systems. Extensive knowledge of and experience with drilling mud properties and interaction with various types of formations are essential. Problems with rotary drilling methods occur when lost circulation zones are encountered; large quantities of the mud are lost in highly porous formations, caverns, and fractures and become difficult to remove during development. Sidewall pressure is not maintained under these circumstances, thereby allowing the potential for the collapse of the borehole. A thick filter cake may also reduce upward velocity of the drilling fluid and may interfere with the movement of the drill pipe and the installation of casing. Use of drilling mud requires a significantly larger site to stage, service, and condition the mud. Potable water is also required—a challenge in undeveloped areas. Likewise, the mud must be brought in and hauled out after use—another challenge in many areas. Water samples cannot be secured during drilling operations. Use of packers (single or straddle packers) is needed during or at completion of the drilling operation to isolate potential water production zones.

Extensive development is required for mud rotary drilling operations because of the introduction of drilling mud into permeable formations (the ones that are most likely to yield water). The inability to remove the mud will reduce aquifer yield. Unfortunately, the lack of well development is a common problem with well drilling. Monitoring the mud, mud quality, and drill string is one of the reasons that more technical expertise is required on the drill rig for rotary drilling methods.

Caution must be exercised when planning for the use of the mud rotary method in an active well field. In such a situation, the borehole may intercept the cone of influence of nearby producing wells. This can cause drilling mud to migrate from the borehole being drilled toward the active wells, resulting in contamination of these wells and a reduction in the yield from the aquifer. As a precaution, it is recommended that active wells be taken out of service during the drilling of the new wells. Preplanning and coordination with the facility affected by shutting down the production wells will result in smoother field operations.

A variation of the mud rotary method is the air rotary method. Air rotary drilling uses air as the drilling fluid instead of mud. Air rotary systems have higher operating costs because of the amount of power required to force air into the borehole with enough force to dislodge the formation materials. Over time, these costs have decreased. Pressures are 40–50 lbs psi. Annular velocity is 2,000–5,000 ft per min for dry air. The annular velocity is based on borehole size, drill pipe size, and air compressor capacity. Although air rotary drilling can result in high drilling costs, this drilling process can only be used where borehole stability is not a concern.

Introduced in the late 1970s, air rotary drilling with foam is also an option. Foam is created by aerating the drilling fluids. It increases penetration rates when compared to mud rotary drilling and helps stabilize the borehole to some degree. The foaming agents must be a biodegradable liquid mixture of anionic surfactant, which is added to fresh, hard water. The slow-moving foam has greater capacity for carrying cuttings to the surface compared to conventional air rotary drilling.

Other advantages include:

- reduced air volume,
- reduced pressure requirements,
- increased well cleaning capabilities,
- reduced hydrostatic head, and
- reduced loss of drilling mud.

REVERSE-CIRCULATION ROTARY

The reverse-circulation rotary method is virtually identical to the hydraulic rotary method of drilling except the drilling fluid circulates in the opposite direction. Reverse-circulation drilling is best for large-diameter, high-capacity wells. Reverse circulation is also less expensive than some other methods. Lehr et al. (1988) notes that many wells have diameters of at least 24 in., with some having diameters of more than 60 in. The reverse-circulation method typically uses clear water with no mud additives. A pit is constructed so that the drilling fluid will flow down the annular space between the borehole and the drill pipe and return through the bit and the inside of the drill pipe to the surface, carrying the cuttings with it. A high-capacity pump is attached to the drill pipe to keep the fluid moving at high velocity.

The borehole is stabilized by the hydrostatic pressure of the fluid in the borehole. The fluid level is maintained at the ground surface. Keeping the borehole open requires a large volume of water to maintain a head above the natural static water level, which results in a flow into the formation. This higher head prevents the wall from caving. Significant damage or collapse of the borehole may occur if fluid circulation is lost. Caving may result from movement of fluid down the borehole when formation materials will not accept water. If the formation is highly permeable, the required head will be difficult to achieve. In such cases, processed clays may be added to create a filter cake similar to the straight rotary drilling process. However, this defeats the primary advantage of the reverse-circulation method of construction. In wet clay soils, collapse is a problem. Caustic soda is used to raise the pH. Another option is to increase the amount of sodium silicate in the mud. Bentonite mud can also be used to prevent collapse of clay formations. In some instances, 20–50 gallons per minute (gpm) of make-up water may be needed in highly permeable soils. This is in addition to the 500-gpm circulation rate of the mud.

Advantages of the reverse-circulation method include its ease of use in situations where artificial-gravel-pack wells are specified because less mud cake forms on the face of the borehole. Less development time is required because mud does not intrude into porous formations. Test wells can be drilled and abandoned at a minimal expense by plugging. The reverse-circulation method is also generally faster than the cable-tool method for drilling larger-diameter wells (>18 in.). However, the reverse-circulation method may prove problematic in soft, loose, unconsolidated materials, such as dune sand and quicksand, because the water pressure cannot keep the hole open.

REVERSE-AIR CIRCULATION DRILLING

Reverse-air drilling has many of the characteristics of hydraulic rotary drilling except that drilling mud is not used; air is. With reverse-air drilling, a small steel or PVC air line connected to an air compressor is run inside the drill pipe. The formation cuttings and formation water are lifted outside the drill pipe (Figure 4-28). Reverse-air drilling is capable of drilling large-diameter wells in many formation types. The wells can be drilled quickly, and high-quality lithologic samples can be gathered. High-quality water samples can also be obtained because there are no drilling fluids recirculated into the borehole.

Advantages of reverse-air drilling include speed of drilling and the ability to drill a wide variety of formation types. Reverse-air circulation drilling is highly effective in cavernous and karstic formations. Uncased formations from reverse-air drilling permit excellent geophysical logs to be obtained. Sampling of water and obtaining high-quality cuttings are major reasons for pursuing reverse-air circulation drilling methods. The disadvantages of this method include the potential for bit plugging, cross-contamination of the aquifer, and intermixing of unconsolidated sediments in the borehole. Stability of the borehole in deep, unconsolidated formations may be a problem as no mud cake forms on the borehole.

DUAL-TUBE (REVERSE-AIR) METHOD

Dual-tube/reverse-air circulation is a variation of the reverse-circulation drilling method. The drill bit and stem are rotated by a top head hydraulically-driven drive. Reverse-air, dual-tube drilling methods are similar to reverse-air circulation. Instead of an air line inside the borehole, a dual-walled drill pipe is used (Figure 4-29). As shown in Figure 4-30, the air channel delivers the air to the base of the borehole through the bit. Figure 4-31 shows a typical dual-tube drill rig and

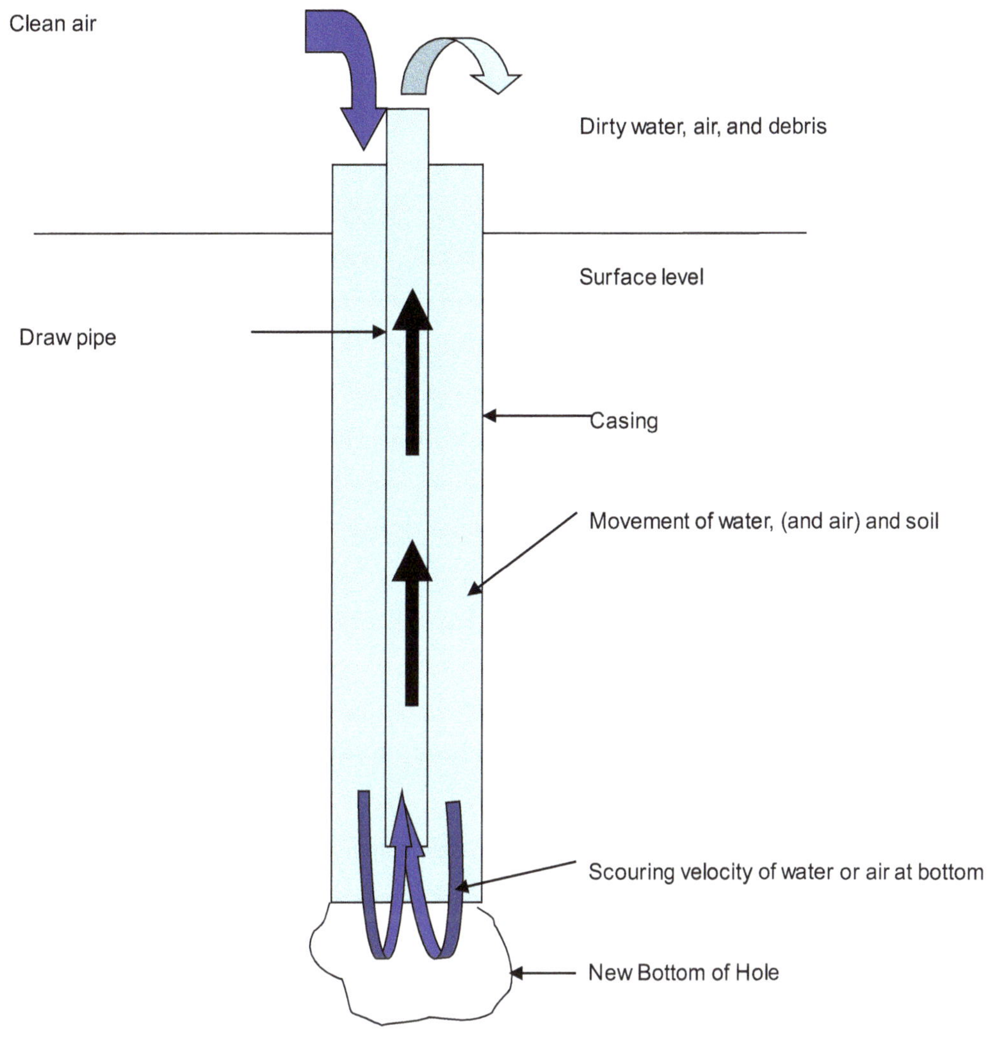

Figure 4-28 Reverse-air drilling method

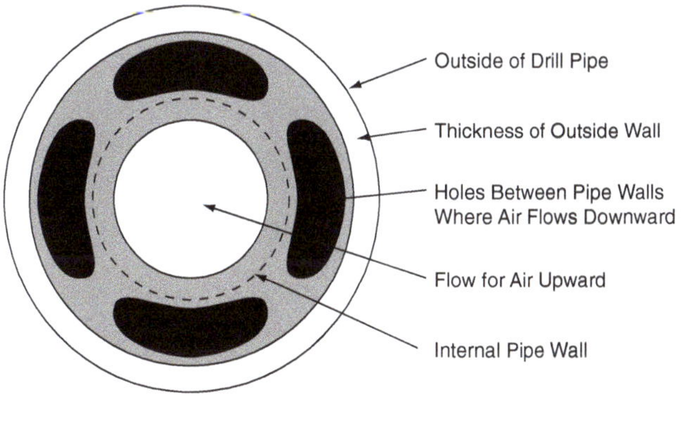

Figure 4-29 Dual-walled drill pipe

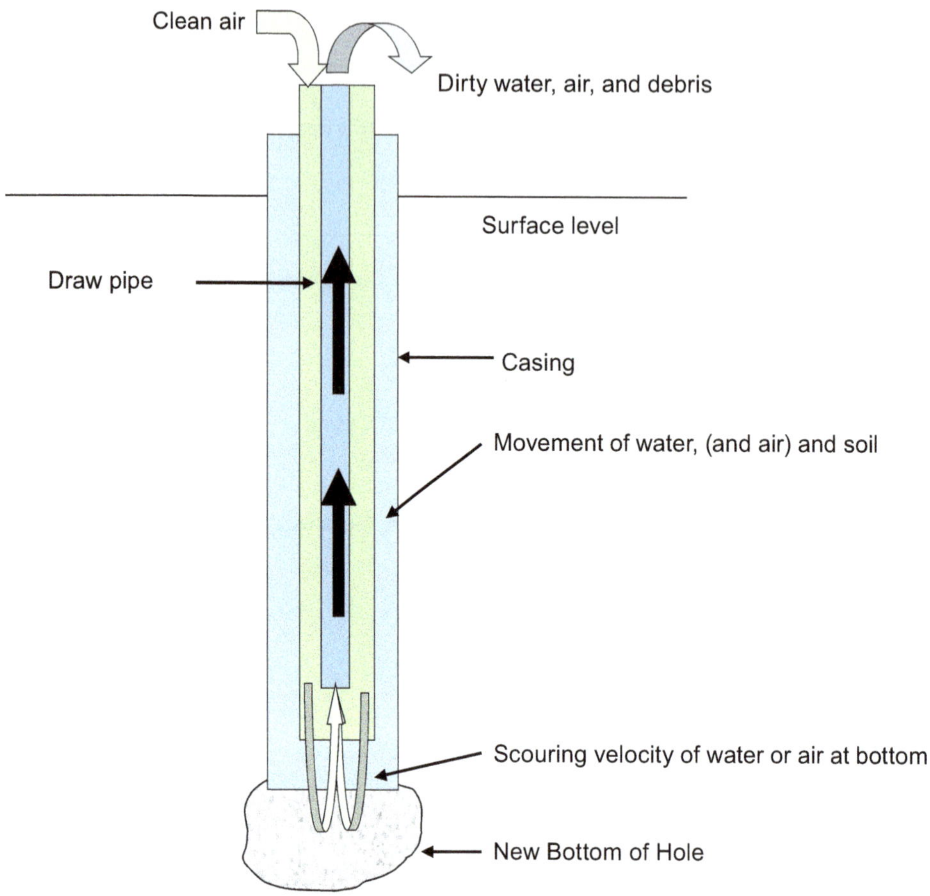

Figure 4-30 Dual-tube circulation system

Source: Courtesy of John Largey

Figure 4-31 Typical reverse-air, dual-tube drill rig with cyclone

Source: Courtesy of John Largey

Figure 4-32 Sample from cyclone on typical reverse-air, dual-tube drill rig

how samples come out of the sampling cyclone. The air and cuttings return to the surface through the inner tube to a sample cyclone where lithologic samples are recovered (Figure 4-32). Significant cores of the formation are recoverable with reverse-air, dual-tube methods. Excellent quality lithologic samples are obtained by this method. A small-diameter well can be installed through the inner tube as well. As a result, the dual-tube method is also successfully used for test wells. Additional advantages of the dual-tube method include:

- fast drilling rate and minimal impact to aquifers,
- well installation can be accomplished on small sites,
- accurate, continuous water and formation samples can be recovered,
- a clean borehole when drilling is complete.
- the loss of circulation is minimized, and
- reduced potential of cross-contamination.

The method is commonly used when high-quality wells with high-quality water quality and lithologic samples are needed. However, because of the precision required for reverse-air, dual-tube methods, the cost is high, and there are limitations to the depth and size of the well. Other disadvantages include:

- limited availability of equipment,
- large quantities of pressurized air are required,
- sloughing of the formation and water saturated sands may cause the drill pipe to bind and become stuck,
- a high volume of formation fluids is generated as a result of the drilling method,
- dual-tube is not applicable to certain types of formations, such as clays, and
- plugging of the formation is possible, so extensive development of the well may be needed.

DOWN-THE-HOLE-HAMMER

As with many other industries, well drilling has seen the introduction of new methods in recent years. One of the more popular, and useful, is the down-the-hole-hammer (DTHH) method (Figure 4-33). This drilling method combines some elements of hydraulic drilling with the percussion action of the cable-tool method. A pneumatic drill is located at the bottom of the drill string. Air delivered from an adequately sized compressor actuates a downhole air piston, transmitting blows from the bit to the formation. The repeated blows are similar to those delivered by a cable-tool rig. The air used to actuate the hammer also removes the cuttings from the borehole similar to reverse-air rotary drilling. Occasionally, as in air rotary drilling, foaming additives are added to aid in removal of the cuttings. This eliminates the need to remove the bit from the borehole and recover the cuttings by using a bailer. As the blows are delivered, the drill pipe is slowly rotated by either a top-head drive or a rotary table. The rotating motion and the removal of the cuttings allow the bit to deliver each successive blow to a different, clean surface instead of repeatedly striking the same surface or previously cut rock fragments. DTHH works well in situations where hard formations are anticipated, and borehole stability is not an issue. Wells drilled using this method are commonly 6 in. in diameter, although hammer sizes range to about 17 in. in diameter.

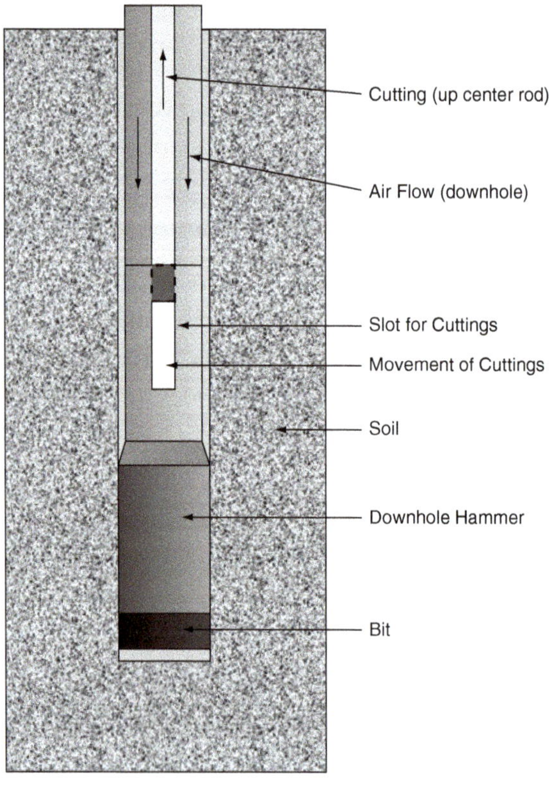

Figure 4-33 Down-the-hole-hammer method

Advantages of the DTHH method include:

- rapid removal of cuttings,
- drilling mud not used,
- high penetration rates in resistant formations,
- high quality water and lithologic samples recovered,
- excellent geophysical logs obtained, and
- yield estimates obtained at selected depths during drilling.

Disadvantages of the DTHH method include:

- restricted to semi-consolidated to consolidated formations,
- high volumes of air needed to activate the pneumatic hammer, and
- high volumes of water produced during drilling.

CORING

Coring, which is not a primary method of drilling, is used to collect high-quality lithologic samples. Cores are cylinders of rock recovered from the formation (Figure 4-34 and Figure 4-35). Various tests may be conducted on cores, including hydraulic conductivity, porosity, permeability, and compressive strength. This information will help professionals determine the appropriate factors to include in designing the well pumps and determining the safe yield of the well.

To recover a core, the following procedure is used.

- Drill a pilot hole to the top of the formation where the core is to be recovered.
- Trip out drill bit and drill pipe.
- Trip in core barrel and drill pipe (see Figure 4-36).
- Cut the core out of the formation where the core is desired. It is essential that the core be cut, not pushed into the barrel.

Source: Courtesy of John Largey

Figure 4-34 Core sample

Source: Courtesy of John Largey

Figure 4-35 Boxed core

- Trip out the core barrel and drill pipe.
- Recover the core.
- The core should be placed in a box and labeled with core recovery date, depth, well, and relevant observations at the time of collection (see Figure 4-35). For core recovery, it is essential that an experienced driller be used.
- Drill to the next point where a core is desired.

During drilling of an exploratory well and any subsequent supply well, it is often useful to collect geologic samples and rock cores of formations penetrated during drilling. Formation cores should be taken at selected intervals. The cores are typically 10–30 ft long and 4 in. in diameter. The recovered core should be described by a qualified geologist, and the data, samples, and descriptions should be submitted as directed by the regulatory agencies.

In addition to the description of the geologic strata and structural features, vertical and horizontal porosity tests should be performed on the samples at appropriate core intervals (i.e., the formations of concern). Porosity, compressive strength, modulus of elasticity, and specific gravity should be recorded. When the pilot hole is completed, geophysical logs and a survey involving the lowering of a camera down the well should be conducted on the borehole.

DRILLING METHOD SELECTION

When preparing to drill a well, the appropriate drilling method must be selected; a combination of methods may be required. A professional engineer, professional geologist, hydrogeologist, or a combination should specify the methods appropriate for a particular project. For example, bucket-type drilling may be used in

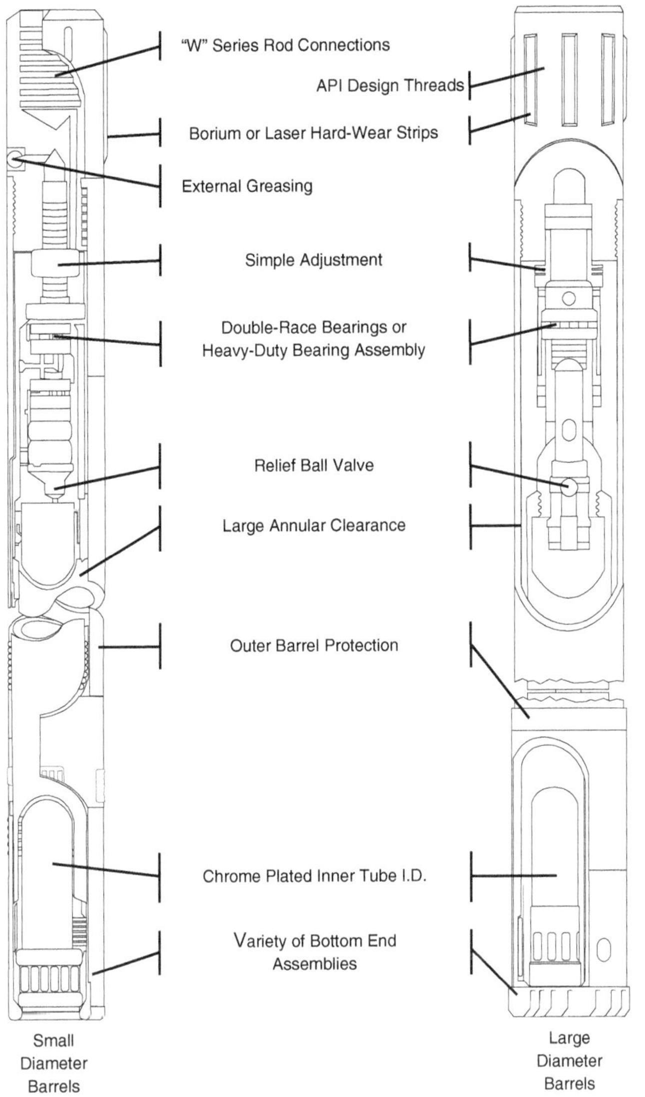

Source: Adapted from Christensen Products

Figure 4-36 Core barrels

setting relatively shallow starter casings for large-diameter water wells, and mud rotary methods may then be used to drill to a predetermined depth above the production zone. After the second string of casing is set and cemented into place, the drilling mud will be displaced with potable water. Reverse-air methods may be used through the production zone. During the drilling, cuttings should be collected and analyzed by a hydrogeologist. Composite samples are used to determine formation characteristics over a predetermined interval of the formation. Grab samples may be used for specific analysis. However, in either case, knowing the depth in the borehole where the sample was recovered is important. Without knowing the depth, the information derived from the samples is of limited value.

In addition to determining the borehole data, planning is needed for the proposed site for the well construction activities. Overhead and buried utilities must be located and accounted for before drilling commences. Large well diameters and deep wells require large drill rigs. Drill rigs for large wells may be very tall, requiring Federal Aviation Administration clearance near airports. Casing must be delivered to the location and staged before installation. If geophysical logging is required, additional areas will be required. Utilities and potable water supplies are useful for the drilling project, and access to them should be considered during drilling operations.

The type of drilling method used will impact the amount of area needed for the drilling process. Hydraulic rotary systems with mud will require more lay-down area for mud separation than reverse-air, dual-tube drilling. Table 4-1 outlines the benefits and issues with the methods discussed.

Table 4-1 Summary of drilling methods, benefits, and limitations

Drilling Method	Uses	Applications	Benefits	Limitations
Driving	Small, shallow wells	Dewatering	Simple and easy	Small wells
Jetting	Small, shallow wells	Dewatering and single family homes	Simple	
Cable tool	Up to 18-in. diameter and 1,000 ft in depth	Water supply	Minimal aquifer impact, favorable sampling, applicable to most formations, relatively simple	Slow, casing must be advanced with drill bit, pulverizes formation
Hollow stem auger	Up to 8-in. diameter and 300 ft in depth	Water supply	Good sampling, auger keeps borehole clean, high quality sampling, does not introduce drilling fluids	Limited to unconsolidated formations, limited to depth and size, running sand limits
Hydraulic rotary	Deep and large diameter wells	Water supply	Fast, most common, no formation limitations, multiple drill bits, good quality cuttings, high quality borehole for logging for uncased boreholes	Requires fluid (mud), large bit sizes needed, possible need for fluids, mud disposal, samples not of high quality, aquifer plugging
Reverse air	Large diameter, depth can vary	Water supply	Aquifer testing possible during drilling, high water quality and formation sampling, fast, excellent logs, supports telescoping	Borehole may collapse, possible cross contamination of aquifers
Reverse air dual-tube	Small diameter, limited depth	Water supply and test holes	Aquifer testing possible during drilling, high water quality and formation sampling, fast, excellent logs, supports telescoping	Cost, volume of cuttings, plugging while drilling

Disposal of water, mud, and cuttings is also an issue that must be planned. Waste cannot be placed on the surface or disposed of on the ground. For instance, saltwater usually must be trucked off-site for treatment. Sewers are not often available where wells are being drilled; if they are, the utility may not want drilling fluids in the sewer system. Federal, state, and local regulations should be consulted before initiating a drilling project.

REFERENCES

American Water Works Association (AWWA). 2014 (4th ed.). M21, *Groundwater*. Denver: AWWA.

Lehr, J., S. Hurlburt, B. Gallagher, and J. Voytek. 1988. *Design and Construction of Water Wells: A Guide for Engineers*. National Water Well Association. New York: Van Nostrand Reinhold.

Scalf, M.J., W. McNabb, R. Dunlap, R. Crosby, and J. Fryberger. 1980. *Manual for Ground Water Sampling Procedures*. Ada, Okla.: R.S. Kerr Environmental Research Laboratory, Office of Research and Development.

CHAPTER 5
Well Design

Typical steps or phases for construction of wells include predesign, determining regulatory criteria, field testing, design, construction, and testing and operations. Each step includes strategies for addressing technical and/or regulatory concerns. For example, there are regulatory conditions that should be addressed early in the project, such as estimation of future demands and the potential for impacts of the additional demands on existing users, water levels in the region, and environmental impacts. As noted in previous chapters, these tasks will include locating the well sites and gathering available data from local wells and existing legal users. The field testing is usually intended to allow collection of site-specific data for use in the final design of wells. A test well may be advisable, especially if it can be converted to a functional well upon completion of testing. Results from testing provide information needed to define the underground conditions (e.g., lithology, water production, and water quality).

PREDESIGN

Design and construction of water supply wells provide the opportunity to gather raw water quality if favorable subsurface conditions exist. These conditions include a formation that is areally extensive and porous enough to permit cost-effective water flow to the well. Initial data to be collected at a potential well site were discussed in previous chapters. After the initial investigation and confirmation of the presence of water-bearing formations, the quantity of water that can be sustainably withdrawn can be estimated.

Important aquifer parameters for predesign purposes are porosity, transmissivity, specific yield, and specific retention. These parameters are used to determine hydraulic conductivity/transmissivity and storage coefficient and assist in describing rock formations and well yield. Hydraulic head and gradient are also important as they provide insight on water movement and represent the mechanics of horizontal and vertical recharge to a well being pumped. Head and gradient are also used to analyze the transport of pollutants that may migrate to a well (AWWA 2014). Hydraulic conductivity and transmissivity indicate how easily water will move in the formation. Hydraulic conductivity is the basic, three-dimensional parameter required for modeling purposes. Aquifer characteristics allow professionals to calculate the extent of the cone of influence, flow, and specific yield.

Specific well design should include calculations for other criteria such as uphole velocity, screen size, well losses, and casing burst strength. Well screens add friction and hence reduce well capacity. Formation information will also help to identify if well screens are needed and, if so, slot sizing.

Porosity

The ratio of openings (voids) to the total volume of a soil or rock is referred to as porosity. Porosity is expressed either as a decimal fraction or as a percentage. Thus,

$$n = \frac{V_t - V_s}{V_t} = \frac{V_v}{V_t} \qquad \text{(Eq 5-1)}$$

Where:

n = porosity, as a decimal fraction
V_t = the total volume of a soil or rock sample
V_s = the volume of solids in the sample
V_v = the volume of openings (voids)

If the porosity determined using Eq 5-1 is multiplied by 100, the result is porosity expressed as a percentage.

Table 5-1 outlines the porosity of various formation materials. Surface topsoils are highly porous materials that are caused by loose soil particles, root holes, and animal burrows, but they rarely make good water supply sources. The porosity of unconsolidated sand and gravel depends on the range in grain size, the degree of sorting, and the shape of the rock particles. Fine-grained materials tend to be better sorted and have the highest porosity values. Clay has a high percentage of voids; however, clay transmits virtually no water because the voids are so small.

Specific Yield and Specific Retention

Specific yield is the portion of water in a formation that will drain under the influence of gravity. Specific yield is important for determining the amount of water that can be withdrawn from a formation and should be checked on an ongoing basis throughout the life of the well. Specific yield is calculated as follows:

$$n = S_y + S_r \qquad \text{(Eq 5-2)}$$

Table 5-1 Values of porosity for various geologic materials

Material	Porosity (%)
Sedimentary	
Gravel, coarse	24–36
Gravel, fine	25–38
Sand, coarse	31–46
Sand, fine	26–53
Silt	34–61
Clay	34–60
Sedimentary rocks	
Sandstone	5–30
Siltstone	21–41
Limestone, dolomite	0–20
Karst limestone	5–50
Shale	0–10
Crystalline rocks	
Fractured crystalline rocks	0–10
Dense crystalline rocks	0–5
Basalt	3–35
Weathered granite	34–57
Weathered gabbro	42–45

Source: Davis 1969; Johnson and Morris 1962

$$S_y = \frac{V_d}{V_t} \qquad \text{(Eq 5-3)}$$

Where:

n = porosity
S_y = specific yield
S_r = specific retention
V_d = the volume of water that drains from a total volume of Vt
V_t = total volume of a soil or rock sample

Specific retention is the opposite of specific yield. Specific retention is the water that is retained as a film on rock surfaces and in very small openings that is not likely to be recovered in wells. The physical forces that control specific retention are the same forces controlling the thickness and moisture content of the capillary fringe, as expressed in the following equation:

$$S_r = \frac{V_r}{V_t} \qquad \text{(Eq 5-4)}$$

Where:

V_r = the volume of water retained in a total volume of V_t

Table 5-2 lists general values of specific yield for various geologic materials.

Hydraulic Head and Gradient

Potentiometric head is the water surface in a confined aquifer. As noted in Chapter 1, this is the point above the confining unit to which the water level rises. In either case, the highest head (i.e., water level) occurs where the aquifer is recharged. The change in head over the distance between the point of recharge and the point where a well is proposed to be installed is referred to as the gradient (dz/dL). Water levels should be established from a fixed measuring point. Typically, the referenced standard is the National Geodetic Vertical Datum (NGVD) of 1929 or, preferably, the North American Vertical Datum of 1988 (NAVD88), a newer elevation reference point. Both have the point 0.0 commonly referred to as *sea level*, but in reality those reference points may be 1.5 ft apart. Use the newer datum.

Table 5-2 Values of specific yield for various geologic materials

Material	Specific Yield in Percent (%)
Gravel, coarse	23
Gravel, medium	24
Gravel, fine	25
Sand, coarse	27
Sand, medium	28
Sand, fine	23
Silt	8
Clay	3
Sandstone, fine-grained	21
Sandstone, medium-grained	27
Limestone	14
Dune sand	38
Loess	18
Peat	44
Siltstone	12
Till, predominantly silt	6
Till, predominantly sand	16
Till, predominantly gravel	61
Tuff	21

Source: Adapted from Johnson 1967

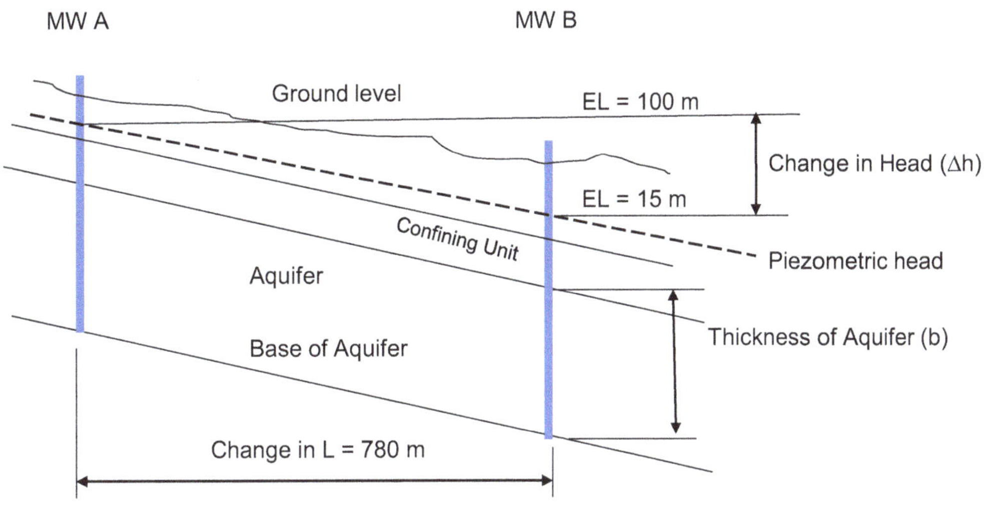

Figure 5-1 Piezometric head

The gradient between any two observation points (Figure 5-1) is found as follows:

$$\frac{\Delta h}{L} = \frac{(100 \text{ m} - 15 \text{ m})}{780 \text{ m}} = \frac{85 \text{ m}}{780 \text{ m}} \tag{Eq 5-5}$$

Where:

h = elevation of the water table or the potentiometric head, whichever applies

L = distance between measuring wells 1 and 2

The hydraulic gradient can be determined if the following data are available for three wells located in any triangular arrangement:

- relative geographic position of the wells,
- distance between the wells, and
- total head at each well.

Total head is the sum of elevation head, pressure head, and velocity head. Because groundwater moves relatively slowly, velocity head can generally be ignored. Therefore, the total head at an observation well involves only two components: elevation head and pressure head. Groundwater moves in the direction of decreasing total head, which may or may not be in the direction of decreasing pressure head.

The equation for total head h_t is:

$$h_t = z + h_p \tag{Eq 5-6}$$

Where:

z = elevation head, the distance from the datum plane to the point where the pressure head h_p is determined.

Flow and Hydraulic Conductivity

The factors controlling groundwater movement are defined by Darcy's law as follows:

$$Q = KA\left(\frac{dh}{dl}\right) \quad \text{(Eq 5-7)}$$

Where:

Q = the quantity of water per unit of time

K = the hydraulic conductivity, which depends on the size and arrangement of the water-transmitting openings (pores and fractures) and on the dynamic characteristics of the fluid (water), such as kinematic viscosity, density, and the strength of the gravitational field (also referred to as the *coefficient of permeability*)

A = the cross-sectional area, at a right angle to the flow direction, through which the flow occurs

dh/dl = the hydraulic gradient

Unlike rivers and streams, groundwater tends to move slowly. As a result, groundwater generally flows under laminar conditions, which means that the individual water particles tend to follow discrete streamlines and not mix with particles in adjacent streamlines. As a result, the quantity of water, Q, is directly proportional to the hydraulic gradient, dh/dl.

If Eq 5-7 is rearranged to solve for K, the following is obtained:

$$K = \frac{Qdl}{Adh} = \frac{(m^3/d)(m)}{(m^2)(m)} = \frac{m}{d} \quad \text{(Eq 5-8)}$$

Thus, the units of hydraulic conductivity are those of velocity (or distance divided by time). In Eq 5-8, the factors involved in the definition of hydraulic conductivity include the volume of water, Q, that will move in a unit of time (commonly, one day) under a unit hydraulic gradient (such as a meter per kilometer) through a unit area (such as a square meter). Expressing hydraulic conductivity in terms of a unit gradient rather than an actual gradient at some place in an aquifer allows for comparison of hydraulic conductivity values for different rock types.

The units of hydraulic conductivity are those of velocity (or distance divided by time). However, the velocity units are less obvious in Darcy's law because the definition of hydraulic conductivity includes the volume of water, Q, that will move in a unit of time (gpd) assuming a unit hydraulic gradient (such as ft/mile) through a unit area (such as ft^2).

Figure 5-2 outlines the ranges of hydraulic conductivity through 12 orders of magnitude. Hydraulic conductivity will vary by type of rock and likely will be different from place to place in the same rock (Table 5-3). If the hydraulic conductivity is essentially the same throughout an area, the aquifer is considered to be homogeneous. If the hydraulic conductivity differs from one part of the aquifer to another, the aquifer is considered to be heterogeneous.

Water typically flows preferentially in one direction. As a result, the hydraulic conductivity may be different along different axes in an aquifer. If it varies by direction, such as differences between conductivity in the vertical and horizontal directions, the aquifer is anisotropic. If the hydraulic conductivity is essentially the same in all directions, the aquifer is isotropic. While convenient to assume that aquifers are both homogeneous and isotropic, these aquifers are rare.

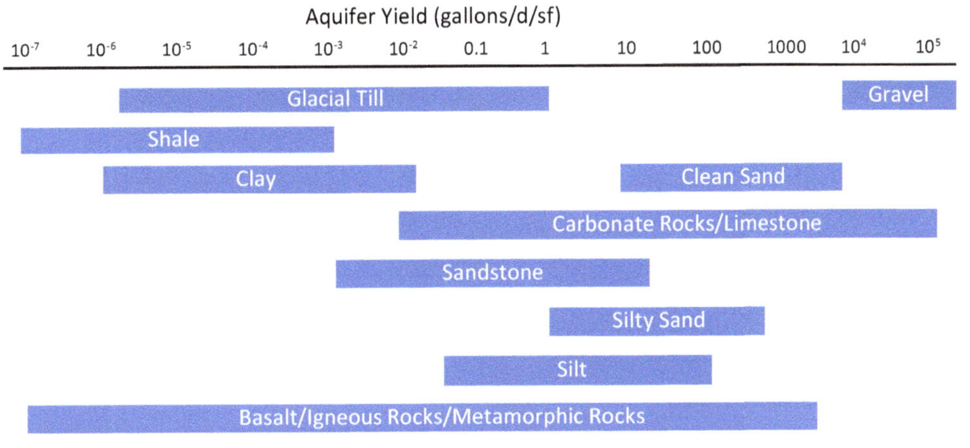

Note: Fractured formations tend to have higher flow values (right side of basalt) compared to unfractured formations.

Source: Adapted from Bear, et al., 1968

Figure 5-2 Typical values of hydraulic conductivity and permeability

Table 5-3 Ranges of hydraulic conductivity for various rock types

	Hydraulic conductivity (m/s)		
Sedimentary			
Gravel	3×10^{-4}	to	3×10^{-2}
Coarse sand	9×10^{-7}	to	6×10^{-3}
Medium sand	9×10^{-7}	to	5×10^{-4}
Fine sand	2×10^{-7}	to	2×10^{-4}
Silt, loess	1×10^{-9}	to	2×10^{-5}
Till	1×10^{-12}	to	2×10^{-6}
Clay	1×10^{-11}	to	5×10^{-9}
Unweathered marine clay	8×10^{-13}	to	2×10^{-9}

(continued)

Table 5-3 Ranges of hydraulic conductivity for various rock types (*continued*)

	Hydraulic conductivity (m/s)		
Sedimentary rocks			
Karst and reef limestone	1×10^{-6}	to	2×10^{-2}
Limestone, dolomite	1×10^{-9}	to	6×10^{-6}
Sandstone	3×10^{-10}	to	6×10^{-6}
Siltstone	1×10^{-11}	to	1×10^{-8}
Salt	1×10^{-12}	to	1×10^{-10}
Anhydrite	4×10^{-13}	to	2×10^{-8}
Shale	1×10^{-13}	to	2×10^{-9}
Crystalline rocks			
Permeable basalt	4×10^{-7}	to	2×10^{-2}
Fractured igneous and metamorphic rock	8×10^{-9}	to	3×10^{-4}
Weathered granite	3×10^{-6}	to	5×10^{-5}
Weathered gabbro	6×10^{-7}	to	4×10^{-6}
Basalt	2×10^{-11}	to	4×10^{-7}
Unfractured igneous and metamorphic rocks	3×10^{-14}	to	2×10^{-10}

Source: Adapted from Domenico and Schwartz 1990. Reprinted by permission of John Wiley & Sons, Inc.

Transmissivity is related to hydraulic conductivity. The transmissivity, *T*, of an aquifer is equal to the hydraulic conductivity of the aquifer multiplied by the saturated thickness of the aquifer as shown in the following equation:

$$T = Kb \qquad \text{(Eq 5-9)}$$

Where:

T = transmissivity
K = hydraulic conductivity
B = aquifer thickness

As with hydraulic conductivity, transmissivity is also defined in terms of a unit hydraulic gradient.

Storage Coefficient

The storage coefficient is the ability of a formation to store and transmit water. These are the formation's most important hydraulic properties. These properties

Table 5-4 Units of hydraulic parameters of aquifers

Porosity	Unit cube of material	Unit prism of aquifer
Transmissive capacity	Hydraulic conductivity	Transmissivity
Available storage	Specific yield	Storage coefficient

are given either in terms of a unit cube of the material or in terms of a unit prism of an aquifer, depending on the intended use. These abilities, as they relate to the two units of measurement, are shown in Table 5-4 (AWWA 2014).

The storage coefficient S is defined as the volume of water an aquifer releases or stores per unit surface area of the aquifer per unit change in head. The storage coefficient is a dimensionless unit, as Eq 5-10 shows, in which the units in the numerator and the denominator cancel.

$$S = \frac{\text{volume of water}}{(\text{unit area})(\text{unit head change})} = \frac{\text{m}^3}{(\text{m}^2)(\text{m})} = \frac{\text{m}^3}{\text{m}^3} \qquad \text{(Eq 5-10)}$$

The size of the storage coefficient depends on whether the aquifer is confined or unconfined. In a confined aquifer having a porosity of 0.2 and containing water at a temperature of about 59°F (15°C), expansion of the water alone releases about 3 x 10^{-7} m^3 of water per cubic meter of aquifer per meter of decline in head. To determine the storage coefficient of an aquifer caused by expansion of the water, the aquifer thickness must be multiplied by 3 x 10^{-7}. If only the expansion of water is considered, the storage coefficient of an aquifer 100-m thick would be 3 x 10^{-5}. The storage coefficient of most confined aquifers ranges from about 10^{-3} to 10^{-5}. The difference between these values and the value caused by expansion of the water is attributed to compression of the aquifer (AWWA 2014).

Capillarity and Unsaturated Flow

Most recharge of surficial groundwater systems occurs during the percolation of water through the unsaturated zone of soil. This movement of water is controlled by both gravitational and capillary forces. The capillarity forces result from the mutual attraction (cohesion) between water molecules and the molecular attraction (adhesion) between water and different solid materials (AWWA 2014). Because most pores in granular materials are of capillary size, water is pulled upward into a capillary fringe above the water table to a height, h_c, above the water level (Table 5-5). In a steady-state condition or conditions in which the moisture content remains constant, flow of water in the unsaturated zone can be determined from a modified form of Darcy's law, as shown in the following equation:

$$Q = K_e A\left(\frac{h_c - z}{z}\right) \pm \left(\frac{dh}{dl}\right) \qquad \text{(Eq 5-11)}$$

Table 5-5 Approximate height of capillary rise h_c in granular materials

Material	Rise, *in.*
Sand	
Coarse	05
Medium	10
Fine	15
Silt	40

Where:

Q = the quantity of water
K_e = the hydraulic conductivity under the degree of saturation existing in the unsaturated zone
A = the cross-sectional area through which flow occurs
$(h_c - z)/z$ = the gradient due to capillary (surface tension) forces
dh/dl = the gradient due to gravity

The +/– sign accounts for the direction of movement: plus for downward and minus for upward. For movement in a vertical direction, either up or down, the gradient caused by gravity is 1. For lateral (horizontal) movement in the unsaturated zone, the gravitational gradient can be eliminated.

Because transmissivity depends on both K and b, its value differs between aquifers and from place to place in the same aquifer. Estimated values of transmissivity for principal aquifers range from less than 1 gpd for some fractured sedimentary and igneous rocks to more than 1,000,000 gpd for cavernous limestones and lava flows (AWWA 2014).

In a related issue regarding transmissivity, Nuzman (1989) accepted that the field coefficient of permeability represents the limit of laminar flow through the formation at a given temperature and viscosity of the water (AWWA 2014). The limit of laminar flow through the borehole wall is defined by the following equation:

$$Q = \pi dLk \qquad \text{(Eq 5-12)}$$

Where:

k = permeability
L = length of screen (ft)
d = mean grain diameter (in.)

This equation assumes vertical flow that does not actually occur in wells. It has been found by field experience that the beginning of turbulent flow through the formation borehole was approximately 2.35 times the laminar flow limit (AWWA 2014). Williams (1985) defined the point where flow transition from predominantly turbulent flow to predominantly laminar flow as:

$$r_e = 0.9587(Q/L)d/\theta \qquad \text{(Eq 5-13)}$$

Where:

r_e = critical radius (in.)
Q/L = specific aquifer discharge (gpm/ft)
Q = discharge rate (gpm)
L = length of screen (ft)
d = mean grain diameter (in.)
θ = effective porosity

It should be noted that Williams (1985) defined the Reynolds number at the point where the flow regime changes from laminar to turbulent as being 30.

Flow volumes can be derived for design purposes from monitoring wells and drawdown information for equilibrium conditions from one of two formulas developed by Theim. For water table aquifers:

$$Q = K\left(H^2 - h^2\right)/\left(1{,}055\log(R/r)\right) \qquad \text{(Eq 5-14)}$$

Where:

Q = discharge rate (gpm)
r = distance to observation well (ft)
R = borehole diameter (ft)
K = hydraulic conductivity (gpd/ft)
H = drawdown at the borehole
h = drawdown at observation well (ft)

For confined aquifers:

$$Q = Kb(H - h)/\left(528\log(R/r)\right) \qquad \text{(Eq 5-15)}$$

Where:

Q = discharge rate (gpm)
r = distance to observation well (ft)
R = borehole diameter (ft)
K = hydraulic conductivity (gpd/ft)
b = formation thickness (ft)
H = drawdown at the borehole
h = drawdown at observation well (ft)

WELLFIELD DESIGN

The proper design of wells and wellfields is possible through measurable field data (Bloetscher et al. 2005). A small-diameter test well is usually constructed to ascertain the depth and thickness of the aquifer and formation properties, along with water quality. Once completed, fully functional wells can be tested (i.e., withdrawal or pumping test) to estimate transmissivity/hydraulic conductivity. Water quality can be analyzed to assess treatment strategies. Once the aquifer parameters have been field-tested and predesign assumptions resolved with the field findings, the well spacing (i.e., separation) between wells in a wellfield, effects of new wells on existing wells, and the optimum pumping rates can be estimated. The field-test results and geophysical logging are very useful when making an overall appraisal of the groundwater resources of an area and the potential for future water supply development.

Wellfield Interference

Possible interference between wells should be determined before locating multiple wells in a well field. Determining interference between pumping wells will allow the design engineer and/or hydrogeologist to optimize spacing and pump capacity to determine the most efficient placement pattern and pumping rates. The total interference drawdowns estimated for various pumping rates are shown in Figure 5-3.

Pumping Rates

When numerical values have been assigned to transmissivity and the storage coefficient, the drawdown effects of pumping can be determined. These effects are for any quantity of water at any reasonable distance from the pumping well. A graphic representation should be plotted of water levels against the logarithm of distance from the center of pumping for a given time period. A minimum continuous pumping period of 100 days is usually used as a conservative safety factor.

WELL DESIGN

Once the wellfield spacing and pumping rates are determined, the proper design of wells and specification of materials are made. Decisions regarding materials,

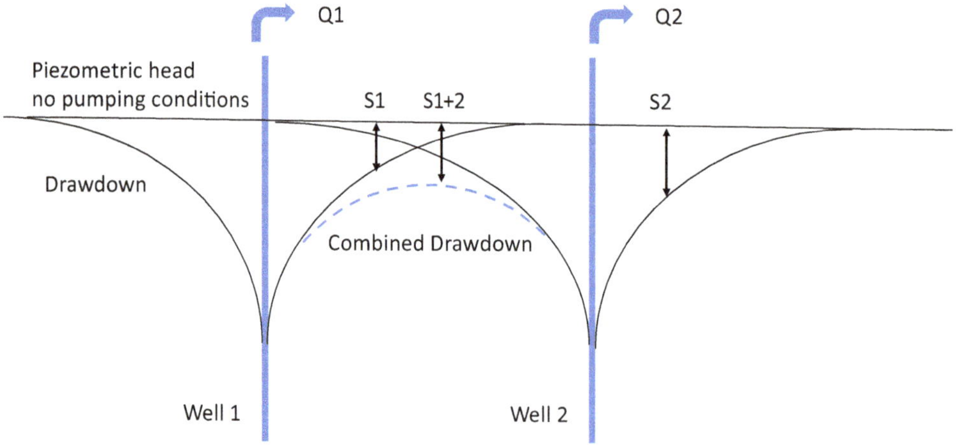

Figure 5-3 Interfering drawdowns between wells

screened interval length, screen openings, casing and column pipe diameter, and the need for sand or silt removal can be made once the field parameters are understood. A series of calculations must be made to ensure the proper size borehole, casing, screen, and screen spacing are designed. Typically, the client has a need for a specific water yield. The designer should use these data to design a cost-effective and efficient well.

Uphole Velocity

The maximum uphole velocity criteria imposed on withdrawal wells is 10 ft/s, which affects casing diameters. However, this velocity is too high for many wells because it tends to entrain sand and other particles if proper screening is not provided (see Entrance Velocity section). High velocity will also exert wear on the casing. Theoretical withdrawal velocities and pressures can be calculated for a maximum day flow rate using the following equation (Heald 1994):

$$V = 0.4085\, Q/d^2 \qquad \text{(Eq 5-16)}$$

Where:

V = velocity of flow, fps
d = inside diameter of the casing, in.
Q = flow rate, gpm

Friction Loss

There are also pressure (head) limitations in a confined aquifer well. The total pressure head consists of three components: (1) friction loss (head loss) through the column pipe; (2) head caused by the formation; and (3) the pump driving pressure (Bloetscher et al. 2005). The loss of pressure and upward buoyant forces decrease radially from the well. Therefore, the greater the distance from the well, the more the pressure head remains in the formation.

Pressure (head) is friction loss through the pipe (in the case of a well, the column pipe). Friction losses through the column pipe are a function of the diameter of the pipe, the rate of flow through the pipe, and the roughness of the pipe referred to as the *friction factor* (or *coefficient of friction*). The head losses through the pipe are most commonly calculated using the Hazen–Williams formula (Heald 1994):

$$h_f = 0.002083 \times L \times \left(\frac{100}{C}\right)^{1.85} \times \left(\frac{gpm^{1.85}}{d^{4.8655}}\right) \qquad \text{(Eq 5-17)}$$

Where:

h_f = head loss due to friction (ft)
L = length of pipe
C = friction factor for Hazen–Williams (dimensionless)
gpm = flow (gpm)
d = inside diameter of the column pipe (in.)

Density Differential

The density differential is calculated using a derivation of the Ghyben-Herzberg principle (Bloetscher et al. 2005). Note the Ghyben-Herzberg relation does not apply for nonequilibrium conditions. As a result, the following equation can be derived for the nonequilibrium condition, which is stated as follows (Fetter 1994):

$$z = \frac{\rho_f}{\rho_s - \rho_f} h_f - \frac{\rho_f}{\rho_s - \rho_f} h_s \qquad \text{(Eq 5-18)}$$

Where:

$z(x,y)$ = depth to the saltwater interface below sea level at location (x,y) (ft)
ρ_f = density of fresh water (g/cm^3)
ρ_s = density of salt water (g/cm^3)
$h(x,y)$ = elevation of the water table above sea level at location (x,y) (ft)

The following equation can be used for the nonequilibrium condition.

$$h_d = (\rho_s - \rho_f) z / \rho_f \qquad \text{(Eq 5-19)}$$

Where:

h_d = head due to density differential (ft)
ρ_s = density of salt water (g/cm^3)
ρ_f = density of fresh water (g/cm^3)
z = length of column of fresh/salt water (ft)

Note that given the density differential in Eq 5-19, this equation is approximately equal to the following:

$$z = 40 h_f \qquad \text{(Eq 5-20)}$$

This equation can be solved for h_f to obtain:

$$h_f = \sqrt{\frac{2(\rho_s - \rho_f) q_v x}{(\rho + (\rho_s - \rho_s))K}} \qquad \text{(Eq 5-21)}$$

where q_v = flow per unit width of the aquifer system and K is hydraulic conductivity as defined previously.

Recalling the Darcy velocity, the following equation results from the differential density problems:

$$v = -\frac{[k]}{\mu}[\nabla p + \rho g \nabla z] \qquad \text{(Eq 5-22)}$$

Where:

ν	=	Darcy velocity
k	=	permeability tensor
μ	=	fluid viscosity.

STEP-DRAWDOWN TESTING

The purpose of step-drawdown testing is to evaluate the performance of the well. The data collected during the step-drawdown tests are used in the evaluation of the performance, efficiency, and specific capacity of wells at the different pumping rates (AWWA 2014). The data are also used for the calculation of the transmissivity and hydraulic conductivity of the aquifer; to quantify the deterioration in well performance over time; and yield information regarding well efficiency, well development, and well screen/borehole clogging. Step-drawdown testing should be performed yearly on each well in conjunction with sand testing. Step-drawdown testing should also be performed before and after well disinfection. This procedure will allow evaluation of the effectiveness of the disinfection procedure. The results also help determine possible rehabilitative procedures and optimum pumping rates.

Step-Drawdown Testing Procedure

Step-drawdown testing involves pumping a well for a predetermined amount of time (approximately 60 min), until water level stabilization is reached, at each of three to five increasing pumping rates. Ideally, the three rates should be at 50, 100, and 150 percent of the design pumping rate. Before each increase in pumping rate, water levels are allowed to recover to static levels for at least the same amount of time as the well was pumped, or at minimum, come to a point of equilibrium. The changes in water levels within the well are measured with an electric water-level probe (M-scope) during both the drawdown and recovery periods. It is imperative that the M-scope be disinfected before use in each well and after use in each well. The time increments for measurements are as follows: 1-min readings for the first 10 min, 2-min readings from 10 to 20 min, and 5-min readings from 20 to 60 min and/or the end of the test.

From a measurement standpoint, the most important test measurements are the first and last water-level measurements. If the well does not stabilize at 60 min, the test readings should be measured until the well is stable. Each successive rate (and recovery period) must be extended out to at least the longest previous time interval, and a measurement must be obtained at the same time as the end of the first drawdown step. The recovery period should be equal to the longest time period in the drawdown step. Example: If step 1 lasts for 60 min, but the second step indicates the well is still drawing down at 60 min, readings should continue until the well is stable (for example, 100 min). Readings will then be obtained at 60 min and at 100 min.

Discharge from the well should be controlled by a gate valve and should be measured using a calibrated flowmeter and/or an orifice manometer assembly. The total drawdown (measured in the field) in a well is a function of the drawdown due to aquifer characteristics and the drawdown due to the loss of efficiency from the well. Total drawdown(s) can be written as the following equation (Dawson and Istok 1991):

$$s = BQ + CQ^2 \qquad \text{(Eq 5-23)}$$

Where:

s = drawdown in the well casing, ft
Q = pumping rate, gpm
C = well loss coefficient, sec^2/ft^5
B = $(264/T)\log[(0.3Tt)/(r2S)]$, gpd/ft

and

T = transmissivity, gpd/ft
t = time, min
r = radius of the well, ft
S = storage coefficient, dimensionless

Because the transmissivity and storage coefficient of an artesian aquifer or a leaky aquifer are constant, the BQ term in the equation does not affect the determination of well loss using Eq 5-23. In a water table aquifer, the transmissivity, storage coefficient, and specific yield values change (decrease) as the aquifer saturated thickness decreases. The drawdown equation does not compensate for the effects of partial penetration of the aquifer. However, if the degree of dewatering is small, changes in transmissivity, storage coefficient, and specific yield are negligible and can be ignored. Assuming that the well is not developing, the total drawdown can be used to determine transmissivity. However, this method gives lower transmissivity values than those calculated without accounting for drawdown caused by well loss.

Specific Capacity Calculation

The productivity (quantity of water produced) of a well can be expressed as specific capacity. The specific capacity of a well is defined as the ratio of the pumping rate to the drawdown at a given time, as illustrated in the following equation:

$$C_s = Q/S \qquad \text{(Eq 5-24)}$$

Where:

C_s = specific capacity of the well, gpd/ft of drawdown at a unit of time
Q = pumping rate, gpm
S = drawdown, ft

Estimating the specific capacity of a well requires determining the drawdown from a static water level to a pumping water level within the well at a known pumping rate after a known span of time. Specific capacity is measured in gallons per minute per foot of drawdown at a given period of time (gpm/ft at a unit of time) to calculate well efficiency. The higher the specific capacity, the more efficient the well—as long as all other factors are equal. Specific capacity changes in a nonlinear fashion with increased pumping rates because a well cannot be 100 percent efficient. Slight decreases in the specific capacity with increased pumping rates are to be expected in wells that have fully stabilized and are no longer developing. If the specific capacity increases at higher pumping rates, the well is still developing.

Well Loss in Feet

Well loss is used to calculate the well efficiency. It is computed in feet using the following equation (Jacob 1946):

$$s_w = CQ^2 \qquad \text{(Eq 5-25)}$$

Where:

s_w = well loss, ft
C = well loss coefficient, sec² per ft⁵
Q = production well discharge, ft³ per sec

Well Efficiency Calculation

Well efficiency can be obtained by dividing the theoretical drawdown by the total drawdown and multiplying by 100 to obtain the percentage.

$$\frac{s_t}{s} x\, 100 - Percent\ Efficiency \qquad \text{(Eq 5-26)}$$

Where:

s_t = theoretical drawdown, ft
s = actual drawdown, ft

For the purposes of this text, the theoretical drawdown is calculated as the total (measured) drawdown minus the well loss in feet. The actual drawdown is the drawdown as measured in the well. The term *well efficiency*, in this context, can be misleading because it does not indicate that the efficiency (productivity) is caused by both the well characteristics (well loss) and aquifer characteristics (theoretical drawdown). Therefore, wells with lower well efficiencies should not be thought of as necessarily inferior to wells with higher well efficiencies. Well efficiencies of greater than 100 percent indicate that the wells are developing.

Bottom-Hole Pressure

As the withdrawal rate increases, the bottom-hole driving pressure (head) from the injection zone decreases. The inverse is also true (Bloetscher et al. 2005). The bottom-hole driving pressure (head) is defined as the change in pressure (head) in the formation caused by the withdrawal of the water. The bottom-hole driving pressure (head) is primarily a function of the pumping rate, assuming the following:

- storage coefficient changes in the aquifer;
- leakage changes in the aquifer; and
- the transmissivity of the aquifer near the borehole is constant and does not change with time as a result of natural and/or man-made phenomena.

The bottom-hole driving pressure would be related to withdrawal rate as follows (Witt and Ameno 1989):

$$\frac{h_A}{Q_A} = \frac{h_B}{Q_B} \quad \text{(Eq 5-27)}$$

Where:

h_A = bottom-hole driving pressure at the withdrawal rate
h_B = bottom-hole driving pressure at the withdrawal rate of QB(psi)
Q_A = withdrawal rate (gpm)
Q_B = withdrawal rate, equation variable (gpm)

Total Dynamic Head

The total dynamic head (pressure) at the well head is calculated by the following equation (Bloetscher et al. 2005):

$$h_T = h_f + h_d + h_B \quad \text{(Eq 5-28)}$$

Where:

h_T = total dynamic head at well head (ft)
h_f = head friction loss (ft)
h_d = head density differential (ft)
h_B = head bottom-hole driving pressure (ft)

The three components were described in the previous three sections.

Head-to-Pressure Conversion

To convert feet of head into psi, the following formula is used (Bloetscher et al. 2005; Heald 1994):

$$psi = \frac{h \, x \, sp.gr.}{2.31} \quad \text{(Eq 5-29)}$$

Where:

psi = pressure (psi)
h = head (ft)
$sp.gr.$ = specific gravity of fluid (water is assumed 1.00)
h = head (ft)

Collapse Strength

In addition to the requirements of velocity and pressure, adequate surge or water hammer protection to the well must be addressed (Bloetscher et al. 2005). The theoretical collapse strength of a well casing is calculated as follows (AWWA 1964):

$$P_a = \left(\frac{2E}{1-\mu^2}\right)\left(\frac{t}{d}\right)^3 \qquad \text{(Eq 5-30)}$$

Where:

Pa = critical collapse pressure (psi)
E = modulus of elasticity for steel pipe (30,000,000)
μ = Poisson's ratio (usually taken as 0.30 for steel)
d = outside diameter of the pipe (in.)
t = wall thickness of the pipe (in.)

Water Hammer Analysis

Adequate surge or water hammer protection must be incorporated into the design of the pumping system (Bloetscher et al. 2005). Pressure surges associated with water hammer can be observed at older or inadequately designed facilities. The potential for water hammer pressures resulting from instantaneous pumping stoppage at maximum rate should be analyzed. Water hammer is calculated using the following equation (AWWA 1964):

$$h = \frac{aV}{g} \qquad \text{(Eq 5-31)}$$

Where:

a = wave velocity (fps)., which can be reduced to:

$$a = \frac{4660}{\sqrt{1 + \left(\frac{d}{100e}\right)}} \qquad \text{(Eq 5-32)}$$

h = pressure rise above normal (feet of water)
d = inside diameter of the pipe (in.)
e = thickness of pipe wall (in.)
V = velocity of flow (fps)
g = acceleration due to gravity (32.2 feet per second per second [ft/s/s])

To use this calculation, if the inside diameter of the injection well tubing (liner) is 9.75 in., and the wall thickness of the injection well tubing is 0.500 in., based on the calculation of pressure wave, the wave velocity (*a*) is equal to 4,263 fps. If a maximum allowed injection velocity is 10 fps, the calculated head is 1,332 ft or a pressure of 577 psi; at an injection velocity of 12 fps, the calculated head is 1,599 ft or 692 psi. At the maximum peak day velocity of 6.86 fps, the calculated head is 913 ft, or 396 psi. Each of the above velocities provides a pressure that is under the 6,634-psi collapse strength of the injection tubing.

Figure 5-4 is an example of the wave that occurs when a well turns on. The pressure here was in excess of 1,000 psi, and it included a weighted check valve to dampen the water hammer.

Well Losses

Drawdown values obtained for a single pumping well using the Theim formulas represent only the head losses suffered by water movement through the formation under laminar flow conditions. The actual pumping level of a particular well cannot be calculated without considering high velocities and turbulence losses during pumping. At and near the well face, fluid velocities may become large enough that turbulent flow conditions exist. The magnitude of turbulence losses varies with each well because of differences in formation characteristics, screen slot sizes required, degree of well development, well diameter, and quantity of water being pumped. There are so many unknown quantities involved in the calculation of these individual factors that they are usually lumped together under the heading of *well losses*.

One method of approximating the well losses for a particular well is to use the step-drawdown equation as defined by Eq 5-23.

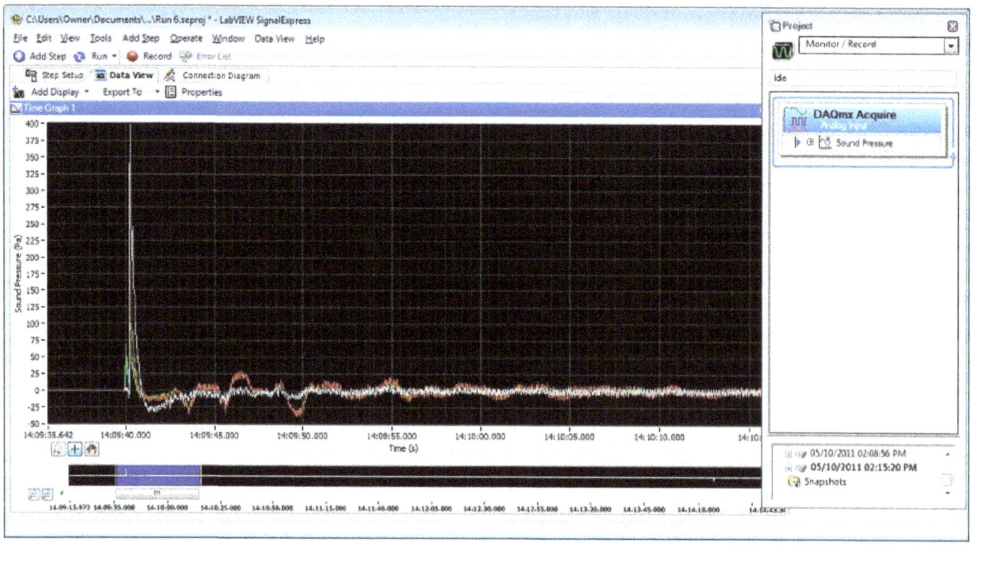

Source: Photo by Fred Bloetscher

Figure 5-4 Water hammer wave

In Eq 5-23, the values of B and C (formation and well losses) may be calculated if proper test data are available. To collect such data, the finished well must be pumped at three to five increasing rates for equal periods of time and the drawdown measured for each pumping rate. When a full-scale aquifer performance test is not conducted, however, a step-drawdown test can differentiate the observed losses in the pumping well. Additionally, this test makes it possible to quickly compare the magnitude of well losses to determine when a well needs cleaning or other repair work. Irregular increasing well losses with increasing pumping rates indicates unsatisfactory development of a new well or deteriorating aquifer or well conditions in an old well. Small regular increases in well loss or decrease in well specific capacity as a result of transition to turbulent flow in the aquifer are normal. A method of approximating the well losses for a particular well has been presented by Rorabaugh (1953).

Entrance Velocity

Water entrance velocities through the screen openings should be between 0.1 and 0.2 ft/s (0.03 and 0.06 m/s). Such velocities will minimize head losses and chemical precipitation. For design of well screens installed in a radial collector well, an average velocity of about 0.033 ft/s (0.01 m/s) is used. Screen entrance velocities are computed by:

$$V = Q/A \qquad \text{(Eq 5-33)}$$

Where:

Q = well capacity, in ft³/s (1 ft³/s = 449 gpm)
A = effective area of screen, in ft²

The effective screen area must be estimated carefully. It is standard practice to assume that 50 percent of the screen slots are plugged by particles after proper well development. The total open area required must be determined by adjusting either the length or diameter of the screen, because the slot size is not arbitrary.

A significant factor in well losses for sand and gravel wells is an open screen area when the percentage of open area is substantially less that the specific yield of the aquifer. Research by Williams (1985) has shown that when the open area of the screen is greater than the specific capacity of the formation, the actual head loss across the well screen is insignificant until the velocity through the screen exceeds 2 ft/s (0.6 m/s). To limit turbulent flow losses around the well borehole, many regulatory agencies have prescribed screen velocities between 0.1 and 0.2 ft/s (0.03 and 0.06 m/s) and a minimum thickness of gravel pack resulting in large-diameter well construction. High velocity turbulent flow through the formation borehole results in higher pumping and maintenance costs. In this case, velocity is a function of quantity and area and is easily approximated in the design stage. For membrane applications where sand may become an operational problem, the velocity should be reduced.

Note the well screen, well diameter, screen length, and other parameters can be adjusted to reduce, entrance velocity, increase water supply and decrease drawdown and as a result must be considered as a part of the design process as will be discussed in the next section.

BASIC DESIGN DECISIONS

Once the previous calculations have been made, the design engineer can evaluate the materials and sizing of the components. In choosing a supply well diameter, the minimum casing and screen diameter should be at least one pipe-size larger than the largest diameter of the pumping equipment to be installed. This gap allows adequate space for pump installation and removal, efficient pump operation, and good hydraulic efficiency of the well. If a shroud needs to be installed around a submersible pump and motor, an appropriate allowance in diameter needs to be made. If additional equipment is to be installed, such as a transducer or water-level controls, then an increase of two pipe-size diameters may be needed.

Because the quantity of water to be pumped from a well, Q, is more correctly established using formation loss and well interference, the open area of screen is the basic parameter to consider. Screen slot size should be selected for accurate sampling and proper sieve analysis. Thus, the screen diameter and length are the two variables in design. Screen-length selection should incorporate more than a casual recollection of the aquifer thickness. The definition of transmissivity, T, incorporates flow through the total thickness of water-bearing material. If less than the total thickness is used, the value of T should be decreased. The Theis equation indicates that as T decreases, the formation drawdown will increase, although not in direct proportion. If the screened portion of the formation is significantly less than one half of the formation thickness (partial penetration), the additional drawdown suffered may be significant. Therefore, it is recommended that as much of the aquifer as practical should be screened to minimize reduction in yield.

If gravel pack construction is used, the borehole should meet minimum thickness requirements of 4 in. (101.6 mm) larger than the screen as specified in AWWA Standard A100-06. If the average hydraulic conductivity, length of well screen and minimum borehole diameter, and the limit of borehole diameter are input, the limit of laminar flow can quickly be calculated. This value may appear to be very low. A flow yield of approximately 4–6 times the laminar flow rate may be cost effective. Some high-capacity wells may be operated in the turbulent flow range, but this type of operation may not be permitted by regulatory agencies. The gravel pack thickness can be increased to the improve the available yield. Unfortunately, in low-permeability aquifers, the maximum practical well borehole diameter will limit the water yield. Other limitations such as saturated thickness, available drawdown, and static water-level depths affect the available yield.

The well casing material and grout used in the construction of each newly drilled well must be designed for the life expectancy of the well. The type of pump installed should provide optimal service over a prescribed number of years, when pumping under specific conditions. Before selecting a pump, different types of pumping arrangements should be investigated to ensure the ultimate needs are met. For instance, vertical turbine pumps should not be used when the treatment involves membranes as the entrainment of oxygen at start-up may encourage fouling of the membranes.

There are several pumping options.

- Piston pumps are low-capacity wells used for hand-pumped wells. They do not meet the needs of public water systems.
- Ejector pumps are small pumps widely used for private home wells. They do not have sufficient capacity to meet the needs of most public water systems.
- Suction pumps work on the principle of creating a vacuum and to pull the water up to the pump level. This type of pump can only be used with relatively shallow wells because the principles of physics limit suction to about 21 ft.
- Turbine well pumps are commonly used for water supply wells. They have a vertical shaft motor located at the ground surface and a long drive shaft extending down the well to operate the pump suspended below the water level. Turbine pumps are available in a wide variety of capacities, can be designed to produce almost any desired pressure, and the motor is easily accessible at the surface for maintenance and repair as well.
- Submersible pumps combine a turbine pump with a waterproof motor. Submersible pumps are located at the bottom of the column-pipe well down the borehole. The pumps require a discharge pipe, power wires, and a lifting cable. Submersible pumps are made in sizes ranging from small pumps used for private home wells to very large units for public water systems. These pumps are very common and provide long useful lives efficiently.

Throttling pumps to match flow demands will cause the systems to run at very low efficiency during low demand periods. To address the problem for continuous operation, a pressure-regulating valve or variable-speed drive should be used to match the pump output with the system demand. The overall cost of equipment and operation should be thoroughly analyzed before adopting such a system. Electric motors are usually selected according to National Electrical Manufacturers Association standards, which include requirements for enclosures and cooling methods.

Pump selection should ensure that overpumping of the aquifer and damage to the casing, pump, formation, and column pipe are minimized. Pumps are available in steel, stainless steel, and bronze. Bronze pumps tend to be resistant to microbiological fouling, while stainless steel and steel are not. Well and pump screens should be installed at the discretion of the hydrogeologist. Additionally, the well must be properly designed and developed before installing the production pump to minimize sand pumping. Often, submersible pumps are used.

A submersible pump eliminates the need for the surface motor, long drive shaft, shaft bearings, and lubrication system of the conventional turbine pump. Submersible pump motors are cooled by water flowing vertically past the motor to the pump intake. The motor is usually longer and of smaller diameter than a surface motor of the same horsepower. When a large-capacity submersible pump is needed, the manufacturer should be consulted for specific design and installation recommendations.

Submersible pump use requires:

- settings that prevent motor burial in sand or silt,
- water temperature and flow past the motor to provide proper cooling,
- use of cable and splices that meet the amperage and voltage requirements,
- pipe tightening to prevent unscrewing by motor-starting torque,
- clamping of cable to delivery pipe,
- proper controls and protections,
- necessary checks before, during, and after installation, and
- adequate electrical power and backup.

The costs for a submersible pump depend on setting depth, required head and capacity, corrosion resistance, and other factors. Operating costs will depend on motor efficiency, column bearing, hydraulic losses, cable losses, setting depth, and similar factors. A thorough analysis of all factors should be performed to compare surface and submersible motor-driven deep-well pumps for a specific installation. Submersible pumps are especially useful for high-head, low-capacity applications, such as domestic water supply.

REFERENCES

American Water Works Association (AWWA). 1964. M11, *Steel Pipe Design and Installation*. Denver: AWWA.

American National Standards Institute (ANSI)/AWWA. 1998. A100-06. Water Wells. Denver: AWWA.

AWWA. 2014. M21, *Groundwater*. Denver: AWWA.

Bear, J., D. Zaslavsky, and S. Irmay. 1968. *Physical Principals of Water Percolation and Seepage*. Paris: United Nations Educational, Scientific and Cultural Organization (UNESCO).

Bloetscher, F., A. Muniz, and G.M. Witt. 2005. *Groundwater Injection: Modeling, Risks and Regulations*. New York: McGraw-Hill.

Davis, S.N. 1969. "Porosity and Permeability in Natural Materials." In *Flow Through Porous Media*. (R.J.M. DeWiest, ed.). New York: Academic Press.

Dawson, K. and J. Istok. 1991. *Aquifer Testing: Design and Analysis of Pumping and Slug Tests*. Chelsea, Mich.: Lewis Publishers.

Domenico, P.A. and F.W. Schwartz. 1990. *Physical and Chemical Hydrogeology*. New York: John Wiley & Sons.

Fetter, C.W., Jr. 1994. *Applied Hydrogeology* (3rd ed.). New York: Macmillan College Publishing.

Heald, C.C. (ed.). 1994. *Cameron Hydraulic Data*. Liberty Corner, N.J.: Ingersoll-Dresser Pumps.

Jacob, C.E. 1946. "Drawdown Test to Determine Effective Radius of Artesian Well." *Transactions of the American Society for Civil Engineers*. 112:1047-1070.

Johnson, A.I. 1967. "Specific Yield—Complication of Specific Yields for Various Materials." US Geological Survey, Water Supply Paper 1662-D. Washington, D.C.: USGS.

Johnson, A.I. and D.A. Morris. 1962. "Physical and Hydrologic Properties of Water-Bearing Deposits from Core Holes in the Los Banos-Kettleman City Area, California." USGS Open File Report. Denver: USGS.

Nuzman, C.E. 1989. "Well Hydraulic Flow Concept." In *Recent Advances in Groundwater Hydrology*. Minneapolis, Minn.: American Institute of Hydrology.

Rorabaugh, M.I. 1953. "Graphical and Theoretical Analysis of the Step-Drawdown Test of Artesian Wells." *Proceedings of the American Society for Civil Engineers.* 79(12):1-23.

Williams, D.E. 1985. "Modern Techniques in Well Design." *Journal AWWA.* 77(9):68-74.

Witt, G.M. and J.J. Ameno. 1989. *City of Plantation Northwest Treatment Plant: R.O. Reject Injection Well Preliminary Design*. Plantation, Fla.: Camp, Dresser & McKee.

CHAPTER 6

Well Construction and Development

INTRODUCTION

This chapter focuses on components of well construction and development. However, the well is not complete until the casings have been set, the surface seals complete, and pumps installed (see Chapter 5 for selection).

There are five primary components common to most wells: the well casing(s), cementing or grouting of the casings, well screens, gravel packs, and sanitary seals at the surface to prevent the migration of surface water into the well. Considerations for these components include material specifications, sizing, and—most importantly—the depth of the casing/screen to allow the appropriate water source into the well. Once the well is completed, it must be developed. Development may be the most overlooked issue in well construction, but good development practices will provide an efficient well with long life and low maintenance costs. In the long term, redevelopment of a well may be necessary to correct challenges that arrive in the future. Data collection during the construction and development stage provides baseline information to be used in the future to assess productivity of the well and help define when rehabilitation or construction of a new well needs to be undertaken.

CASINGS

Purpose of Casings

The well casing is a lining for the drilled hole that maintains the open hole from the land surface to the water-bearing formation. For the casing to be entirely effective, it must be constructed of suitable materials and be properly installed to be watertight for its entire depth. If the formation could likely cave over the full depth of the well, a single casing is usually sufficient. In these situations, the sand

and gravel caves in around the outside of the casing and closes the space between the drill hole and the casing. However, single casings are usually restricted to smaller, surficial wells. In deeper wells, it is preferable to have multiple casings as this makes drilling easier in formations that may collapse. Likewise, using multiple casings may be desirable in aquifers that are corrosive or polluted, in zones that are unproductive, or to seal off zones that the user does not wish to pump. Casings also serve to seal out contaminated water from the land surface and undesirable water from formations above the aquifer.

Figure 6-1 shows an example of a well with a series of casings that seal off various zones above the production zone. This concept is referred to as *telescoping*. In a telescoped well, there are several types of casings: pit, surface, intermediate, final, and tubing. The final casing string is the one that seals off all formations except the one where the water is desired. A pit casing is the initial casing installed at the surface to prevent the introduction of contaminants from the surface and provide containment for the drilling operation at the surface. The pit casing should be steel and extend through the surface soils. The surface casing is the next casing installed. The surface casing typically seals the surficial formation from the rest of the well. A surface casing may not be used for water table aquifer wells because the surface formation is the aquifer. Intermediate casings may be installed to seal off successive formations where the water is not desired or not available. Each successive casing is typically 6 in. diameter smaller than the prior casing. The final casing will be filled with the water to be withdrawn. The column pipe and pump are installed inside the final casing.

If one or more outer casings are installed in a well, the annular space between the casings is filled with cement grout, often containing bentonite, to limit shrinkage and cracking in the grout (and thereby creating a seal). With this type of installation, the outer casing may be either left in place or withdrawn completely. If withdrawn, the grout is placed as the temporary casing is removed. Each outer casing is generally one pipe-size larger in diameter than the outside diameter of the couplings or the protective casing. This type of grouted installation may also be used where the water-bearing formation underlies clay, hardpan, or other stable formations.

In fractured formations, care should be taken to identify where there are connections to poorer quality water sources than those desired for water supply or polluted water that flows as a result of vugs and fractures in the formation. It is preferable to use such formations only where a competent layer of low-permeability rock overlays the aquifer. The well can be protected if it is watertight to a depth greater than that of the deepest existing well of questionable construction in the area, and substantially below the lowest anticipated water level. The watertight construction is achieved by drilling the hole in the fractured rock 2 in. (50 mm) larger than the outside diameter of the casing couplings and filling the annular space between the drill hole and the outside of the casing with bentonite grout. In some areas, such construction may not be realistic because available water is cased-off. Other methods of assuring adequate water quality protection may be necessary.

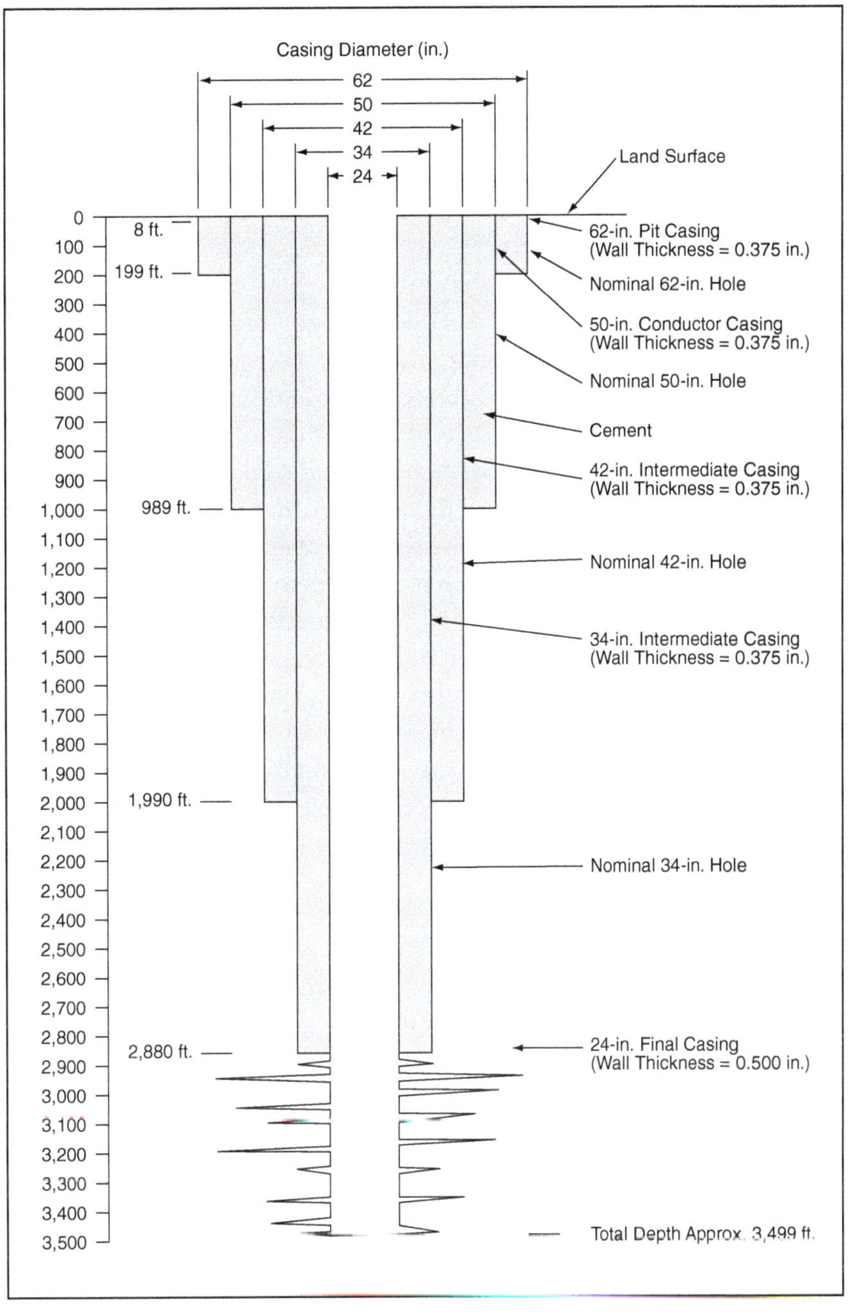

Source: Hazen and Sawyer, P.C., Boca Raton, Fla.

Figure 6-1 Multiple casings in a telescoped well

Environment (Factoring in Conditions Such as Inland, On Coast, Corrosive Areas)

Regardless of construction method, water well casings and screens must be adequate to withstand the stresses of installation and the stresses applied during well completion, development, and use. The tensile strength of the material must address the forces of collapse on the casing (see Chapter 5 and Appendix C). Note that hydrostatic stresses may be greater than the negligible stresses on small diameter wells in consolidated formation, and the difficult-to-determine stresses from unconsolidated formations. The selection of well casing wall thickness depends on the tensile (ultimate) strength of caused by installation and the compression load due to the hydrostatic and formation pressures. This value is equivalent to the yield strength.

Well casing thickness impacts durability, whereby small increases in the wall thickness of ordinary low carbon steel casings not only improve strength, but extend well life, from a corrosion standpoint, by a factor greater than the percentage thickness difference (Roscoe Moss ND). Casings and screens should be designed to reduce the potential for corrosion damage, biofilm formation, wear from water flow and sand, and damage from installation of removal of pumps, while facilitating future rehabilitation and repair. The next section on Casing Types/Materials of Construction outlines considerations for well casing specification.

The importance of water quality is relevant to the issues associated with corrosion. Saltwater obviously increases the potential for corrosion of steel and stainless steel casings in the column pipe. Bicarbonate retards corrosion in wells, while sulfate and nitrate are neutral and chlorides, typical of deeper wells, islands and coastal areas, accelerates corrosion (Roscoe Moss ND). Concentrations of carbon dioxide of more than 50 mg/liter are usually corrosive as it indicates a lower pH. In the presence of corrosion, damage to stainless steel may be accelerated. Selection depends on the anticipated life of the well; a 30-year life is a reasonable expectation.

Casing Types/Materials of Construction

Casings are usually one of four materials: carbon steel, stainless steel, fiberglass, or PVC. Fiberglass and PVC have been used extensively in recent years for installations in shallow wells or where corrosion and/or bacteria may be an issue. Ingot iron is used in constructing gravel-wall wells or other large-diameter wells. In selecting a suitable material, the stress that the casing experiences during installation and the corrosiveness of the water and soil must be considered. All casings will provide satisfactory service given the correct groundwater and stress environment.

Many grades of casings are available, so specifying casings must include more than the nominal diameter of the casing. Tables 6-1 through 6-5 show examples of the standard casing sizes for steel, fiberglass, and PVC as outlined in ANSI/AWWA Standard A100-06. Carbon and stainless steel are the most common casing materials (Figure 6-2). Carbon steel has several benefits that make it useful: ability to weld, high yield and tensile strength, and high burst strength. Carbon steel provides the most amount of protection against borehole collapse because the strength of the material is greater than fiberglass and PVC. However, carbon steel is subject to corrosion from galvanic and microbial contamination. Therefore, careful consideration to the materials used in the well must occur to limit galvanic action. For example, it may not be appropriate to use steel casings in aquifers with high microbial populations or high chlorides.

Table 6-1 Wall thickness for steel casing

Nominal Size (in.)	Casing Diameter External (in.)	Internal (in.)	Wall Thickness (in.)	Weight (lb/ft) Ends	Collars
2	2.375	2.067	0.154	3.56	3.71
2.5	2.875	2.469	0.203	5.79	5.88
3	3.5	3.068	0.216	7.58	7.67
3.5	4	3.548	0.226	9.11	9.27
4	4.5	4.026	0.237	10.79	11.01
5	5.563	5.047	0.25	14.62	14.9
6	6.625	6.065	0.25	18.97	19.33
8	8.625	8.071	0.25	27.7	25.4
10	10.75	10.192	0.279	31.2	32.2
12	13.75	12.09	0.33	43.77	45.4
14	14	13.25	0.375	54.57	55.8
16	16	15.25	0.375	62.58	64.08
18	18	17.25	0.375	70.59	72.37
20	20	19.25	0.375	78.6	80.7

Source: ANSI/AWWA A100-06

Table 6-2 Steel well casings fabricated from standard plates

Diameter (in.)	Standard Plate		Well Casing Sheets	
	Thickness (in.)	Gauge	Thickness (In.)	Gauge
6	0.1046	12	0.1094	12
8	0.1046	12	0.1094	12
10	0.1046	12	0.1094	12
12	0.1345	10	0.1406	10
14	0.1644	10	0.1406	10
16	0.1644	8	0.1719	8
18	0.1644	8	0.1719	8
20	0.1644	8	0.1719	8
22	0.25	—	—	10*
24	0.25	—	—	10*
30	0.25	—	—	8*

*Double thick

Table 6-3 Fiberglass casing sizes

Nominal Size (in.)	Casing Diameter External (in.)	Internal (in.)	Wall Thickness (in.)	Weight (lb/ft)	Internal Pressure (psi)
Future Pipe Industries, Inc., Red Box 1500					
4	3.75	3.33	0.21	2.3	1,500
5½	4.96	4.42	0.27	3.8	1,500
6½	6.1	5.43	0.34	5.7	1,500
7	6.97	6.21	0.38	6.9	1,500
10½	9.94	8.85	0.54	15.3	1,500
13½	13.29	11.97	0.66	23.7	1,500
16	16.08	14.48	0.8	35	1,500
Burgess Well Company, Inc. "EON" Fiberglass Column Pipe					
4	5	4.5	0.25	4.5	400
6	6.625	6	0.31	5.8	400
8	8.625	8	0.31	7.8	300
10	12.75	10	0.375	10.5	250
12	12.75	12	0.375	12.6	225

Table 6-4 Small diameter PVC casing sizes—SCH 80

Nominal Size (in.)	Outside Diameter (in.)	Inside Diameter (in.)	Minimum Wall Thickness (in.)
1.5	1.9	1.72	0.09
2	2.375	2.149	0.113
2.5	2.875	2.601	0.137
3	3.5	3.166	0.177
4	4.5	4.072	0.214

Table 6-5 Large diameter PVC casing sizes—SCH 40

Nominal Size (in.)	Outside Diameter (in.)	Inside Diameter (in.)	Minimum Wall Thickness (in.)
5	5.563	5.047	0.258
6	6.625	6.065	0.28
8	8.625	7.981	0.322
10	10.75	10.02	0.365
12	12.75	11.938	0.406

Source: Photo by Fred Bloetscher

Figure 6-2 Steel casing materials

Stainless steel has the same benefits as carbon steel. Stainless steel is an upgrade that attempts to solve the corrosion problems but is still subject to both galvanic and microbial contamination (in some cases, the stainless steel may be more susceptible to microbial contamination). It is also significantly more expensive than the other options.

Fiberglass is lightweight and corrosion resistant (Figure 6-3)—neither galvanic nor microbiological activity will damage it. Fiberglass is less expensive than steel. However, fiberglass has less burst, tensile, and yield strength than steel, so deeper wells are unlikely to use fiberglass casings (column pipes may also be fiberglass).

PVC is the least costly and lightest weight of the potential casing options (Figure 6-4). Its use is becoming more common in shallow wells as it is not subject to corrosion or microbiological attack. Galvanic activity is not a factor. PVC has less burst, tensile, or yield strength than steel or fiberglass. Care must be taken during construction of wells with PVC, as the heat created during the grout curing operation may buckle PVC casings.

The lighter materials (PVC and fiberglass) may be used for test wells or temporary casings. Temporary casings may be used as forms when a grout seal is placed around the outside of the permanent casing. The temporary casing is withdrawn as the grout seal is placed. Joints for permanent casings should have threaded couplings or should be welded (in the case of steel—see Figure 6-5) to ensure watertightness from the bottom of the casing to a point above grade. This precaution will prevent surface contamination or undesirable groundwater from entering the well from formations above the water-bearing formation through the casing.

Source: John Largey

Figure 6-3 Fiberglass casing materials

Source: John Largey

Figure 6-4 PVC casing being installed

Source: Hazen and Sawyer, P.C., Boca Raton, Fla.

Figure 6-5 Welding a casing pipe

Measuring Plumbness

The appendix to ANSI/AWWA A100-06 includes a section on testing for plumbness of the borehole. While there are gyroscopic and laser methods for design plumbness, there is an alternate-alignment tolerance method that is straightforward, elementary, and easy to perform in the field. The procedure involves lowering a cylindrical plummet into the well to the specified depth. The plummet is a rigid spindle with round plates at both ends (Figure 6-6). For normal plummets, the outer diameter of the end plates is 0.5 in. (13 mm) smaller than the inside diameter of that part of the casing or hole being tested. The distance between end plates is approximately 1.25 times the diameter of that part of the casing or hole being tested, and they are heavy enough to keep the plumb line taut. The plumb line is attached to the plummet at the exact center of the top end plate (Figure 6-7).

Per ANSI/AWWA A100-97, "Plumbness and alignment are determined by lowering the plummet a maximum of 10 ft (3.05 m) at a time, or more frequently when approaching the allowable maximum, and measuring the horizontal deflection of the plumb line from the center of the top of the casing or hole at each interval. The horizontal deflection shall be measured in two planes 90° from each other."

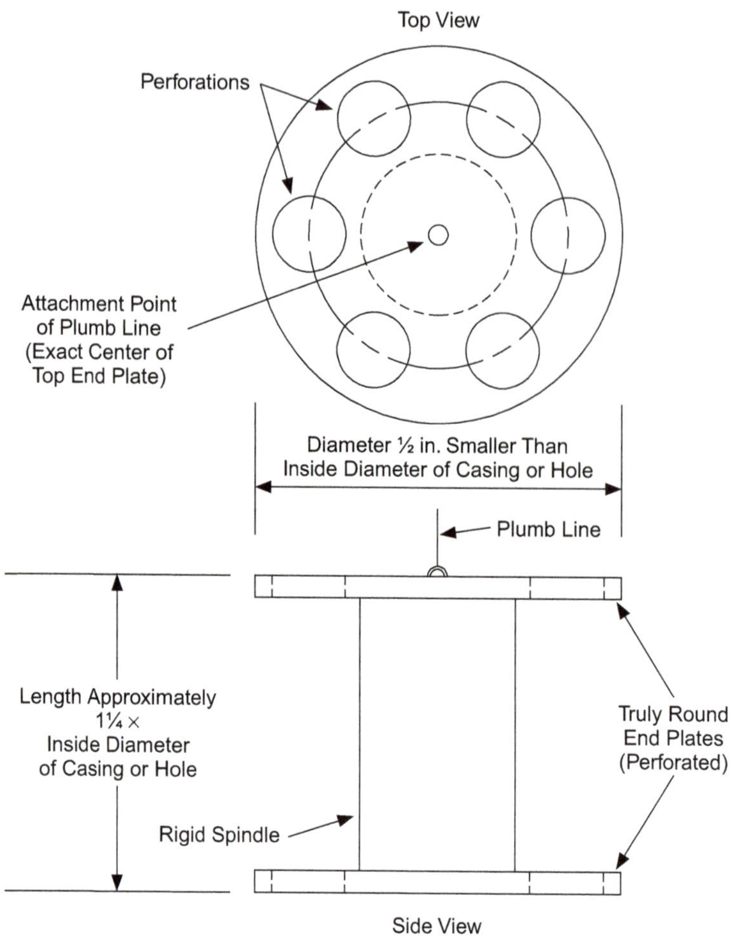

Source: ANSI/AWWA A100-97

Figure 6-6 Plummet

The drift (horizontal deviation) of the casing or hole at each recorded depth is calculated by using the following formula:

$$\text{drift} = \text{deflection (height + depth)/height}$$

Where:

- Drift is calculated horizontal deviation of casing or hole from the vertical, inches (millimeters).
- Deflection is measured horizontal deflection of the plumb line from center of the top of casing or hole, inches (millimeters).
- Height is the height of apex above the top of casing or hole, in feet (meters).
- Depth is the depth of plummet below the top of casing or hole, in feet (meters).

The calculated drift of the casing or hole at the depth intervals recorded is part of the testing procedure for well plumbness and alignment:

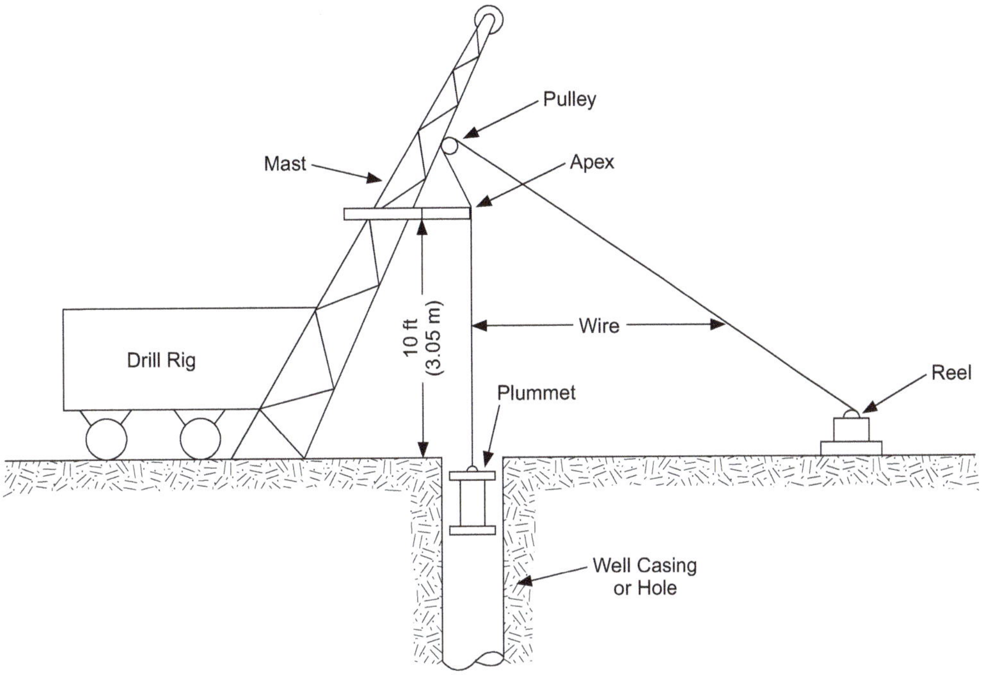

Source: ANSI/AWWA A100-97

Figure 6-7 Use of drill rig and plummet to measure plumbness

1. Test sheet—written statement covering details of the plumbness and alignment test data
2. Well diagram—longitudinal projections of actual well centerline and proposed pump centerline
3. Plumbness graph—calculated drift of the well-casing centerline from vertical
4. Alignment graph—horizontal deviations of actual well-casing centerline from proposed pump centerline
5. Diagram—a diagram showing the effective well diameter and the pump

The point is to determine if the pump can be placed in the borehole. Placing or replacing a long casing string (Figure 6-8) necessitates plumbness of the borehole, or there will be a challenge with pump installation and removal.

INSTALLATION METHODS

Setting the Casing

How the casing is installed is related to the drilling method used to construct the borehole (see Chapter 4), the formation type and the location of the water source (AWWA 2003). There are five basic methods for setting the casing:

Source: Photo by Fred Bloetscher

Figure 6-8 Example of a long column pipe string and pump that will be reinstalled

- driving,
- vibrating,
- cable tool,
- dual tube, and
- rotary.

Driving the casing has the benefit of not requiring grouting, but it is limited to softer rock and soil formations and limited depth. Jetted wells or well-points are such examples. This method is generally only used for low volume wells and will not be discussed further, as it is generally not applicable for most utilities who drill wells.

When drilling a well using the cable-tool method, the casing should be driven when the ground formation could begin caving. A drive shoe, attached to the lower end of the casing, keeps the hole from collapsing. Drive shoes are threaded or machined to fit the pipe or casing, and the inside shoulder of the shoe butts against the end of the pipe. Drive shoes are forged of high-carbon steel, without welds, and are hardened at the cutting edge to withstand hard driving. Some regulatory agencies do not accept casings that are not cemented in place because they do not positively seal the borehole. Table 6-6 shows casing sizes and pump types for small wells.

Table 6-6 Casing sizes for small wells based on yield

Yield*	Recommended Casing Diameter (in.)	Jet	Drilling Method	
			Double Jet	Submersible
Less than 8 gpm	2	✓	✓	
	3	✓	✓	✓
	4	✓	✓	✓
	5		✓	✓
	6			✓
8 to 16.5 gpm	2	✓	✓	
	3	✓	✓	✓
	4	✓	✓	✓
	5		✓	✓
	6			✓
Greater than 16.5 gpm	3	✓		
	4	✓	✓	✓
	5		✓	✓
	6			✓

*Yield at 50 ft of drawdown

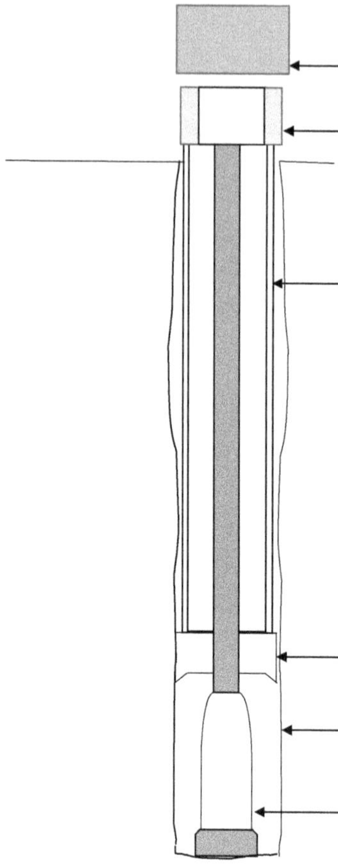

Figure 6-9 Drive shoe at base of casing being driven in cable tool well with drive head on top

Casings are driven using drilling tools, drive clamps, and a drive head. Where the well penetrates water-bearing rock underlying unconsolidated material, the casing is driven into the rock to obtain a good seal (Figure 6-9). Unfortunately, a tight seal that will prevent pollution or unconsolidated material from entering the well from above is not guaranteed, so a grouted seal is usually required. A length of casing is attached to the casing previously installed by threaded coupling or welding. A drive head is then attached to the upper end of the casing to protect it from the driving blows of the drive clamp, which is attached to the drill stem. When the drill is lowered into the length of casing and subsequently raised and lowered, the action of the dropping clamp on the drive head forces the casing into the drill hole. The concept is similar to drilling the well and uses the same drilling rigs. Additional protection can be gained by driving the casing down to stable rock and under-reaming the borehole beneath the casing to a diameter 2 in. (50 mm) larger than the outside diameter of the shoe for a depth of 10 ft (3 m) below the casing.

The under-reamed portion of the drill hole is filled with cement grout, and the casing is driven to the bottom of the hole. Before drilling is resumed, the cement grout is allowed to set for several days, providing a good seal. Once drilling is restarted, the cement inside the casing is drilled out (Figure 6-10). An open,

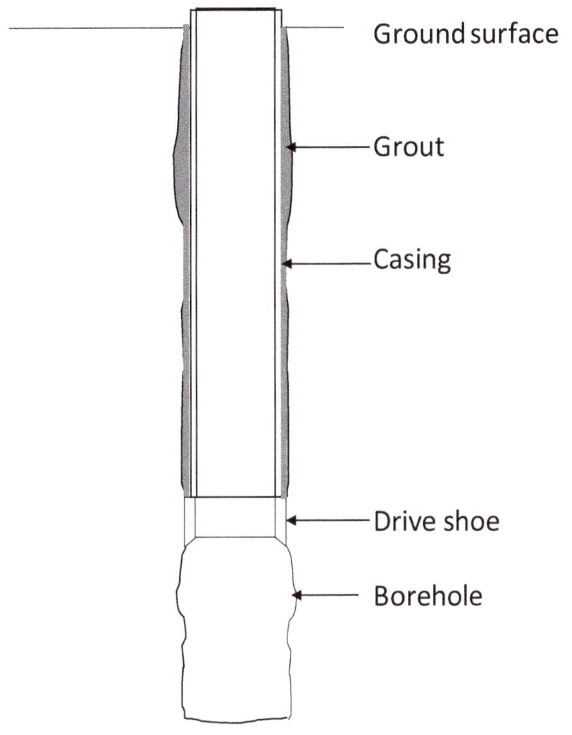

Figure 6-10 Grouted cable tool well

uncased hole is constructed in the water-bearing rock below this point. Note that cable-tool methods limit data collection on the formation because the casing is immediately installed.

Vibrating the casing into the aquifer is another option for unconsolidated formations. Well casings that are vibrated into place normally do not require cementing; however, the same limitations that apply to driven casings also apply to vibrating casings into place. Vibration methods limit data collection on the formation because the casing is immediately installed, just like cable tool wells.

To get water quality data, dual tube and rotary methods are used because they minimize disruption of the aquifer when setting the casing. For collection of geophysical data, rotary methods are the preferred method for setting the casing. Wells constructed using rotary methods are not usually cased until drilling is completed. Because the casing is smaller than the drilled hole, no driving is required. In some instances, a casing is installed concurrently with drilling, such as with the use of dual-rotary drilling methods.

WELL COMPLETION

Screens

Once the casing is placed and grouted in place, the column pipe and screen, if required, are placed. Screens are needed in most wells, especially where sand and

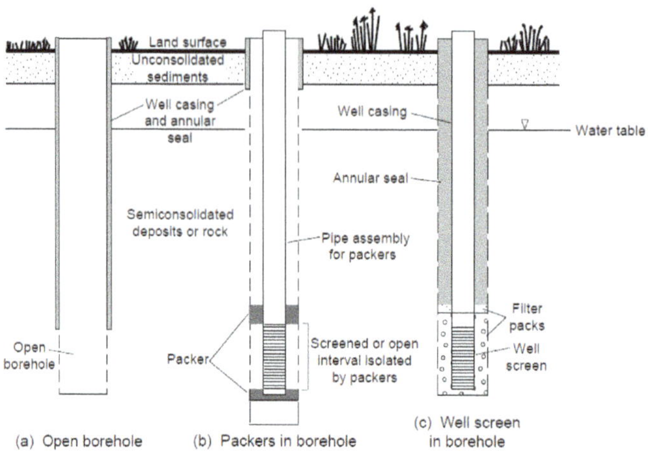

Figure 6-11 Construction of screened well

Source: Photo by Fred Bloetscher

Figure 6-12 Fiberglass screen with centralizers to center the screen in the borehole. Note the small size of the packing gravel.

fine materials may enter the borehole (Figure 6-11). The screen is designed to eliminate fine particulates that may damage downstream pumps and treatment equipment, while allowing the maximum amount of water from the aquifer to enter the well with a minimum of resistance. An example is shown in Figure 6-12. Some screens, especially in California, may have a vertical orientation (Figure 6-13).

Source: Photo by Fred Bloetscher

Figure 6-13 Some well screens, especially those in California, have vertical slots.

Generally, wells completed in unconsolidated formations, such as sands and gravels, are equipped with screens. In competent rock that will not release fines, such as limestone or granite, screens may not be required. Screens are sometimes installed in fractured formations that may collapse into the borehole and trap equipment.

Although a screen prevents sand from entering the well during pumping, a screen may allow fine formation particles to enter the well during the development process so they may be removed by bailing. At the same time, the large particles of sand are held back, forming a permeable, graded natural-gravel pack around the well screen itself. In this way, the hydraulic conductivity of the water-bearing formation around the well screen is greatly increased, resulting in lower velocity head loss and higher capacity per foot of drawdown.

Proper screen selection is extremely important in the design of a well drawing from unconsolidated aquifers. Selection is often a complicated matter that demands a highly specialized knowledge of well construction and operation. The size of screen openings, or the slot number, is usually expressed in thousandths of an inch. Screens have many sizes of openings, as shown in Figure 6-14. The proper screen slot size is determined through:

- collection and analysis of representative samples of the formation to be used for water production;
- identification of the lithologic properties of the formation; and
- laboratory analysis of grain-size gradation.

The width of the slot, or slot size, is best determined using a mechanical sieve analysis of a sample from the water-bearing formation. Representative samples of the formation must be selected for mechanical grain-size analyses. The largest slot

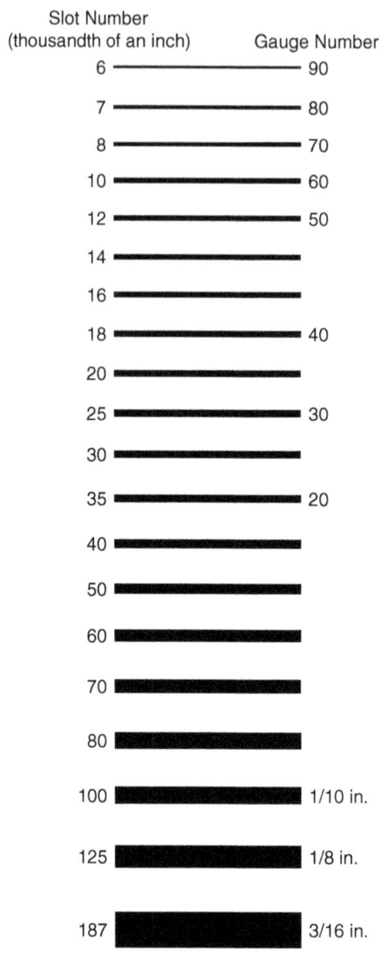

Source: AWWA 2014

Figure 6-14 Well screen size chart

opening practical is normally specified to maximize the amount of water that can enter the column pipe, while correcting for screen entrance velocity limitations. Hence, the tighter the screen, the less available flow area, which means the volume must be reduced, subject to meeting the goals of maximizing the amount of water withdrawn while minimizing screen losses and the introduction of fines. Depending on the type of well construction, the slot size is selected to permit a percentage of the formation material to pass through it. For naturally developed wells, this amount usually ranges between 35 and 65 percent, depending on uniformity of the material and the overlying formation (AWWA 2014). The design criteria for water entering through the screen opening that has been adopted by many regulatory agencies for well construction is between 0.1 and 0.2 ft/sec (0.03 and 0.06 m/sec). The very low screen velocity criteria promoted the use of large-diameter well screens and more efficient well construction. However, research indicates that the actual head loss across the screen is minimized as long as the thickness of the well screen and the percentage of open area in the screen is equal to or greater than the specific yield of the aquifer (Williams 1985). Therefore, the most important factor is the degree of turbulent flow that may be generated in the water flow through the formation and gravel-pack material surrounding the well screen.

Turbulent flow head losses around the borehole increase by the velocity squared. In laminar flow conditions, the head loss is linear with the velocity. In properly constructed and properly developed wells of high capacity, the well loss in head can be quite significant because of the turbulent flow in the well screen. Turbulent flow causes movement of sand particles, mechanical plugging of the gravel pack, and mechanical blockage and chemical precipitation of minerals around the outside of the well screen.

Gravel Packs

All gravel-packed wells have screens. A gravel pack is included to act as a filter to permit the use of larger slot sizes in the well screen than would be possible if the area surrounding the screen were not gravel-packed. When a well screen is surrounded by an artificial gravel wall, the size of the openings is controlled by the size of gravel used and by the types of openings.

A gravel-wall well must be carefully designed. Table 6-7 outlines the typical grain sizes used in gravel packs. Selection of the gravel pack material depends on the aquifer formation. The material used in the gravel filter must be clean, washed gravel composed of well-rounded particles. Like the screen, the intent of the gravel pack is to prevent fines from plugging the screen or entering the borehole in large quantities. Gravel packs can consist of pea gravel, sand, or other rock. The filter size depends on the size of the natural formation and the intended slot openings of the well screen. Without proper gravel size, fine sand will not be prevented from entering the well, and the yield of the well will be reduced. The size of individual grains of gravel filter material should be four to six times larger than the median size of the natural material. At the same time, the uniformity coefficient of the gravel treatment should be similar to that of the formation material. The slot size for the screen should retain 90 percent of the pack material. An artificial gravel-pack filter can also be installed around the lateral well screens in a radial collector well to match finer-grained formation materials.

The gravel pack is placed between the outside of the well screen and the borehole. After the outer casing is in place, the screen is lowered to the bottom of the well and centered. Selected gravel is added to the annular space between the screen and the casing through a small-diameter tremie pipe. The gravel is placed evenly around the screen in 2- to 4-ft (0.6- to 1.2-m) layers. As the gravel is added, the casing and tremie are slowly raised. The procedure continues until the entire screen is surrounded with gravel and the pack extends several feet (0.5–1 m) above the top of the screen. The outer casing is pulled back high enough to expose the entire screen section. As a rule, the screen is attached to an inner casing, extending to the land surface, into which the pump is placed. About 25 ft (8 m) of the outer casing is required to provide a seal against contamination by surface water. After the gravel filter has been placed, a pipe is often installed in the finished pump base or foundation to allow additional filter materials to be added if the gravel filter settles as a result of normal pumping operations, well development processes, or well rehabilitation procedures (AWWA 2014). If the entire casing is removed, the gravel pack must not extend to the land surface.

Table 6-7 Various size grade scale in common use

Size (mm)	U-W Values	Udden Wentworth	German (Atterberg)	USDA/ Soil Science Society of America	US Army Corps of Engineers/ BOR	Size (inches)
			Cobbles			
200 mm						
	-6	Cobbles	Gravel	Cobble (80 mm)	Boulders	
64 mm						10
		Pebbles	Gravel	Gravel	Cobbles	
20 mm						3
	-2	Pebbles	Gravel	Gravel	Gravel	
4 mm						4 mesh
	-1	Granules	Sand	Gravel	Coarse sand	
2 mm						10 mesh
	0	Very coarse sand	Sand	Coarse sand		
1 mm						
	1	Coarse sand	Sand	Medium sand		
0.5 mm						
	2	Medium sand	Sand		Medium sand	
.25 mm						40 mesh
	3	Fine sand	Sand	Fine sand	Fine sand	
.125 mm						
	4	Very fine sand	Sand	Very fine sand		
.0625 mm						200 mesh
		Silt	Silt	Silt	Fines	
.002 mm						
	8	Clay	Clay	Clay		

Sealing the Well

Once the casing has been placed in the borehole, it must be sealed in place to fill the annular space and hold the casing in place. The annular space is the space between the casing and the native formation. The annular space is the most likely source of contamination from the surface if not properly sealed. In loose caving formations, such as sand, the opening may be self-sealing, but in stable formations, this space must be cemented to prevent contamination from the land surface or porous formations connecting with the surface. In addition to sealing against contamination from the surface, wells are cemented, or grouted, and sealed for the following reasons:

- to seal out water of an unsatisfactory chemical quality;
- to increase the life of the well by protecting the casing against exterior corrosion;
- to stabilize soil or rock formations of a caving nature; and
- to protect the casing from potential deterioration from microbiological activity.

Grout material should be placed uniformly into the annular space after water or drilling fluids have been circulated sufficiently to ensure any obstructions in the annular space have been cleared. In shallow wells, grout can be placed by dropping it from the surface (*rude and crude*—see Figure 6-15) or a bailer. The rude-and-crude methods should only be employed when the interval to be grouted can be seen from the surface. This generally limits usage of this method to intervals of less than 30 ft.

The dump-bailer method is perhaps the simplest method for grout placement. The cement grout is lowered in a dump bailer that discharges its load when it reaches the bottom of the hole. The bailer is placed in the annular space 1 ft above the bottom of the hole. After the grout is placed in the well, the casing is pulled up so that the shoe is above the grout. A plug is placed in the bottom of the casing, which is then driven to the bottom of the hole, displacing the grout into the annular space around the outside of the casing. Bailer methods permit the grout to be placed in stages. The elapsed time between dumps should not be more than 10 min.

Source: Photo by John Largey

Figure 6-15 Dump methods for grouting

Source: Photo by John Largey

Figure 6-16 Tremie pipe

However, in deeper wells, grouting the casing in place can be a challenge. If the annular space outside the casing is large enough to accommodate a grout (tremie) pipe, an air- or water-pressure drive is used. The tremie pipe should extend from the surface to the bottom of the annular space. Grout is then pumped into the tremie pipe (Figures 6-16 and 6-17). As the grout is placed, the pipe is slowly withdrawn to the surface, circulating around the casing to ensure a smooth and consistent pour all the way around the casing. The tremie pour should be continuous, and the tremie pipe discharge should be submerged in the grout at all times. The tremie method requires a minimum annular space of 3 in. (7.6 cm) between casings. The minimum tremie pipe diameter is generally 2 in. (5.1 cm), although concrete grout tremie pipes should be a minimum 3 in. (7.6 cm) to prevent clogging. Grout material placed using the tremie method should occur after water or drilling fluid have been circulated to clear obstructions.

A variety of pumping methods can be used but may be limited to site-specific applications. Pressure grouting involves forcing grout into the annular space. Grout pumping methods begin with the installation of a pipe inside the casing. The casing is suspended slightly above the bottom of the borehole, and a suitable packer connection is provided at the bottom of the casing. The packer allows removal of the grouting pipe and prevents grout leakage into the interior of the casing.

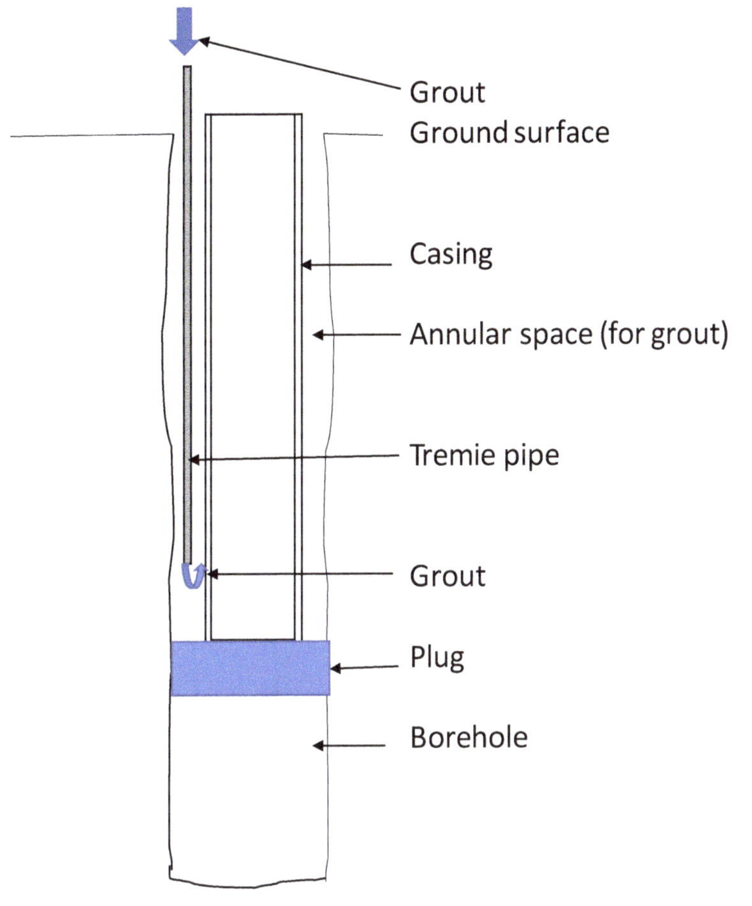

Figure 6-17 Tremie pipe in the borehole

The continuous injection method requires the grout to be placed with a float shoe with a back-pressure valve. Tubing is run to a float shoe to carry the grout. When the annular space is deemed clean and free of obstructions, the grout is pumped down the tube into the bottom of the annular space (Figure 6-18). When the space is filled, the grout pipe is removed. Work on the well is not resumed for at least 72 hours, after which time the packer connection and plug are drilled out (AWWA 2014). Pumping should be continuous until the entire annular space is filled with grout. Concrete grout cannot be used with this method. Pressure grout has the potential problem of exceeding the burst strength of the casing pipe if not monitored. After grouting, an acoustic sonic log (i.e., cement bond log) should be run in the well to determine the competency of the cement bond to the casing and formation (or second casing, whichever is appropriate). The log should be run from the top of the casing to the bottom at least 72 hours after the grouting operation but before further construction commences.

Cement Grouting Materials

Three materials are commonly used for grouting: concrete, sand cement, and neat cement grout. Concrete grout is a mixture of Portland cement (ASTM C150), sand, coarse aggregate, and water in the proportion of at least five bags of cement (94 lbs/cf) per cubic yard of concrete, and not more than 7 gal of water per sack

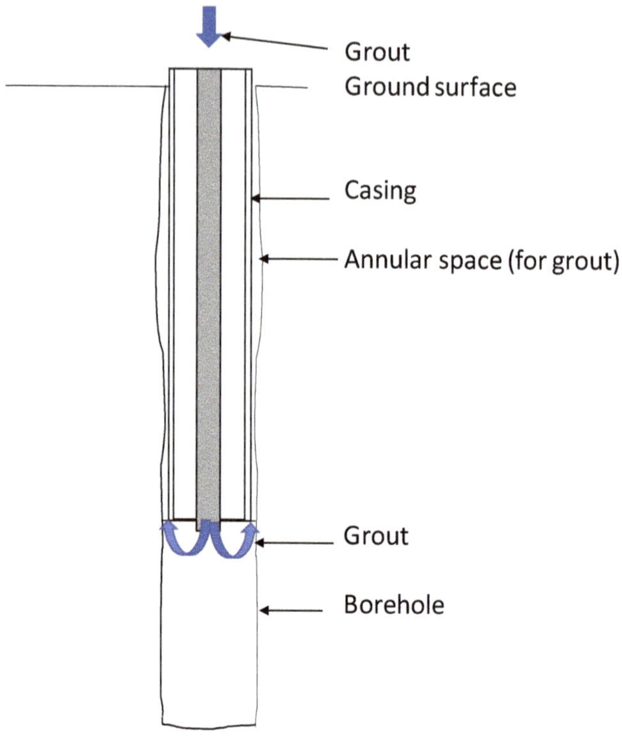

Figure 6-18 Pumping grout (below-ground method)

of cement. Bentonite and other admixtures (ASTM C494) are commonly used to reduce shrinkage, increase viscosity, and reduce permeability (Lehr et al. 1988). A minimum of 2 percent and a not-to-exceed maximum of 12 percent, by weight, of bentonite clay should be added to neat cement grout to compensate for shrinkage. Regardless of the materials used, cement, additives, and water must be mixed thoroughly.

Sand cement grout consists of sand, Portland cement, and water. The sand to cement ratio is 2:1. Water content remains the same as other admixtures. Neat cement is made of only Portland cement and water in the ratio of 7 gal of water per sack of cement (94 lbs/cf in each bag). Admixtures are similar to those used in concrete grout (Lehr et al. 1988). Curing time before further construction is based on the type of Portland cement used. Type I cement has a minimum curing time of 72 hours. Type II Portland cement has a curing time of 36 hours and is preferred for many installations as a result.

When formations located below the depth of the protective casing are known to yield water of an unsatisfactory chemical quality, these formations may be sealed off with liners set in cement grout for their entire length, which may be several hundred feet deep. When a casing is extended to a consolidated formation lying below an unconsolidated formation, the most effective way to prevent sand or silt from entering the well at the bottom of the casing is cementing. The casing exterior is protected against corrosion by encasing it in cement grout, as described earlier in the section on casing installation. A minimum of a 2-in.

(50-mm) thickness of grout is recommended; more may be required by some regulatory agencies.

The grout should be applied in one continuous operation if possible; however, it often must be placed in stages to ensure a satisfactory seal and be entirely in place. The grout must always be introduced at the bottom of the space to be grouted to avoid segregation of materials, inclusion of foreign materials, or bridging of the grout mixture, and if above the fluid level, to avoid leaving large packets of air in the annulus. An air pressure test (i.e., 7–10 psi) should be applied to the grout seal for a period not less than 1 hour to determine if any leakage exists. If the pressure drops during the 1-hour test, the necessary repairs and resealing of the grout should be made and the new seal retested.

Well Completion

Ultimately the wells can be constructed using a variety of configurations. ANSI/AWWA A100-97 identified 12 configurations that are shown in Figures 6-19 to 6-31. These include options with and without gravel packs, with and without screens, single and multiple aquifer zones, and differing formation types.

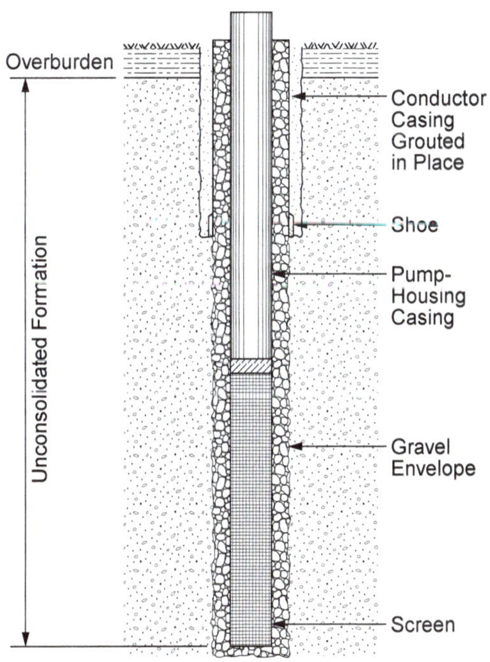

Source: ANSI/AWWA A100-97

Figure 6-19 Gravel-packed well with conductor casing grouted in place and gravel envelope extending to the surface

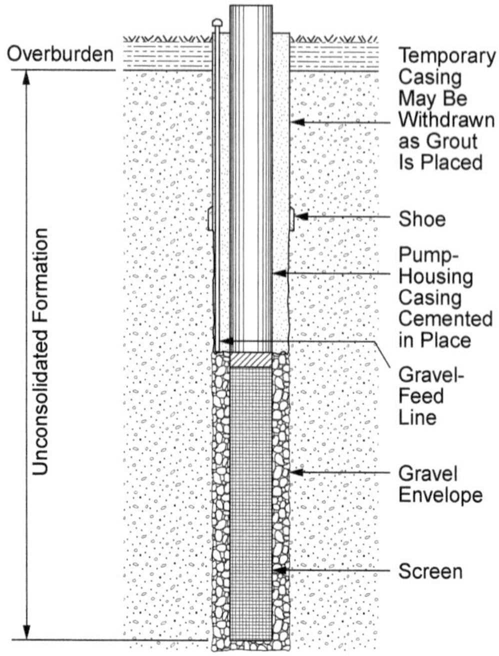

Source: ANSI/AWWA A100-97

Figure 6-20 Gravel-packed well with well casing cemented in place and gravel envelope terminated above the top of the screen with gravel feed line (tremie pipe)

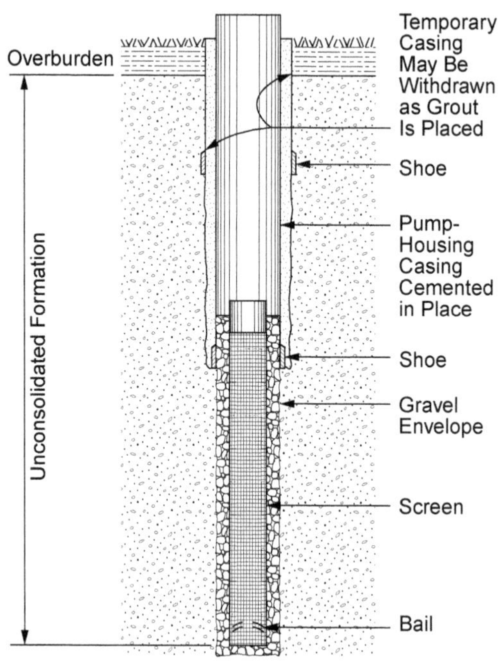

Source: ANSI/AWWA A100-97

Figure 6-21 Gravel-packed well with telescoped screen, well casing cemented into place, and gravel envelope terminated above the top of screen

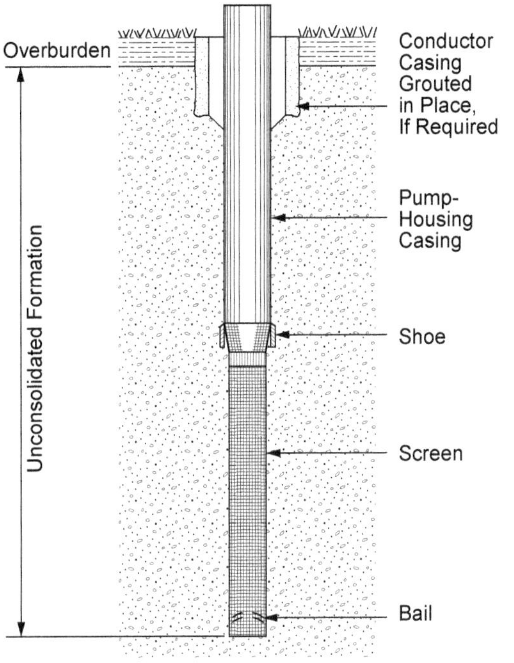

Source: ANSI/AWWA A100-97

Figure 6-22 Naturally developed well with telescoped screen, well pump housing casing driven or jacked into place, and conductor sealed as locally required

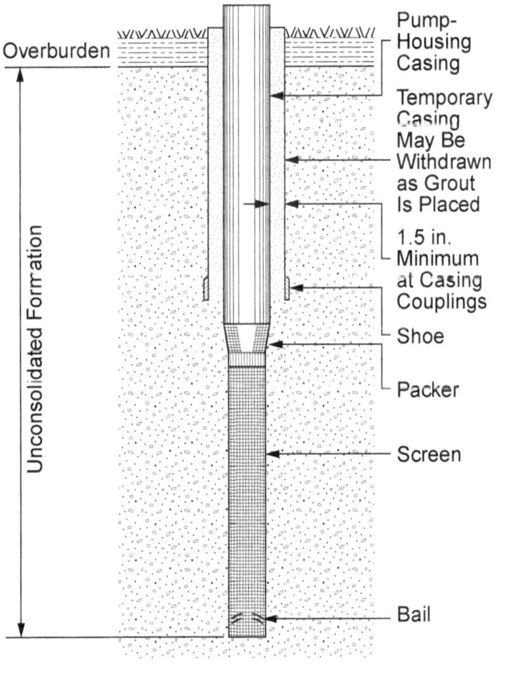

Source: ANSI/AWWA A100-97

Figure 6-23 Naturally developed well with telescoped screen, temporary casing driven or jacked into place, and the pump housing casing conductor sealed in to prevent contamination

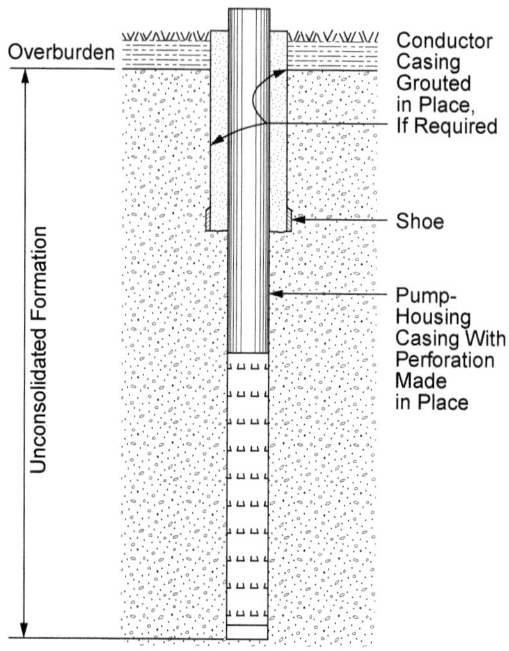

Source: ANSI/AWWA A100-97

Figure 6-24 Naturally developed well with well casing advanced by driving or jacking and perforated in place (California well)

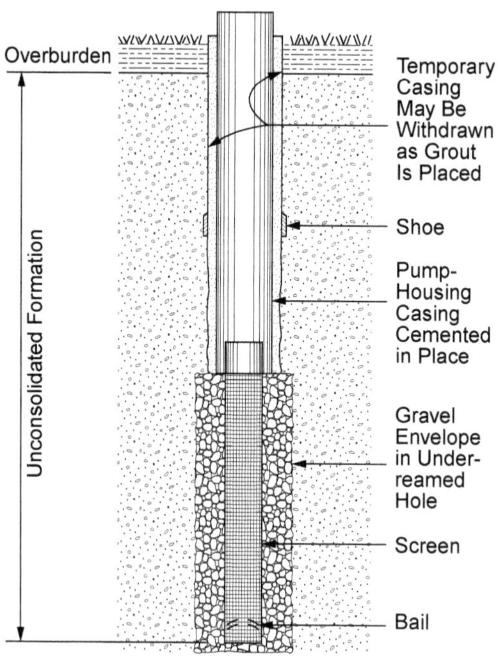

Source: ANSI/AWWA A100-97

Figure 6-25 Gravel-packed well with under-reamed borehole for screen and pump-housing casing cemented in place

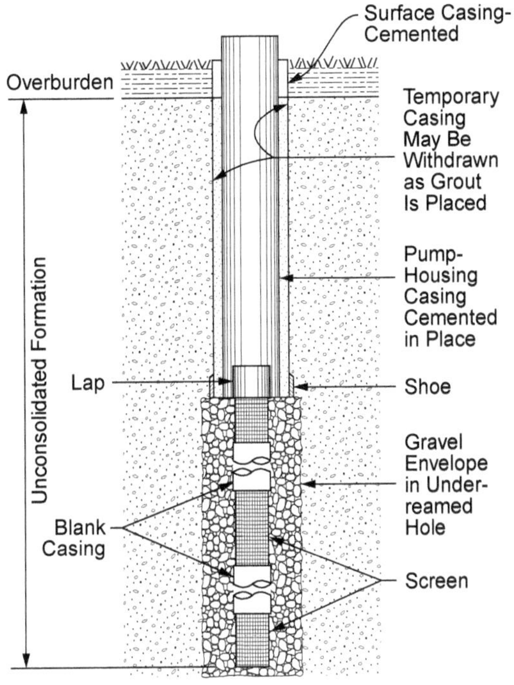

Source: ANSI/AWWA A100-97

Figure 6-26 Gravel-packed well with under-reamed borehole for screens in multiple unconsolidated aquifers

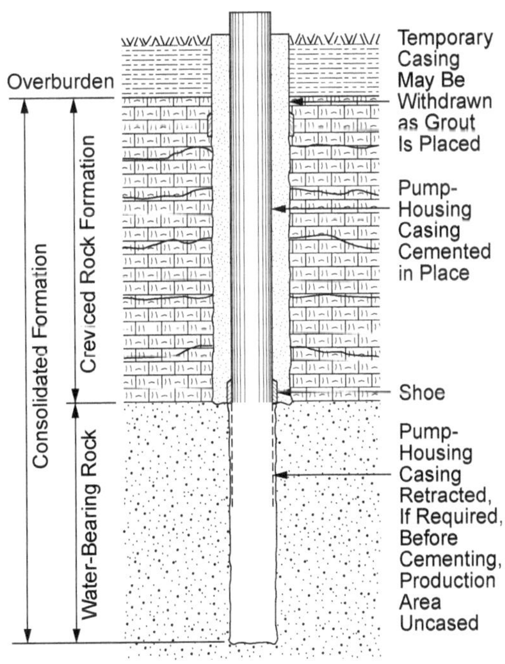

Source: ANSI/AWWA A100-97

Figure 6-27 Well with open hole completion in consolidated rock and well casing cemented in place

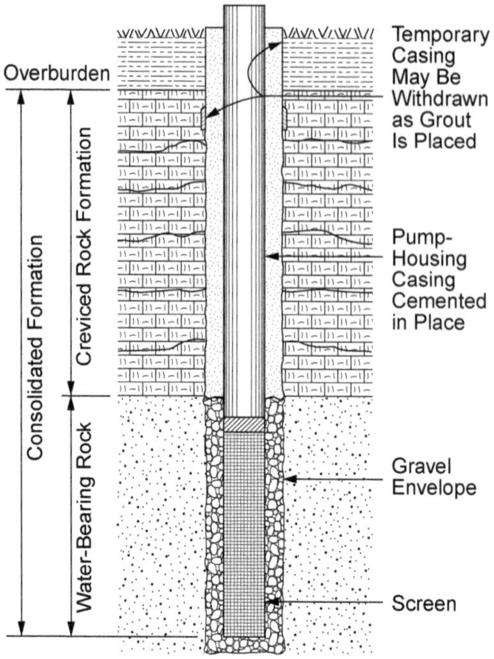

Source: ANSI/AWWA A100-97

Figure 6-28 Gravel-packed well completed in consolidated rock with well casing cemented into place

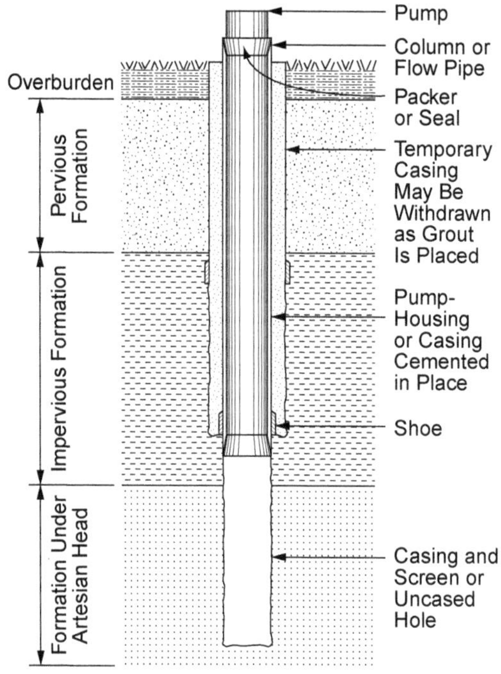

Source: ANSI/AWWA A100-97

Figure 6-29 Open hole or screened well completion in artesian aquifer where piezometric level is above the ground elevation

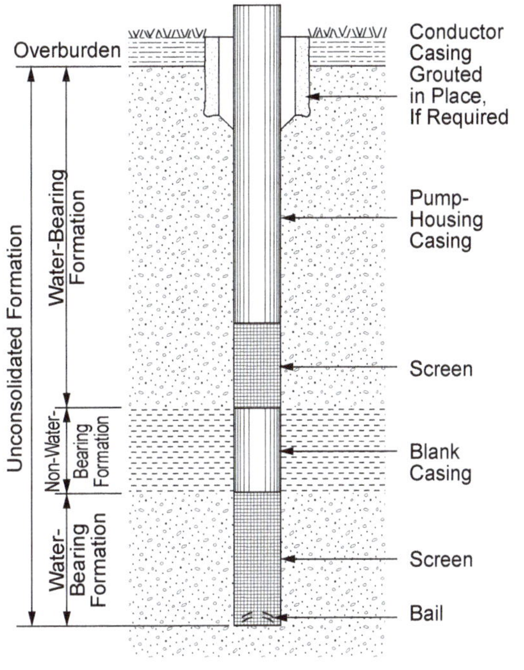

Source: ANSI/AWWA A100-97

Figure 6-30 Naturally developed well with screen and casing installed in place in an open hole. Blank casing in non-water-bearing formation is optimal.

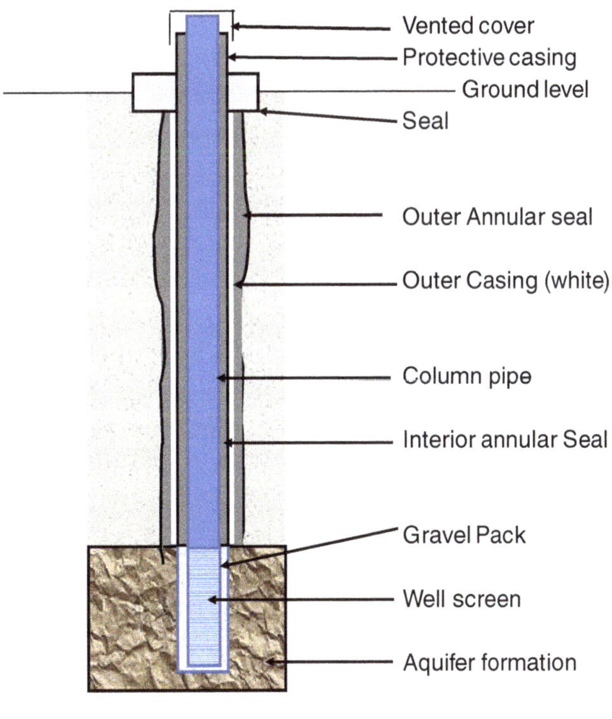

Figure 6-31 Typical water table well with primary and secondary filter pack, screen, and casing with annular seal

WELL DEVELOPMENT METHODS, PROTOCOL, AND MONITORING

Well development principles are designed to maximize the efficiency and specific capacity of the well efficiently. The well development protocol should be modified based on well type, aquifer type, grain size distribution, available equipment, well construction methods, and installed materials and equipment.

There are a variety of methods for well development.

- Over-pumping
- Raw-hiding
- Pump surging
- High-velocity jetting
- Air surging and pumping
- Mechanical surge blocks
- Double-flanged swabbing
- Chemical additives

Each method has its advantages and disadvantages. Over-pumping is the easiest method to use and may be satisfactory in many cases. This method pumps the water at much higher rates than the anticipated production rate to draw fines, drill cuttings, and other contaminants out of the formation. This takes time, especially when drilling mud is introduced into the production zone. Over-pumping has the tendency to preferentially develop the more permeable zones at the time of well development. If a very permeable zone has been significantly intruded with drilling mud, the well may not be developed properly. However, there are risks to over-pumping in loose formations or where fine sands may exist. The fines may be removed, and the grains, in poorly consolidated formations, reoriented. In either case, there is a potential to damage the formation around the borehole, collapse the borehole, or cause bridging of the fines outside the screen, effectively sealing off the well. Over-pumping should be used with care in situations with unconsolidated formations and sand.

Raw-hiding is a variation of over-pumping that involves placement of the pump near the screen. The pump is turned on and run until the water is clear. The water is then reversed so it pushes into the formation, loosening adjacent particles. The process is repeated until no turbidity is found in the water. The method is inexact and not appropriate for most installations. It requires a deeper aquifer to get enough back-siphonage to dislodge formation materials.

Pump surging involves turning the pump on and off for short periods. This develops a mild water hammer in the well to dislodge the cuttings and other materials that need to be removed. Surging may also involve pumping water into the formation and then reversing the pump to draw it out. This method solves two problems with over-pumping—it breaks up the bridging of small particles and is less likely to pull sands and fines to the well. However, surging does create pressure and may damage friable formations, creating more fines and sediments than anticipated. Fines will continue to migrate toward the well once pumping starts so development may be incomplete.

Air purging and mechanical block surging are essentially the same concept using different materials. With air purging, the air is pumped into the borehole through a drop pipe. With mechanical blocks, specific parts of the borehole are purged as opposed to the entire borehole. The same benefits and concerns of pump surging apply to air purging and mechanical block surging.

High-velocity jetting is a means to scour the borehole and remove particles. The jetting occurs as the well is being pumped, so loose materials are moved to the surface. Jetting has advantages in hard formations where it may quickly remove materials on the borehole wall and screen. Soft formations should not be subjected to jetting because the formation could be damaged.

Double-flanged swabbing can be used to develop the well and to scrub the well to remove materials on the borehole wall. A typical swab consists of two rubber discs sandwiched between three wood or steel discs. The swab is constructed so that the outside diameter of the rubber disc is equal to the inside diameter of the screen, fitting closely to the inner surfaces of the well screen and casing. The swab is mechanically raised and lowered along the casing and well screen to draw drilling fluids and fines through the gravel pack and into the borehole. An air-lift may be used in conjunction with the swab to clear the borehole, or the swab may be fitted with a one-way valve allowing removal of development water and fines.

A variety of chemicals can be used to help in well development. Acidization is common to remove materials that cannot be swabbed or scoured off. Carbon dioxide and freezing may have value. Dispersing agents can also be used, but most chemical action is unnecessary for initial development. Chemical use is more appropriate for well maintenance (see Chapter 7).

Well Development Protocol

Records of well development should be maintained. Establishing an efficient and timely protocol for development will ensure a smooth operation in the field (Bloetscher et al. 2005). Such a protocol might be as follows:

1. Measure and record flow.
2. Measure and record pumping distance to water level.
3. Turn pump off and wait 5 min.
4. Record static water level.
5. Calculate drawdown (pumping water level minus static level).
6. Calculate specific capacity.
7. Develop the well by surging or air lifting.
8. Repeat steps 1–7.
9. If the well specific capacity increases significantly (at least 25 percent), disinfect the well and place it back into service. Otherwise, keep repeating steps 1–7.

Any development effort, whether for plugging or not, will create a wastewater product that may contain chemicals, silt, sand, or other debris. The quality of this wastewater may require treatment. In each case, the waste stream characteristics must be identified, including:

- pH of the water,
- chloride level,
- toxic substances,
- silt,
- the quantity of the water to be discharged,
- the time element for which the discharge will occur (i.e., a relatively consistent flow over a period of time or surges),
- the new water quality of the wells, and
- the uptake of metals, SOCs, or VOCs that might violate air or water standards.

There may be a potential for environmental problems if chemicals are used in the redevelopment process. Otherwise, the major concern will be the potential for flooding areas near the well as the redevelopment water is discharged to the ground. This discharge will contain silt, sand, and other debris. If highly turbid, this water may require treatment or capture in an overflow basin, filter sock, or other sedimentation-control device. Regulatory agencies that may be involved in any discharge may include the USEPA, state agencies, and local environmental agencies. Discharges to a sanitary sewer system will involve local utilities and is generally not recommended.

Redevelopment

Periodically, the well will need to be redeveloped to remove accumulated precipitants in the screen, biological masses, and sediment buildup. Redevelopment will restore much of the initial aquifer efficiency. The same basic procedure is used to redevelop a well as was used to initially develop it.

Clearing Fines from the Borehole

Removing fine-grained materials from the borehole involves removing the clay-sized particles that either naturally exist in the aquifer or are introduced or created as a result of the drilling process. The benefit is a reduction of wear on the mechanical parts of the well, as well as limiting future blockage of the screen and gravel pack. Insufficient development will permit migration of near-borehole fines to the screen or gravel pack. For certain types of treatment process, such as membranes, the introduction of small particles may have adverse effects on the treatment process.

Clean and Stabilize the Formation

The intent of cleaning the formation is to create a zone of increased porosity adjacent to the borehole. If the near-borehole formation is cleared of fines, sand, and debris, the screens and gravel pack are less likely to clog. Less clogging will keep the specific capacity of the well high and limit potential mechanical damage on the system. Removing fines and debris will also stabilize the formation and prevent collapses from above. This assumes the proper seals are in place and that competent rock exists on top of the aquifer.

There are several things that can lead to a reduction in aquifer porosity and permeability near the well. Excessive pumping or pumping that removes substantial amounts of fines (because the screen size is incorrect) may cause the aquifer to compress, thereby narrowing flow paths or possibly collapsing the borehole. In addition, in alluvial or sand formations, the compaction may occur because of reorientation of the grains as a result of pumping. Figure 6-32 shows a cube with a series of similar-sized grains in a cubic arrangement. The porosity of the cube exceeds 40 percent. However, the grains reorient in a rhombohedral fashion, which would reduce porosity to less than 25 percent (Figure 6-33).

Sediments composed of well-sorted grain size will maintain initial porosity even if different layers have different grain sizes, but if the grain sizes vary, the porosity can drop to less than 10 percent. The well may subside at the surface as a result of formation collapse (Figure 6-34). In all these cases, if the voids in the formation are clogged during the drilling process, the permeability and efficiency of the aquifer will diminish.

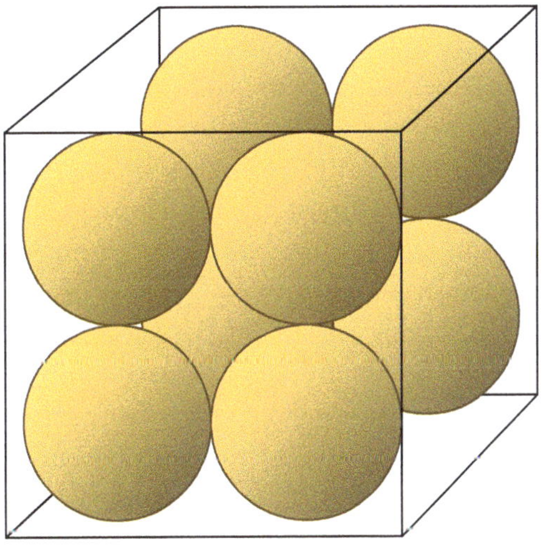

Figure 6-32 Cubic packing >40% porosity

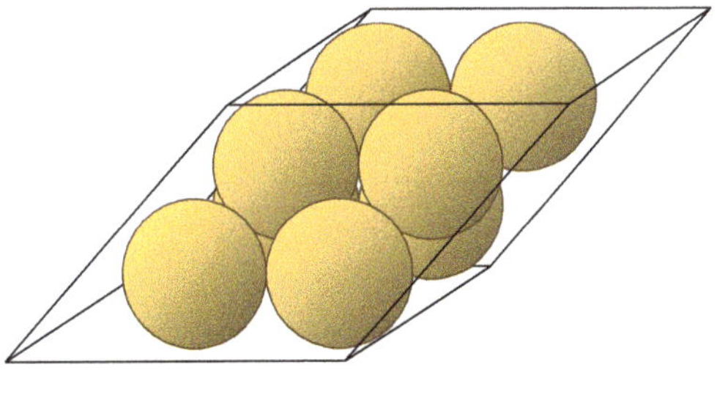

Figure 6-33 Rhombohedral packing <25% porosity

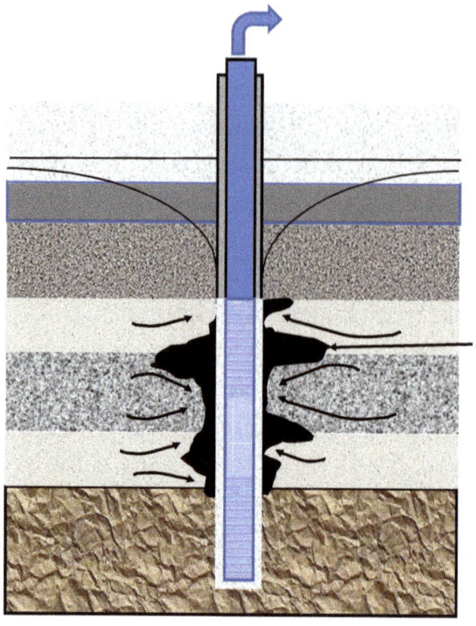

Figure 6-34 Formation collapse

Unconsolidated sediments, such as sand, gravel, and alluvial formations, have significant potential to have compaction, grain reorientation, and clogging through the movement of fines. If the materials are significantly compacted, the well may subside at the surface. Limestone, sandstone, and dolomite are unlikely to have grain reorientation but still may suffer from migration of fines into the aquifer and some compaction if over-pumped.

A third benefit of the casing is to identify construction parameters (i.e., well diameter, screen length, etc.) that are used to establish baseline data for determining the optimum pumping rates given other aquifer parameters and logging results. This initial information is used to establish baseline conditions for future evaluation of pump and well performance.

DISINFECTION/SANITARY PROTECTION

Well Disinfection

During the process of well drilling and construction, the borehole is subject to contamination from the land surface. Contamination can also be introduced by tools, drilling mud (in the case of the rotary method), the casing and column pipes, and the screen. While extended pumping may rid the well of this contamination, disinfecting the well with chlorine is faster and provides more assurance. Most regulatory agencies will require disinfection of the well and a period of clearance testing from fecal coliforms prior to the well being put into use for water supply purposes (AWWA 2003; Bloetscher et al. 2005).

Many disinfection methods are available, although most involve chlorine in some form (typically, hypochlorite). Disinfection is achieved by pumping chlorine into the well casing and producing a mix by alternately starting and stopping the pump. As a general rule, a concentration of at least 50 mg/L of chlorine must be present in the well after introduction of the disinfection fluids. The disinfectant pumped into the well should be thoroughly mixed with the water in the well casing and must come in contact with the pump and discharge piping.

A gravel-pack well may prove difficult to disinfect. The material used for gravel treatment, even though washed and clean, still carries contamination. To resolve the problem, a tablet or powdered calcium hypochlorite can be occasionally added by hand to the gravel-filling tube as the gravel is placed.

Even with disinfection, the water pumped from a well may still show evidence of contamination. Under such circumstances, a chlorinator can be installed at the well to treat all the water discharged to the system. In time (perhaps as long as three or four months), normal pumping will usually rid the well of contamination. During this period, a free chlorine residual will make it possible to use the water. Additional information on disinfection is available in ANSI/AWWA Standard A100-06.

Improved Disinfection

Clean boreholes will have a minimum amount of surface contamination and a limited amount of colonization of native or surface bacteria in the well. As a result, disinfection of the well can be accomplished more easily and more quickly than in wells that are not clean or that have not removed the biological components in the aquifer.

SURFACE FACILITIES

Wellhead

The final piece of well construction is the installation of the surface equipment. Surface equipment for wells is generally limited to some piping, a meter, and some sample taps or probes to monitor water quality. Figure 6-35 shows an engineering drawing for a wellhead that includes the well (top only), an air release valve, meter, butterfly valves for flow control, a check valve to prevent backflow, piping, and pipe supports. This installation is above ground and is shown in different perspectives. Figure 6-36 shows an example of an installed wellhead similar to the drawing shown in Figure 6-35. Surface equipment should be designed by a competent professional engineer. The design should include the following:

- lightning and transient voltage surges protection, including lightning arrestors, surge capacitors, or other similar protection devices, and phase protection;
- access for repair and maintenance purposes that will not cause interrupted operation;

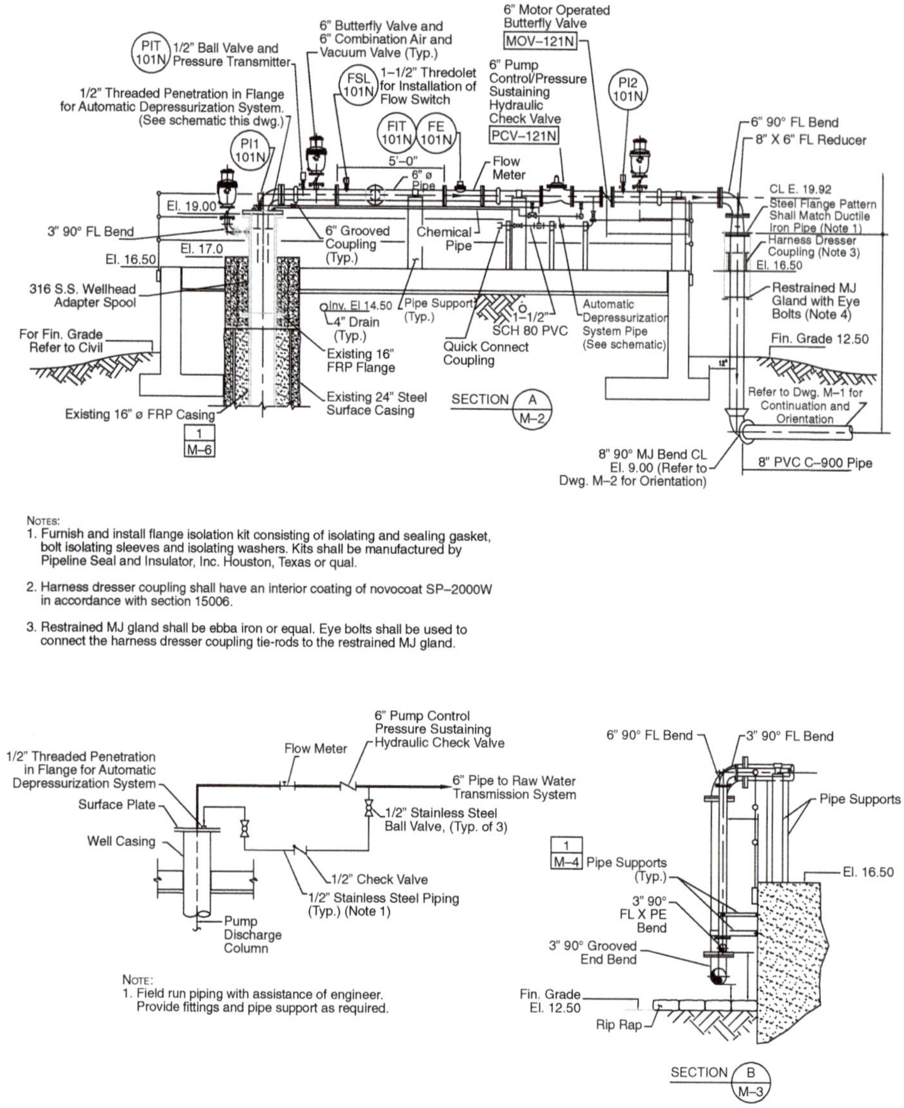

Source: Hazen and Sawyer, P.C., Boca Raton, Fla.

Figure 6-35 Drawing of wellhead and piping

- protection against surge and water hammer to protect the integrity of the well system;
- operational reliability and flexibility in the event of damage to or failure of the pipeline or a well;
- access to the well for geophysical logging without disruption of operations;
- necessary screening for floatable solids prior to withdrawal to avoid plugging of the injection horizon; and
- vandalism protection.

Source: Hazen and Sawyer, P.C., Boca Raton, Fla.

Figure 6-36 Photograph of wellhead and piping

Sanitary Protection

Once the well has been constructed and developed fully, the final sanitary protection should be provided along with disinfection of the borehole, pump, casing, pipe, and fittings. Where the minimum depth for withdrawal varies with soil formations and surrounding conditions, the well casing should extend at least to the depth where protection from surface contaminants is anticipated. The screen should be set below that point. The sanitary seal prevents contamination from migrating downward to the screen from the surface.

The sanitary seal is usually constructed of neat cement. Every well casing should be grout sealed from land surface to the full depth. Many regulatory agencies require a minimum of a 6 ft^2 concrete pad around the well casing, sloped at 1 in. per yard away from the well. Where a well is installed to a depth less than where protection can be assured, the well needs to be located in a wellhead protection zone as defined in Chapter 2. In the immediate vicinity of the well, an impervious layer of well-compacted clay or neat cement at least 2-ft (0.6-m) deep should be placed on the land surface around the well. This barrier will minimize percolation from surface water to the withdrawal point. Once installed, the wellhead can be finalized with a connection from the column pipe to external piping (Figure 6-37). Note the raised pad that serves as a part of the sanitary seal of the well. This one is outdoors.

Another means of protection is to submerge the well screen below the pumping level of water in the well. The pumping level of the well should never be allowed below the top of the screen, as aeration of the well screen may promote aerobic bacteriological activity. The cascading water causes air entrainment and possible cavitation to the pump.

Source: Photo by Fred Bloetscher

Figure 6-37 Final well head after pump installed

Source: Photo by Fred Bloetscher

Figure 6-38 Looking down the casing pipe of the well

Pumps

Once the well casing is set in place, the well pump will be installed. Figure 6-38 is a look down a well with a PVC casing pipe (from the surface). The sanitary seal will be installed on the top (Figure 6-39) and the pumps and column pipe installed (Figure 6-40). Like the grout, the sanitary seal is designed to prevent stormwater from entering the well.

MONITORING

Testing During Development and Redevelopment

ANSI/AWWA A100-97 noted that there are a series of tests that should be undertaken as a part of the development and redevelopment process:

> #### "1 Step-Drawdown Test Method
>
> Step-drawdown tests shall be conducted to characterize well performance at varying rates and to determine the general parameters for a constant-rate pumping test. The well shall be pumped at a minimum of at least three progressively increasing rates, and the length of each discharge step shall be long enough to indicate a straight-line trend on a plot of drawdown versus logarithm of time since pumping began.

Source: Photo by Fred Bloetscher

Figure 6-39 Sanitary seal connection for pump
Note: The seal is the concrete box.

Source: Photo by Fred Bloetscher

Figure 6-40 Submersible pumps being dropped down the casing pipe

2 Constant-Rate Tests

After the step-drawdown test, the well shall be allowed to recover until water levels return to approximately static conditions. After recovery, a constant-rate test shall be conducted at a designated capacity to determine the trend of drawdown versus prolonged time of pumping at the pumped well and any observation wells.

2.1 *Pumping well.* The pumping well shall be pumped at a constant rate until a straight-line trend is observed on a plot of water level versus the logarithm of time since pumping began.

2.2 *Recovery time.* Recovery time of the pumping well and any observation wells to be used in the test should be such that a straight-line trend is observed in all of the wells on a plot of water level versus the logarithm of time since pumping stopped.

3 Water-Level Measurements

Water-level measurements shall be obtained before, during, and after the pumping test in order to acquire background information (static water levels), the effects of pumping (pumping water levels), and a profile of the recovery of the water level from the pumping level to the original state. The measurement frequency of water levels during pumping shall be such that adequate definition of the drawdown trend is achieved.

4 Air-Line Method

A tube, free of air leaks, shall be installed in the well with the development pump, terminating at least 5 ft (1.5 m) above the pump intake. The tube shall have an accurate altitude gauge and an air valve attached to it at the surface. The vertical distance from the bottom of the air line to the center of the gauge shall be recorded. The line shall then be charged with air under pressure until the gauge will read no higher. The water level in the well shall be computed by subtracting the altitude registered on the gauge from the known length of the line.

5 Electric-Sounder Method

If the level cannot be measured accurately due to cascading water, a 0.5-in. (13-mm) or larger diameter pipe shall be installed in the well from the surface to 2 ft (0.61 m) above the pump bowl to permit the installation and operation of an electric sounder.

6 Measurement of Sand Content

Measurement of sand content may be accomplished by the installation of the Rossum centrifugal sand sampler. The sample line shall tap the discharge pipe as close to the pump head as possible to ensure that flow is sufficiently turbulent to keep the sand uniformly distributed in the stream. The purchaser may specify other methods acceptable to local regulatory agencies."

Well development may be the most important part of the well drilling process and is often underestimated with regard to the time required to properly develop the well. As a result, many wells suffer from incomplete development from the start, which makes them less efficient and less productive than they were designed to be. In the long term, this causes additional client time and both capital and operating expense that are unnecessary. The goals of well development are:

- to clear fine materials from the face of the borehole;
- to clean and stabilize the formation by removing drilling mud, sand, and other foreign materials that are pushed into the formation by the drilling process (thereby improving porosity);
- to correct damage caused by the drilling process; and
- to improve ease of well disinfection.

The well development procedure includes all steps necessary to accomplish these goals, including subjecting the aquifer to high levels of energy and pressure to dislodge and remove materials that may clog the formation and reduce well efficiency. Not all wells are developed in exactly the same way. Gravel-pack wells and open-hole wells require different approaches, and the material moved in the open-hole well are far different than those of the gravel-pack. The rotary and cable-tool drilling methods have different impacts on the aquifer during construction, so well development strategies may be slightly different.

PERMITS

Permits are normally required for any well drilling operation. Obtaining permits is usually the responsibility of the well driller or the engineer or hydrogeologist in charge of the project for the owner. A hydrogeologist normally supervises the work in the field, including the procurement of well construction materials, well logging and interpreting logs (Chapter 3), and well drilling (Chapter 4), while an engineer supervises well design (Chapter 5) and certifies as-built well construction drawings after well completion and development (Chapter 6). Reports that document field work can be very valuable in later phases of groundwater development or protection. Many utilities are required to report data to regulatory agencies. For states with permit requirements (such as Florida) to states operating under the Prior Appropriation Doctrine, data on water use must be filed. Regulatory agencies are concerned about well construction (and closure of a well) because of the possibility of cross contamination between shallow zones and deeper high-quality aquifers. Chapter 7 covers general reporting requirements.

REFERENCES

American Water Works Association (AWWA). 2003 (3rd ed.). M21, *Groundwater.* Denver: AWWA.

AWWA 2014 (4th ed.). M21, *Groundwater*. Denver: AWWA.

American National Standards Institute (ANSI)/AWWA. 2006. A100-06. *Water Wells*. Denver: AWWA.

ANSI/AWWA. 1997. A100-97. *Water Wells*. Denver: AWWA.

ASTM. 2008. ASTM C494/C494M-08. Standard Specification for Chemical Admixtures for Concrete. West Conshohocken, Pa.: ASTM.

ASTM. 2018. ASTM C150-18. Standard Specification for Portland Cement. West Conshohocken, Pa.: ASTM.

Bloetscher, F., A. Muniz, and G.M. Witt. 2005. *Groundwater Injection: Modeling, Risks and Regulations*. New York: McGraw-Hill.

Lehr, J., S. Hurlburt, B. Gallagher, and J. Voytek. 1988. *Design and Construction of Water Wells: A Guide for Engineers.* National Water Well Association. New York: Van Nostrand Reinhold.

National Groundwater Association. 1998. *Manual of Water Well Construction Processes*. Westerville, Ohio: National Groundwater Association.

Roscoe Moss. ND. *A Guide to Water Well Casing and Screen Selection.* https://roscoemoss.com/wp-content/uploads/publications/ggwc.pdf.

Williams, D.E. 1985. "Modern Techniques in Well Design." *Journal AWWA*. 77(9):68–74.

CHAPTER 7

Operation, Monitoring, and Maintenance

The intent of this chapter is to outline basics for well maintenance and operations. The goal is to provide the drillers and operators with a common point where the well has been installed, developed, and placed into service and has become the operator's responsibility. However, much of the data the driller has developed should be used as baseline information by the operators.

WHAT DATA OPERATORS SHOULD MAINTAIN

Once the well is designed and constructed, two issues must be monitored on a regular basis: well performance and water quality. Well-performance issues include the routine maintenance records for pumps and motors, work orders for repairs, tracking specific capacity with time, and recording well withdrawal volumes. Monitoring specific capacity can help identify potential fouling in the well. Certain water quality tests (e.g., bacteria and sand) may indicate issues associated with damage to the well. Ongoing issues with fouling and microbiological activity appear to be common with wells, so utilities should be cognizant of the type of deterioration that is occurring and applying appropriate solutions to limit long-term damage to assets. As a result, topics will be covered herein in more detail relating to fouling and microbiologic activity in wells.

AS-BUILT WELL DATA

Data that should be collected and compiled relating to well design and construction include (also see prior chapters):

- detailed individual well (geologic) logs,
- well diameters and depth for each string of casing and screens,

- casing specifications (e.g., wall thickness, couplings),
- total well depth and diameter,
- position of the screens (or portion of the open hole if constructed in rock),
- slot sizes of the screen,
- gravel-pack material (if used),
- method of construction and materials,
- driller's log of the materials encountered during drilling,
- detailed individual well lithologic logs,
- water quality analyses,
- static (nonpumping) water levels in the aquifer,
- design pump discharge pressures, and
- other data developed during the design and construction phase (e.g., well completion report).

The above data, along with as-built drawings, should be included in well completion reports. Well completion reports should be saved in a safe place with the utility. The well completion report is often required by regulatory agencies as well. They are valuable sources of data for future well exploration or well challenges.

ONGOING RECORDKEEPING

One of the major issues that is often neglected with well systems is operating records (Bloetscher 2019). Gaining useful information on well maintenance requires some form of work order tracking tool to be in place, maintenance schedules that match O&M manuals, documentation of the parts and maintenance needs, and training. This is a critical preventive maintenance function for water utilities. For consistency of operations between shifts, there should be a standard reporting process. Computerization is preferred so that computer-tracked data can be compiled along with operator input data. Some of this data is needed for reporting to regulatory agencies, some for maintenance tracking, and some for budget justifications. The lack of data is one of the biggest issues that prevents operations staffs from getting needed budget allocations. Documentation using work orders for maintenance work is key to justify funding. A log tracking the dates and time for work is recommended.

PUMP AND MOTOR MAINTENANCE

AWWA's Manual M21, *Groundwater* (AWWA 2014), suggests that well owners maintain records regarding the following:

- type (and make) of the pump installed,
- type and horsepower of the motor (driver),
- pump setting (depth to the pump intake),
- setting of the air line or other device for measuring the water level in the well,

- notation for the point (and reference elevation) used for measurement of the water level, and
- information provided by the pump and motor manufacturer, such as capacity and efficiency data.

Most of the information provided by pump and motor manufacturers is contained in the operations and maintenance (O&M) manuals that should be provided to the utility upon completion of the well construction and testing. The O&M manuals will contain the recommended maintenance activities for the pumps and motors. These should be followed as failure to do so may void warranties on the equipment. The manufacturer's information and the data from the O&M manuals should be kept in a safe place, updated as necessary when parts and/or pumps are replaced, and used for ordering parts when they fail or are recommended to be replaced before failure (Bloetscher et al. 2005). As operations commence, personnel need to track issues related to the pump equipment such as:

- initial cost of the well and equipment,
- expected equipment life,
- records on repairs (and their costs),
- cost of operation, and
- cost of maintenance.

The issues that should be tracked include the expected operating conditions, such as the ambient air and water temperature ranges, pressures, flow, corrosive and abrasive factors, power supply variation, duty cycle, and protective devices; these must be checked against the pump manufacturers' information to ensure reliable operation. The pump must be installed according to the manufacturer's instructions; warranties rely on the equipment remaining within manufacturer-designed tolerances.

Continuous operation of a pump is generally preferable to intermittent operation, but varying water demands and the desire to operate all wells evenly, usually requires some combination of off- and on-time. For improved well performance and pump life, system components and storage capacity should be designed to minimize the number of starts and stops per day. At the same time, the pump must be sized and set so that it will never run for even a few minutes at "no delivery," as this will cause damage to the pump bearings by the overheating and failure of the submersible motor. If the well drawdown or the delivery system causes the pump to run at little or no delivery, protection should be provided to the pump. This protection could include a well-level switch that would shut off the pump or sound an alarm if the flow or water level dropped below a safe minimum level.

Well maintenance activities should also be recorded. This data can be used to predict times when maintenance needs to be performed, identify possible causes of well decline, and plan for annual budgets for well maintenance when compared to the initial test data. These records should include:

- dates that maintenance was performed,
- results of pre- and post-maintenance pumping tests,

- methods (and materials) used in the maintenance procedures, and
- other factors such as the coloration of the pumped water, amounts of sand removed, odors, and water quality analyses.

Proper selection of system components can ensure system performance, but changing conditions sometimes justify altering or reselecting components to maintain economical operation. Routine maintenance is often specified by the manufacturer and will include greasing bearings, polarity checks, and alignment checks.

OPERATING RECORDS

As part of the proper operation of a groundwater system, gathering, compiling, and recording of a wide variety of data must be performed to document the operating history of the wells. These data are used to detect a loss of production efficiency and possibly the cause of a loss, to schedule maintenance at opportune times to avoid breakdowns, evaluate the cost of water production, and schedule capital improvements. The forms used for recordkeeping are not critically important; the key is that the records must be collected and maintained in a logical fashion regardless of the form that is used.

To accomplish water volume reporting, the first item that is needed is a meter that can be read. Figure 7-1 is an example. However, large meters tend to lose accuracy quickly, so they need to be calibrated regularly. Having an outside person

Source: Photo by Fred Bloetscher

Figure 7-1 Water meter for recording flow from the well

PC CONTROLS INC

P.C. Controls Inc
132 NE 30th Street, Wilton Manors, FL 33334
Phone: (954) 568-9663, Fax: (954) 563-0919
E-mail: pccontrolsinc@bellsouth.net

Certified Flow Meter Test Record

Customer: City of Dania Beach

Consumptive Use Permit # 06-00187-W

Test Site Influent to Membranes Flow transmitter, FIT-1929

District ID# ________

Contact Nate Costa   Number 954 924-3616

Meter Model Rosemount 8750WA   Serial# 0011439

Meter Accessories: Mag

Pipe Information at test site

Material Stainless Steel Schedule 10

Outside Diameter 10.75" OD   Pipe Thickness .165" WT

Transducer Spacing 7.78" V-Mount   Distance from Meter installation is 5 feet

Test Information using Fuji Porta-Flow Portable Ultrasonic Flowmeter

Test performed @ ___777.5__________ GPM

Initial Meter Reading at Calibration   Start: _1955916032_X _1_ End: _1955924096___X_1___

___8064_____Gallons Totalized in _631____Seconds   Average _766.8__ GPM

Test Meter Reading at Calibration   Start: ____0______X_1__ End: __8008_______X_1___

___8008_____Gallons Totalized in _618____Seconds   Average _777.5__ GPM

Meter Tested at _99.9_ % Accuracy   Last Meter Test _6/20_____   Last Accuracy_100.1%_

Comments: Prior to calibration meter was reading 99.9% of test meter – Site meter read 766.7 GPM and test meter read 777.5 GPM.

Performed By: R Felicien   Date: 6/3/21

This is to certify that the above listed equipment has been calibrated and meets or exceeds all published specifications. Calibration was performed using standards and/or instrumentation whose accuracy is traceable to the N.I.S.T.

Signature :

Source: Courtesy of City of Dania Beach (Fla.)

Figure 7-2 Meter calibration report

calibrate the meters provides a degree of confidence for both the operations staff and regulatory staff. Figure 7-2 is an example of a calibration report for a meter.

The total volume pumped from each well is generally required by permit to be recorded daily and reported monthly on operating reports submitted to regulatory agencies. These numbers should be graphed to illustrate the seasonal and yearly production rates. An example is included on Figure 7-3. This data can be used for future projection of water withdrawal rates and to monitor the actual

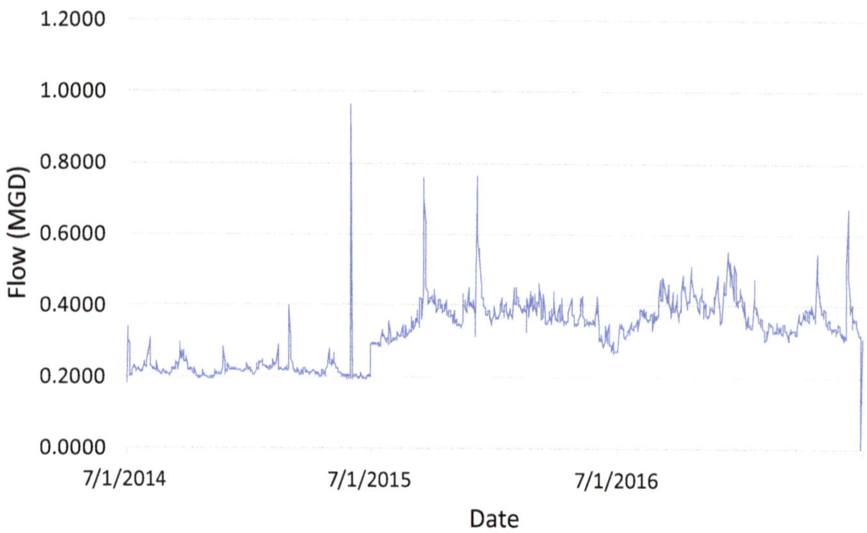

Figure 7-3 Graphic of water use over time

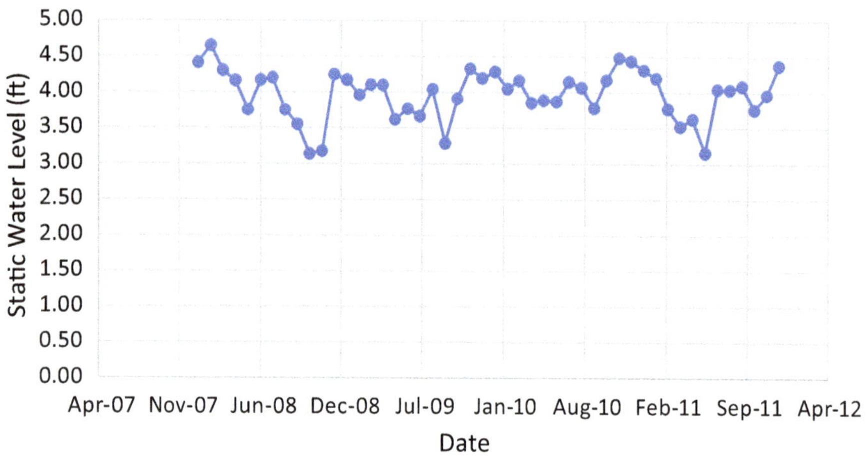

Figure 7-4 Record of static water level over time

volume of water produced from each well. Data can usually be recorded from a totalizer on flowmeters installed in the discharge piping for each well.

Water levels in the well during periods of non-use (static) and during pumping should be recorded to provide a baseline for determining the amount of drawdown. The static levels can identify changes in the amount of water that may be available in the aquifer with time or at any given time. An example is shown in Figure 7-4.

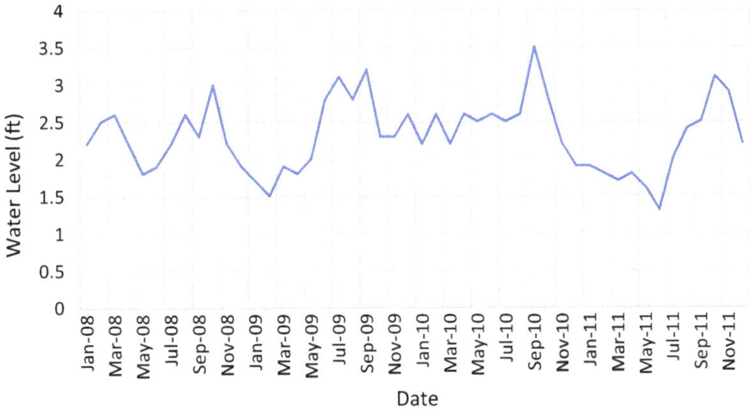

Figure 7-5 Monitoring well levels with time

Monitoring wells can illustrate aquifer levels; they can also be used to monitor the production zone and to monitor adjacent zones. Figure 7-5 is an example of a monitoring well's water levels with time. This is a surficial aquifer, so the aquifer level would be expected to vary with rainfall. Confined aquifers would not be expected to have such fluctuations. Daily reports can capture much of what happens in the system. Operators should take measurements and note issues that arise on the daily log. Figure 7-6 is an example of a daily log. Note that much of this information may be downloaded to a monthly report submitted to regulatory agencies. Computers can be used to track alarms that might indicate issues that arise with the wells, mechanical equipment, and/or transmission system.

ASSESSING WELL PERFORMANCE ISSUES

Operations personnel should evaluate any well failure or long-term decline in performance to determine if physical or mechanical problems are causing the decline. Specific capacity is a method to monitor well performance; it is the ratio of the yield of each well to its drawdown during pumping. The easiest thing to do is plot the operational trend of each well. The specific capacity of a well should be calculated annually to identify the potential need for maintenance, plugging problems, or water supply concerns.

The apparent specific capacity or ratio of the yield of each well to its drawdown, expressed in gallons per minute per foot of drawdown (gpm/ft), is used to plot the operational trend of each well. The apparent specific capacity for a well is determined by dividing the pumping rate (gpm) by the observed drawdown (ft) in the well measured at any specific time. Specific capacity depends not only on the transmissivity of the aquifer but also on well construction factors, such as screen type, well diameter, degree of aquifer penetration, and degree of well development. In general, when a well is new or new screens have been installed,

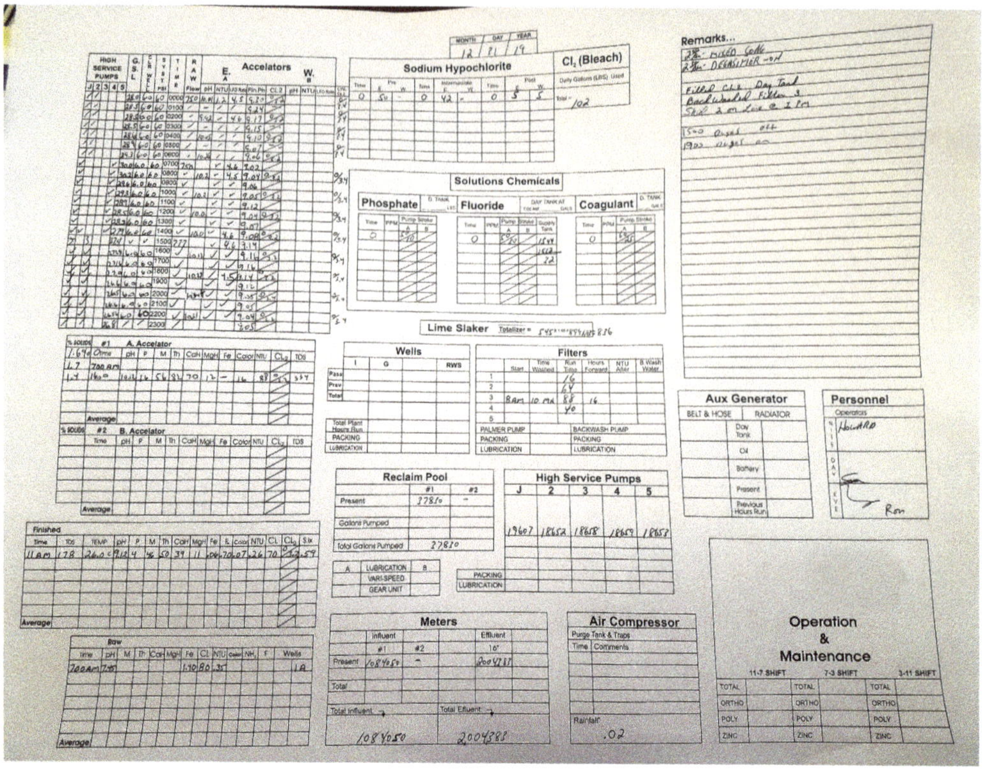

Source: Courtesy of City of Dania Beach (Fla.)

Figure 7-6 Example of a daily log form

the specific capacity is expected to fluctuate with pumping rate, river stage, temperature, aquifer water levels, and degree of screen efficiency.

Well operators should track data on specific capacity on an ongoing basis. As noted in Chapter 5, the formula for specific capacity is shown below and usually reported in units of gpm / foot:

$$\text{Specific Capacity} = \text{Flow} / \text{Drawdown}$$

The utility should track specific capacity on a regular basis. Monthly may be too often in stable environments. Yearly is probably too infrequent in any environment. Specific capacity indicates that there are changes in the aquifer, screen, or pump that should be checked. Figure 7-7 is an example of a utility that tracked specific capacity with time. When the specific capacity reaches a given number, such as 70% or the original specific capacity, the well should be taken out of service and appropriate action taken to return as much specific capacity as possible to the well.

If specific capacity decreases, it may be the result of a drop in pumping water levels or a reduction in pumping yield caused by microbiological fouling, chemical precipitation, formation, well screen or gravel-pack plugging, pump corrosion, or biofouling. Water-level declines can be caused by regional water-level declines or reduced hydraulic efficiency in the well—most commonly plugging

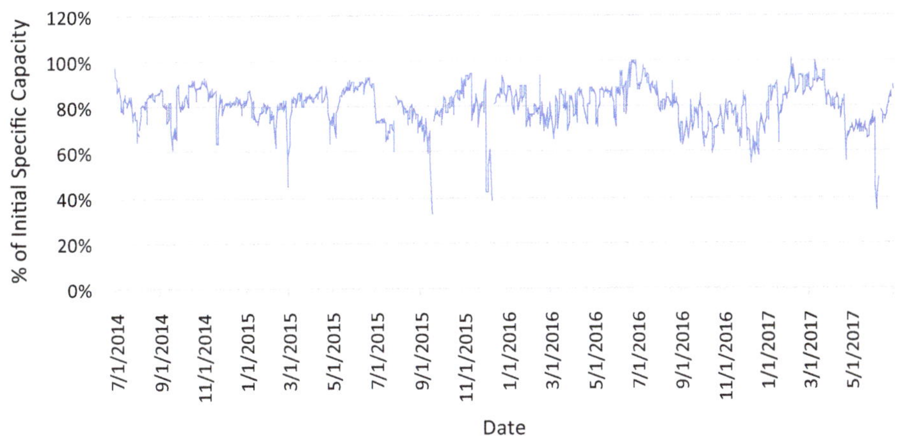

Figure 7-7 Tracking of specific capacity with time

or encrustation of the borehole, screen, or gravel pack. Other specific capacity problems may relate to:

- changes in the water-bearing zone,
- insufficient development of the well at time of drilling,
- pump wear, and
- impeller detachment from the shaft.

All the above may require further investigation to determine maintenance and repair needs. Comparison to "like new" condition should be performed to identify how far capacity has diminished. Utilities should establish a point where once specific capacity diminishes to that point (e.g., to 70–80 percent of original conditions), actions are taken to identify potential concerns. This will include the pulling of the pump and motor. At this time, comparative data pertaining to the physical condition of the pump unit should also be collected.

Design choices create several operational conditions:

- over-pumping (which results in lowering of the water table by overestimating the amount of water that can be pumped from the well);
- clogging or collapse of a screen due to underestimation of the amount of sand in the formation or suction on the screen;
- perforation of a screen section due to wear;
- excessive entrance velocities;
- poor material choices;
- corrosion due to material choices in the formation (steels may pose a challenge in the presences of major concentrations of groundwater bacteria); and
- encrustation due to bacteria and mobile metals, among others.

Other problems include incorrect specification of pumps and poor construction (casing damage, breaches in the grout, misplacement of screens and/or gravel pack, and misalignment of the pump and motor).

Over-pumping can damage the well by reducing the storage and production capacity of a groundwater system. In granular formations, the water-bearing formation may consolidate. Where this occurs, it results in a lower water table, less water storage space, reduced yield from individual wells, and potential collapse of the well casing. Wear in the screens or pumps may be a result of entrance velocity, as water passes through the well screen (or the edge of the formation depending on the type of well). As the entrance velocity increases, sand, silt, and colloidal matter can enter the flow stream.

Other problems include suction breaks and electrical surges. No pump should operate at a rate at which it breaks suction as this condition may cause severe damage to both the pump and the aquifer. Water hammer resulting from surging could be a concern. Surging in the well may collapse the well if it was not properly stabilized. Surging can stimulate sand, silt, and colloidal activity or dislodge corrosion and precipitates. Air bubbles may be entrained into the wells, which can damage the distribution system piping by causing air pockets. The solution is to remove a bowl or slow the motor speed, not for operators to partially close the valves to reduce pumping. Lightning strikes and poor grounding may cause electrical surges that damage motors and pumps. Appropriate lightning attenuation should be installed to avoid surges that can damage pumps and motors.

Any time that a well is removed from service for maintenance, it is recommended that pump tests be run to measure specific capacity. The data collected should include:

- water level measurements made before, during, and after the (drawdown) pumping test,
- a record of the pumping rate,
- hydrographs generated during the test, and
- any other raw data collected.

The results should include manual and computer-generated information. Updates to aquifer computer models should be documented, and model results should be included in the collected information.

WATER QUALITY MONITORING

As operations personnel review changes in static and drawdown levels to identify any trends, they should be cognizant of changes in specific capacity and water levels in the aquifer. Long-term reductions in water availability in the aquifer or limitations in specific capacity of the mechanical system caused by pump wear, clogged screens, or formation of bacterial fouling may be indicated. Repairs to correct mechanical problems should be scheduled before they become serious, but problems that are not mechanical may be more challenging.

Regulations require periodic monitoring of microbiological and chemical quality. Intensive water quality monitoring should occur after the well has been completed to establish a baseline that can permit the water system operator to reduce the frequency of groundwater sampling and indicate if long-term changes in water quality are occurring. Fortunately, groundwater quality in many locations does not change significantly with time because the movement of groundwater is generally very slow compared with surface water. When changes do occur, potentially serious problems could be present.

Where contamination risks are high, sentinel monitor wells should be installed for this purpose. Sentinel wells, located at various depths, will provide definition for the initial groundwater assessment. Sentinel wells also serve as an early-warning system to detect changes in water quality and water elevations before they affect the water-supply wells.

All water quality data should be routinely graphed to identify patterns in the quality of water from the well. Important changes in water quality over time are not limited to changes in concentrations of drinking-water contaminants but can include changes in the general characteristics of the water, such as temperature, dissolved oxygen, pH, alkalinity, metals, and dissolved-solids concentrations. Table 7-1 shows the typical components that should be analyzed; they should be analyzed for each well and as combined average. Because temperature is often indicative of changes in flow regimes in aquifers, the groundwater temperature should be recorded and plotted. As the temperature of the groundwater varies, the capacity of the well fluctuates as a result of the viscosity of the water. In projects where recharge to the aquifer may come from infiltration of surface water, the temperature of the adjacent surface water body should also be recorded. Figure 7-8 shows a comparison of water temperatures in an aquifer that is not subject to frost or major changes and an aquifer that is impacted by adjacent surface waters. Note the temperature changes on the latter but not on the aquifer that is not impacted by surface waters. Figures 7-9 and 7-10 show trends over times for metals analyses

Table 7-1 Example of typical water quality analysis for a series of water supply wells used for a utility

Parameter	Well 1	Well 2	Well 3	Well 4	Well 5	Average
pH	7.5	7.5	7.5	7.4	7.6	7.5
Alkalinity	244	228	236	294	288	258
Total hardness	244	252	248	293	300	367
Calcium hardness	236	236	236	261	292	252
Magnesium	30	16	23	53	58	36
Fluoride	0.22	0.16	0.19	1.42	0.028	0.50
Iron	0.72	0.44	0.58	2.09	1.88	1.10
Color (SCU)	53	46	49.5	65	60	55
Chloride	70	70	70	59	46	63

SCU: standard color units

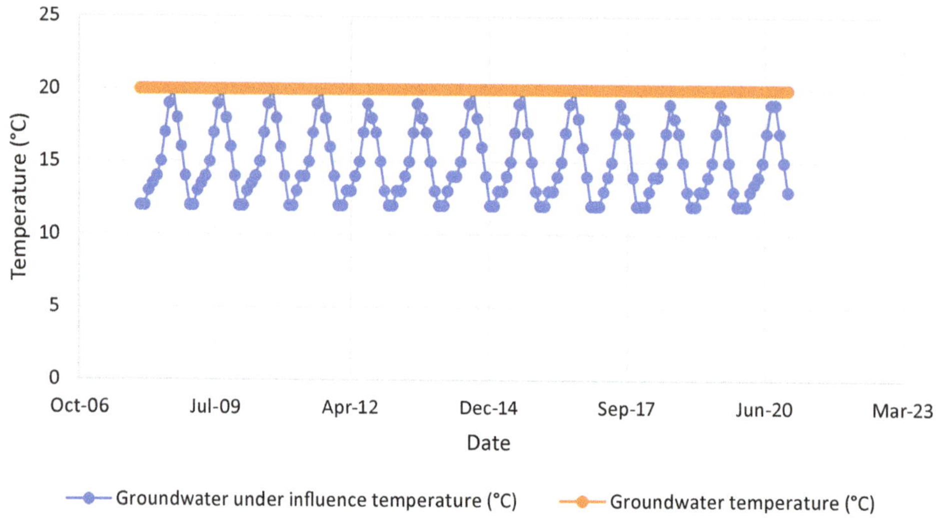

Figure 7-8 Comparison of water temperatures in an aquifer that is not subject to frost with an aquifer impacted by a nearby stream

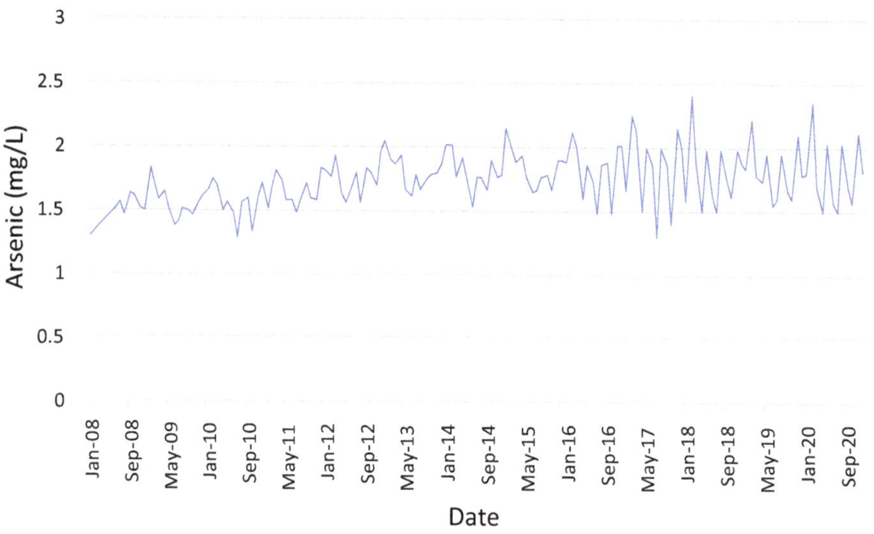

Figure 7-9 Trend analysis for arsenic

for a raw water supply. Operators must monitor this information as treatment may be required or a shutdown of the well may be needed or treatment initiated.

Table 7-2 shows an example of a report where the utility tracked water quality data with time. Ammonia was the example here. Ammonia is usually caused by something other than natural conditions (e.g., fertilizer, septic tanks, industry). As a result, ammonia usually indicates the potential for contamination. Ammonia is also acted upon by bacteria to become nitrate, which is a regulated constituent that causes blue-baby syndrome. Tracking ammonia is therefore important for

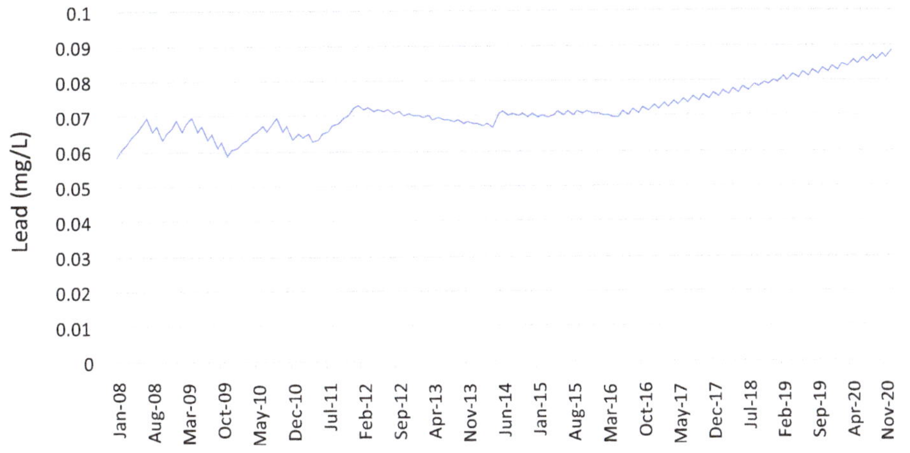

Figure 7-10 Trend analysis for lead

Table 7-2 Example of report tracking water quality data with time of a surface water impacted well

Date	Ammonia (mg/L)	Lead (mg/L)	Temp (°C)
Sep-19	1.67	0.08	20
Oct-19	1.59	0.08	20
Nov-19	2.10	0.08	20
Dec-19	1.79	0.08	20
Jan-20	1.80	0.08	20
Feb-20	2.35	0.08	20
Mar-20	1.69	0.08	20
Apr-20	1.50	0.09	20
May-20	2.04	0.08	20
Jun-20	1.56	0.09	20
Jul-20	1.50	0.09	20
Aug-20	2.04	0.09	20
Sep-20	1.69	0.09	20
Oct-20	1.56	0.09	20
Nov-20	2.11	0.09	20
Dec-20	1.81	0.09	20

operators to understand the nature and ongoing conditions indicated by whatever causes the ammonia and provide an indication if treatment might be needed. For example, ammonia can be removed by chlorine and/or aeration. If levels climb, the utility may consider installing aeration equipment at the head of the plant or on a storage tank. Adding chlorine might be an option, but less so if natural chlorides exist in the water. These are water quality/treatment issues that must be considered.

Likewise, hydrogen sulfide is an indicator of bacterial activity in the formation. Because of the impact hydrogen sulfide can have on metals and concrete, removal of hydrogen sulfide by aeration is often practiced. More discussion of water quality monitoring occurs later in this chapter along with some solutions to address water quality concerns.

Temporal fluctuations in water quality can help identify the presence of preferential flow pathways (i.e., pathways that provide little resistance to flow), which can increase the vulnerability of the well to contamination. Short-term water-quality fluctuations in deep public-supply wells should not be expected in the absence of preferential flow pathways. For example, a well in Modesto, Calif., had higher concentrations of nitrate, uranium, and alkalinity during winter months. This was because the well was functioning as a conduit for the downward movement of contaminants during times of light or no pumping because of a downward hydraulic gradient in the surrounding aquifer. These conditions allowed shallow contaminated water to accumulate at depth around the wellbore, only to be drawn back into the well when heavier pumping resumed. The quality of the water from the well improved once the pumping schedule was updated to include longer pumping periods during winter months.

PATHOGENS

There are more than 100 microorganisms that are human pathogens (Feachem et al. 1981), most of which are introduced into the body via ingestion, inhalation, dermal contact, or entry through wounds or body orifices (Hurst 1996). Infected persons excrete large numbers of these pathogens, which often find their way into surface water and groundwater systems via septic tanks or sewer systems. Each organism has a different dose–response relationship with vastly different threshold doses for infection. Typically, very high quantities of these organisms are required to cause bacterial infections, while with certain viruses one organism may be sufficient to cause infection. Studies indicate that bacteria are generally removed during wastewater treatment and disinfection, but viruses may only experience a 50 percent removal depending on the treatment process employed (Yates et al. 1987).

Microorganisms associated with waterborne disease can be broken into three groups: protozoans, bacteria, and viruses. Each has unique environmental fate and effect characteristics in groundwater. Protozoans and their cysts are common in surface waters and are much larger than either viruses or bacteria. The cyst stage is an encapsulation that protects protozoans from harsh environmental conditions. *Cryptosporidium* and *Giardia lamblia* are the two protozoans most studied because of their presence in drinking water (generally unfiltered surface water), and their recent link to waterborne illness outbreaks (such as the *Cryptosporidium* outbreak that occurred in Milwaukee in 1993).

Giardia lamblia is believed to be the most common protozoan pathogen present in surface waters. Its population appears to remain constant throughout the year in surface water impoundments (Rose and Carnahan 1992). Neither *Cryptosporidium* nor *Giardia lamblia* appears to be a common problem for groundwater except in those groundwater systems under the influence of surface waters, where the surface–groundwater connection allows them to enter the aquifer and potentially contaminate wells. Agricultural operations, sludge, manure fields, and recently turned cropland where manure has been placed are potential sources of this contamination. While *Cryptosporidium* and *Giardia lamblia* are generally believed to be too large to move significant distances in groundwater systems, they are serious issues for surface waters and where groundwaters have direct connection to surface activities. Disinfection is generally not effective in destroying protozoans.

Bacteria are the most widely distributed life form on Earth (Chapelle 1993). Chapelle (1993) notes that bacteria are important to consider in groundwater projects as bacteria inhabit virtually every subsurface environment, producing methane gas and consuming rich organic soils. The key bacteria families responsible for waterborne diseases include gram-negative bacteria such as: *Legionella*, the Pseudomonads, *Klebsiella*, *Escherichia coli*, *Shigella*, *Enterobacter*, *Salmonella*, and *Vibrio cholerae*. Most of these pathogenic bacteria are approximately 0.4–14 µm long and 0.2–12 µm wide, which mean they are much smaller than protozoans, thus making it easier for them to move in the subsurface.

Bacteria have their own enzymes and most are mobile, allowing them to move in the subsurface. Bacteria reproduce by splitting into daughter cells, each of which continues to split, forming additional bacteria and eventually a biomass. The respiration ability of bacteria permits them to survive in soils and aquifers. There are three respiration types:

- bacteria that use inorganic chemicals to serve as electron acceptors such as oxygen, ferric iron, and sulfates;
- bacteria that are aerobic (i.e., require oxygen); and
- bacteria that are facultative anaerobes (i.e., capable of fermentation or using oxygen as electron receptors [Chapelle 1993]).

The respiration mechanism is important because it affects the ability of bacteria to colonize wells and the aquifer; it also affects the growth rate of bacteria indigenous to the aquifer because of constituents introduced by surface activities.

Bacteria will commonly colonize wells because of the nutrients that are brought into the borehole by pumping. One example is *Pseudomonas aeruginosa*, one of the most common opportunistic bacterial pathogens. It has a colonization rate of 2.6–24 percent of the human population (Bloetscher et al. 1997, 2005) and is the most common infection in hospitals. *Pseudomonas aeruginosa* is an extraordinarily versatile organism that will live in nearly any environment. *Pseudomonas aeuoginosa* requires no specific vitamins, growth factors, or amino acids; it is a facultative anaerobe. However, the most important concern about this pathogen is its ability to create a slime matrix that encapsulates other bacteria and protects them from otherwise harsh aquifer conditions (Bloetscher et al. 2005).

Bacteria commonly found in the subsurface include *Gallionella* and *Desulfovibro gallionella*. These are obligate aerobes that obtain energy by oxidizing dissolved ferrous iron to form ferric oxyhydroxides—meaning it will be a problem in wells constructed with steel materials (Chapelle 1993). *Desulfovibro* is a sulfur-reducing bacterium that uses hydrogen or simple organic compounds as an energy source and sulfates as the terminal electron acceptor, which leads to hydrogen sulfide gas formation (Chapelle 1993). Sloughing events may introduce significant quantities of these bacteria into the treatment plant. Other bacteria may also colonize the slime matrix (Bloetscher et al. 1997). Bacteria are generally removed by filtration and destroyed by disinfection, although the Pseudomonads are resistant to chlorine.

Viruses are molecular entities that possess little or no enzymatic capabilities, no energy capability, and no mechanisms for synthesis. They are small, ranging from 20 to 300 nm in size. They cannot reproduce; they require a host cell to multiply. All viruses are composed of nucleic acid and either RNA or DNA (but not both), which allows them to replicate in other cells, including bacteria—where they are called bacteriophages (Chapelle 1993). Pathogenic viruses tend to be smaller than other viruses and can only be seen with an electron microscope. Most are 27–70 nm in size and are symmetrical in shape. Viruses are obligate parasites, always searching for the correct host cell that will allow the virus to multiply. Viruses cannot survive or infect without such a host organism or cell (Chapelle 1993). Most viruses tend to be resistant to chloroform but may be inactivated to various degrees during wastewater treatment processes or by chlorine, bromine, ozone, ultraviolet light, or formaldehyde (Block 1989). Viruses are conserved at –20°C (0°F) (Block 1989).

Human viruses found in natural waters are almost always associated with fecal material eliminated from the bodies of infected individuals. Therefore, virus concentrations in wastewater are high, and groundwater that may be influenced by treated wastewater or septic tanks may be contaminated. Major viruses of concern are: Hepatitis A, Coxsackie, Echo, Norwalk, rota- and reoviruses (Block 1989). While vaccines may be available for some viruses, the wild strains never disappear from the environment (Bouwer 1991).

Communities with poor hygiene and a high proportion of children have provided opportunities for a series of studies attempting to characterize the survival times of various pathogens. The studies confirm that in groundwater, filtration by straining is the most effective method for reducing bacterial presence in an aquifer (Powelson et al. 1993), while virus depletion is affected by adsorption to soil particles (assuming there are no fractures or fissures in the rock that might cause viruses to move significantly further from the source water, making them harder to find).

Viruses and bacteria have been shown to live 28–90 days in groundwater and move 7–30 m routinely (Asano 1991; Teutsch et al. 1991). In tests for viral contamination under sludge land application sites, both Norwalk virus and hepatitis A viruses have been known to move in the groundwater environment (Gerba and Bitton 1984). Vaughn et al. (1983) detected viruses originating in septic tanks that had passed through 3.6 m of unsaturated soil and 67 m of saturated soil (Powelson et al. 1993). Schaub and Sorber (1977) recovered bacteriophage #2 in 47 percent of samples after 72 hours, after it had flowed through 18 m of unsaturated, silty soils (Powelson et al. 1993). Janson et al. (1989) recorded Echovirus 11 m deep, 14 m from a recharge basin for reclaimed water (Powelson et al. 1993). In a Waldo, Fla., migrant camp, the survival of indigenous populations of total coliforms and fecal streptococcus in situ was over 70 days, and a link was shown between the same septic tanks and the presence of Echoviruses 22/23 (Feachem et al. 1981).

Table 7-3 outlines the factors that can be expected to remove bacterial particles successfully over a short distance. Filtration is useful because the bacteria are relatively large, and filter theory indicates that filtration will remove particles successfully up to 0.05 times the pore size (Gerba and Bitton 1984). Therefore, smaller pores equal better removal but faster plugging. Because of this correlation, the physical process of particle removal in saturated soils is fairly well understood and is probably adequate for predicting removal, except in fractured rock formations. Virus removal faces the same problem with rock fractures (Yates et al. 1987). The shape of the bacteria also is a factor in removal via filtration (Gerba and Bitton 1984).

Bacterial survival increases with increased moisture content, increased moisture retention capability of the rock, warmer temperatures, higher pH, and increased nutrient capacity (Teutsch et al, 1991). It is not surprising that bacterial counts are highest in areas having wet, organic-rich soils (Teutsch et al. 1991). Sorption can also play a role in bacterial survival. Clays are fine particles with

Table 7-3 Factors affecting survival of enteric bacteria in soils

Factor	Comments
Moisture content	Survival time increases with moisture and high rainfall moisture
Holding capacity	Survival time less in sandy soils with lower moisture content
Temperature	Survival time longer at lower temperature
pH	Survival time shorter in acidic soils than alkaline soils
Sunlight	Survival time shorter in the presence of sunlight
Organic matter	Increase survival time and rate of regrowth with high amounts of organic matter
Microflora	Increased survival time in sterile soil

Source: Adapted from Yates et al. 1987

negative charges that are very conducive to sorption (Gerba and Bitton 1984). The repulsing factors resulting from these negative charges are reduced when sorption occurs. Cations in solution (Fe^{+2}, Cu^{+2}, and Zn^{+2}) play a role in removal of bacteria, while anions have little effect (Gerba and Bitton 1984). Where soil is absent, the benefit of filtration is eliminated, but the benefits of rock filtration and reduced surface biomass matter may provide better indications of survival times (Bouwer 1991).

While bacteria removal is preferentially accomplished in the soil via filtration, sorption, or biodegradation, virus removal is most efficiently accomplished via soil sorption—although increases in temperature significantly affect viral survivability as well (Gerba and Bitton 1984). Inactivation of viruses also appears to depend on the efficiency of adsorption sites, temperature, and the initial number of viruses discharged. Other factors include soil conditions, pH, moisture content, aerobic or anaerobic conditions, particle size, clay content, organic content, cation-exchange capacity, virus type, and rainfall (Keswick et al. 1982).

True field studies are difficult to conduct because the use of actual viruses injected into the ground is not viewed favorably, and lab studies often cannot replicate the actual aquifer condition. What is known is that the survival of viruses and bacteria in the subsurface is determined in part by their retention on soil particles, which are generally dependent on temperature and rainfall.

Because viruses are sensitive to ultraviolet (UV) light, they are likely to be active longer in groundwater than in surface water because UV rays do not penetrate the aquifer system (Chapelle 1993). Groundwater is assumed to be isothermal (constant temperature) (Yates et al. 1985). However, viruses can survive for years in refrigerators at 41°C (39.2°F), and it appears for every 10°C increase, the virus die-off rate doubles (Gerba and Bitton 1984). Above 30°C (86°F), temperature is the controlling factor for virus survival (Gerba and Bitton 1984). Gerba and Bitton's (1984) multivariate study concluded that temperature could explain 77.5 percent of the variation in die-off rates between samples (Yates et al. 1985).

WATER QUALITY CONSIDERATIONS

There are 14 principal chemical constituents of natural surface and groundwater, and their effects are listed in Table 7-4 . Several of these constituents are regulated under the Safe Drinking Water Act (SDWA). Only small to moderate amounts (micrograms to milligrams) of these substances occur in most freshwater sources. Moderate amounts of dissolved minerals make water more palatable, as mineral-free water tastes flat to most people. Minerals are also important to human health and plant and animal growth. Some minerals reduce the corrosiveness of the water in pipelines and storage tanks.

Chemical and Physical Characteristics

Groundwater possesses chemical and physical characteristics resulting from characteristics of the water and the physical and geochemical setting of the

Table 7-4 The principal natural chemical constituents in water, concentrations, and effects of usability

Constituent	Native water concentrations	Effects of usability in water
Silica (SiO_2)	Ranges generally from 1.0 to 3.0 mg/L, although as much as 100 mg/L is common; as much as 4,000 mg/L is found in brines.	In the presence of calcium and magnesium, silica forms a scale in boilers and on steam turbines that retards heat and fluid flow; the scale is difficult to remove. Silica may be added to soft water to inhibit corrosion of iron pipes.
Iron (Fe)	Groundwater having a pH of less than 8.0 may contain 10 mg/L; rarely as much as 50 mg/L may occur. Acid water from thermal springs, mine wastes, and industrial wastes may contain more than 6,000 mg/L.	More than 0.1 mg/L precipitates after exposure to air; causes turbidity, stains plumbing fixtures, laundry and cooking utensils, and imparts objectionable tastes and colors to foods and drinks. The secondary, aesthetic, standard under the SDWA for iron is 0.3 mg/L. More than 0.2 mg/L is objectionable for most industrial uses.
Manganese (Mn)	Generally 0.20 mg/L or less. Groundwater and acid mine water may contain more than 10 mg/L. Water at the bottom of a stratified reservoir may contain more than 150 mg/L.	More than 0.2 mg/L precipitates on oxidation; causes undesirable tastes, deposits on foods during cooking, stains plumbing fixtures and laundry, and fosters growth in reservoirs, filters, and distribution systems. The secondary, aesthetic, standard under the SDWA for manganese is 0.05 mg/L. Most industrial users object to water containing more than 0.2 mg/L.

(continued)

Table 7-4 The principal natural chemical constituents in water, concentrations, and effects of usability (*continued*)

Constituent	Native water concentrations	Effects of usability in water
Calcium (Ca)	Averages about 15 mg/L in surface water, higher in groundwater. As much as 600 mg/L in some western streams; brines may contain as much as 75,000 mg/L.	Calcium and magnesium combine with bicarbonate, carbonate, sulfate, and silica to form heat-retarding, pipe-clogging scale in boilers and in other heat-exchange equipment. Calcium and magnesium combine with ions of fatty acid in soaps to form soap suds; the more calcium and magnesium, the more soap required to form suds. A high concentration of magnesium has a laxative effect, especially on new users of the supply.
Magnesium (Mg)	As much as several hundred milligrams per liter in some western streams; ocean water contains more than 1,000 mg/L and brines may contain as much as 57,000 mg/L.	
Sodium (Na)	As much as 1,000 mg/L in some western streams; about 10,000 mg/L in sea water; about 25,000 mg/L in brines.	More than 50 mg/L sodium and potassium in the presence of suspended matter causes foaming, which accelerates scale formation and corrosion in boilers. Sodium and potassium carbonate in recirculating cooling water can cause deterioration of wood in cooling towers. More than 65 mg/L of sodium can cause problems in ice manufacture.
Potassium (K)	Generally less than 10 mg/L; as much as 100 mg/L in hot springs; as much as 25,000 mg/L in brines.	
Carbonate (CO_3)	Commonly 0 mg/L in surface water; commonly less than 10 mg/L in groundwater. Water high in sodium may contain as much as 50 mg/L of carbonate.	Upon heating, bicarbonate is changed into steam, carbon dioxide, and carbonate. The carbonate combines with alkaline earths—principally calcium and magnesium—to form a crust-like scale of calcium and magnesium carbonate that retards flow of heat through pipe walls and restricts flow of fluids in pipes. Water containing large amounts of bicarbonate and alkalinity is undesirable in many industries.
Bicarbonate (HCO_3)	Commonly less than 500 mg/L; may exceed 1,000 mg/L in water highly charged with carbon dioxide.	

(*continued*)

Table 7-4 The principal natural chemical constituents in water, concentrations, and effects of usability (*continued*)

Constituent	Native water concentrations	Effects of usability in water
Sulfate (SO_4)	Commonly less than 1,000 mg/L except in streams and wells influenced by acid mine drainage. As much as 200,000 mg/L in brines.	Sulfate combines with calcium to form an adherent, heat-retarding scale. More than 250 mg/L is objectionable in water in some industries. Water containing about 500 mg/L of sulfate tastes bitter; water containing about 1,000 mg/L may be cathartic.
Chloride (Cl)	Commonly less than 10 mg/L in humid regions; tidal streams contain increasing amounts of chloride (as much as 19,000 mg/L) as the bay or ocean is approached. About 19,300 mg/L in sea water; and as much as 200,000 mg/L in brines.	Chloride in excess of 150 mg/L imparts a salty taste. Concentrations greatly in excess of 150 mg/L may cause physiological damage. Food processing industries usually require less than 250 mg/L. The secondary, aesthetic, standard under the SDWA for chlorides is 250 mg/L. Some industries—textile processing, paper manufacturing, and synthetic rubber manufacturing—desire less than 100 mg/L.
Fluoride (F)	Concentrations generally do not exceed 10 mg/L in groundwater or 1.0 mg/L in surface water. Concentrations may be as much as 1,600 mg/L in brines.	Fluoride concentration between 0.6 and 1.7 mg/L in drinking water has a beneficial effect on the structure and resistance to decay of children's teeth. Fluoride in excess of 1.5 mg/L in some areas causes mottled enamel in children's teeth. Fluoride in excess of 6.0 mg/L causes pronounced mottling and disfiguration of teeth. The primary, health-based, standard for fluoride is 4.0 mg/L. The secondary, aesthetic, standard under the SDWA for fluoride is 2.0 mg/L.

(*continued*)

Table 7-4 The principal natural chemical constituents in water, concentrations, and effects of usability (*continued*)

Constituent	Native water concentrations	Effects of usability in water
Nitrate (NO_3)	In surface water not subjected to pollution, concentration of nitrate may be as much as 5.0 mg/L but commonly is less than 1.0 mg/L. In groundwater, the concentration of nitrate may be as much as 1,000 mg/L where polluted, but generally less than 50 mg/L.	Water containing large amounts of nitrate (more than 100 mg/L) is bitter tasting and may cause physiological distress. Water from shallow wells containing more than 45 mg/L has been reported to cause methemoglobinemia in infants. The SDWA standard for nitrate is 10 mg/L. Small amounts of nitrate help reduce cracking of high-pressure boiler steel.
Dissolved solids	The mineral constituents dissolved in water constitute the dissolved solids. Surface water commonly contains less than 3,000 mg/L; streams draining salt beds in arid regions may contain in excess of 15,000 mg/L. Groundwater commonly contains less than 5,000 mg/L, and most of it at shallow depths contains less than 1,000 mg/L; some brines contain as much as 300,000 mg/L.	More than 500 mg/L is undesirable for drinking and many industrial uses. Less than 300 mg/L is desirable for dyeing of textiles and the manufacture of plastics, pulp paper, and rayon. Dissolved solids cause foaming in steam boilers; the maximum permissible content decreases with increases in operating pressure. The secondary, aesthetic, standard under the SDWA for total dissolved solids is 500 mg/L.

Source: Adapted from Durfor and Becker 1964

groundwater. High quantities of minerals mean low-quality water. A significant relationship exists between mineral content and depth of groundwater; the mineral quality of groundwater generally declines with depth. In many sedimentary basins, where the older and deeper sediments were deposited by oceans, mineral content can change abruptly. Poor-quality water can be drawn upward after production begins (i.e., up-coning), even if a production well does not penetrate a saline zone—a common situation in southeastern Broward County, Fla., that affects the cities of Hollywood, Hallandale Beach, and Dania Beach. Similarly, operation of coastal production wells can induce saltwater intrusion into freshwater aquifers (Bloetscher et al. 2005). Currently, many forms of contamination exist. Synthetic and naturally occurring organic compounds, solvents, petroleum products, refined minerals, and heavy metals must be considered when evaluating the development potential of a groundwater resource to meet SDWA requirements. Microbiological substances, especially in membrane treatment applications, are increasingly a concern. In many cases, construction, maintenance, and operation of facilities to remove these substances are more costly than finding a new water source.

OTHER CONSTITUENTS OF CONCERN

Groundwater contamination is a widespread and challenging problem. Contaminants often flow undetected into aquifer formations, migrating through the aquifer until a sizable portion of the aquifer has become degraded. Even where physically and economically practical, rehabilitating a contaminated aquifer is difficult, complicated, and expensive. In arid regions, where water resources are limited, aquifer remediation may be the only option for a reliable, long-term water source. Ongoing monitoring of sentinel wells is the only means to monitor for such contamination.

PESTICIDES

Pesticides are substances used to destroy or inhibit the action of plants or animal pests and include insecticides, herbicides, rodenticides, and nematocides. They are associated with irrigated agriculture, dairy, and livestock activities. Virtually all are toxic to humans to some degree and they vary in biodegradability. Table 7-5 lists some of the regulated pesticides that have been detected in groundwater (Anderson 1990).

NITROSAMINES

Nitrosamines are an emerging contaminant of concern throughout North America and have been found in polluted air and water (Bolton and Nihaela 2000). The major concern has been the occurrence of N-Nitrosodimethylamine (NDMA) in

Table 7-5 Regulated pesticides detected in groundwater

Pesticide	Type	Additional uses / comments
Atrazine	Herbicide	Plant growth regulator used for highway weed control
Bentazon	Herbicide	Food crops
Chlordane	Insecticide	Fumigant
2, 4-D	Herbicide	Defoliant; agriculture and pasture weed killer; fruit crop control
DBCP	Nematocide	Soil fumigant
Endrin	Insecticide	Banned US use and manufacture
Heptachlor	Insecticide	Banned, except for termite control
Heptachlor Epoxide	Insecticide	Heptachlor and chlordane degradation product
Lindane	Insecticide	Livestock, crops, lumber
Methoxychlor	Insecticide	Acaricide; livestock, dairy farms, food crops
Simazine	Herbicide	Algaecide; agriculture, aquatic sites
Silvex	Herbicide	Banned plant growth regulator
Toxaphene	Insecticide	Not recommended for dairy activities

Source: Anderson 1990

potable water systems, first noted in California in 1998 but found throughout the United States and Canada at levels significantly higher than in the past (Yoo et al. 2000). Industrial contamination was initially investigated as an NDMA source in Canada, leading to investigation of potential formation of NDMA in the drinking water treatment process. NDMA is also being found in groundwaters, and current research is expanding to include other nitrosamines (Andrews and Taguchi 2000).

PFAS AND PFOS

The most recent water quality issues are perfluoroalkyl and polyfluoroalkyl substances (PFAS) and Perfluorooctanesulfonic acid (PFOS). They have been in commercial use since 1940 because of their resistance to heat, water, and oil. These were initially used during World War II as a waterproofing coating on Allied tanks; later, they became the basis for Teflon® for pans and many other purposes. DuPont and 3M developed the compounds. They are common in household cookware, lubricants, and the like. They have a history of various negative public health effects including cancer. The USEPA is currently in the process of regulating PFAS and PFOS. Everyone likely has these compounds in their bloodstream.

Over the years, these materials have found their way into groundwater near airports, factories, and military bases. In the past several years, more than 50 bills in Congress were introduced to regulate these substances. Two states have regulatory limits (California 5.1–6.5 ppt and New Hampshire 12–15 ppt).

Groundwater utilities should monitor for these compounds because they may indicate surface contamination from industrial sites. Removal of PFAS and PFOS compounds is also being studied. Treatment or abandonment of contaminated wells are solutions. Utilities will receive more scrutiny of the issue especially in rural areas where oversight in the past may have been lacking.

PHARMACEUTICALLY ACTIVE SUBSTANCES

Pharmaceutically active substances are also known as personal care products (PCPs) and emerging contaminants (ECs), among others. It has been estimated that 70 percent of pharmaceuticals consumed pass through the body unchanged. Further, because concentrations of these substances are typically less than 1 μg/L, the microorganisms in wastewater treatment facilities are not induced to metabolize these substances as an energy source. As a result, current research indicates that many of these substances survive the biodegradation process and are discharged into receiving waters. Those that are altered can revert to their original form or daughter products in the environment (Daughton and Ternes 1999). Recent research also indicates the presence of chemicals in water and wastewater that may disrupt the endocrine system of many species, including humans (Bloetscher and Plummer 2011). Endocrine chemicals are used by organisms to regulate important metabolic activities, such as ion balance, reproduction, basal metabolism, and fight or flight responses through changes in hormones secreted by the thyroid, parathyroid, pituitary, adrenal, sex, and other glands. Because endocrine systems are interconnected, effects on one will affect others as well. Chemicals, whether derived from pharmaceuticals, industrial emissions, or natural sources, that interfere with endocrine systems of humans and wildlife are termed endocrine disruptors, and those that elicit a pharmaceutical response in humans are termed pharmaceutically active substances (PAS). Research has identified more than 60 PAS that impact the endocrine system of animals and humans in the 1 ng/L or lower concentrations in the ecosystem.

SOURCES OF CHEMICAL CONTAMINATION

Potential contaminants are generated by virtually all of society's industrial, agricultural, urban, and rural activities. The principal sources of contaminating substances are shown in Table 7-6.

Contaminants may enter groundwater reservoirs by intent and design, such as deliberate placement in the subsurface through a waste-injection well. Contaminants may also enter a groundwater system inadvertently—for example, by leakage from a ruptured pipeline, as leachate from an inadequately sealed landfill, or by agricultural fertilizer application.

Aquifer contamination occurs because of the downward movement of a contaminant from a land surface source (in this example, a waste lagoon), through the zone of aeration (the zone of rock and soil above the water table; it is unsaturated with water), and then into the aquifer. The recharge mound builds up beneath the source of the contaminant and may propel some of the liquid in an upward direction for relatively short distances, but the dominant direction of movement for contaminant liquids, with densities similar to the native groundwater, is downward in the direction of general groundwater flow, indicated by the direction of slope of the old water table.

The relative densities of incoming fluids and of the receiving water in the aquifer influence the pattern of contaminant movement. High-density organics, such as chloroform, dichloroethane, and tetrachloroethylene, tend to "sink" to the lower part of the aquifer and are commonly called dense nonaqueous phase liquids (D-NAPLS). Comparatively, fluids that are less dense than water, such as gasoline and oil, tend to "float" on the groundwater and are called L-NAPLS.

WELL MAINTENANCE CONCERNS

As operations personnel review changes in static and drawdown levels to identify any trends, they should be cognizant of changes in specific capacity and water levels in the aquifer. Long-term reductions in water availability in the aquifer or limitations in specific capacity of the mechanical system caused by pump wear, clogged screens, or formation of bacterial fouling may be indicated. Repairs to correct mechanical problems should be scheduled.

Table 7-6 Major sources of groundwater contamination

Point Sources	Nonpoint Sources
Landfills	Agriculture
Superfund-type sites	Dairies and feedlots
Leaking underground storage tanks	Seawater intrusion
Wastewater treatment plants	Urban runoff
Oil production and refinery facilities	Oil, sewer, and other pipeline networks
Industrial and manufacturing facilities	Oilfield brine injection
Septic tanks	Acid-mine drainage
Spills and accidents	

PARTICULATE PLUGGING

Plugged screens increase the entrance velocity of the raw water, which can increase particle movement as well as drawdown. Sand, silt, and other particulates may clog the screen, providing less area to draw the water. When this happens, the capacity of the well decreases, the pumps become less efficient, and operations costs for electricity increase. Figure 7-11 shows a borehole with sand entering a pumping well. Sand will also increase wear on pumps and settle in large pipelines. Sand is problematic for membrane processes.

Removal of sand can be a delicate process because it is possible to damage the screen. Water samples from wells developed in sand aquifers should also be periodically inspected for the presence of sand. The presence of sand in a well may be an indication of eventual collapse of the well, collapse of the formation, holes in the casing and, in extreme cases, sinkholes. These problems are generally repairable but require appropriate expertise to review the situation.

In most cases, particulate plugging is caused by poor well design or construction, including insufficient development of the well or inadequate formation sampling leading to poor screen selection and/or location. In some cases, the logging may not have been sensitive to thin layers of sand, silt, or colloidal matter that may be exposed with time. In wells with gravel packs, incomplete development or over-pumping may be indicated by plugging of the gravel pack and the screens. Use of a sand separator (Figure 7-12) is suggested where persistent sand occurs.

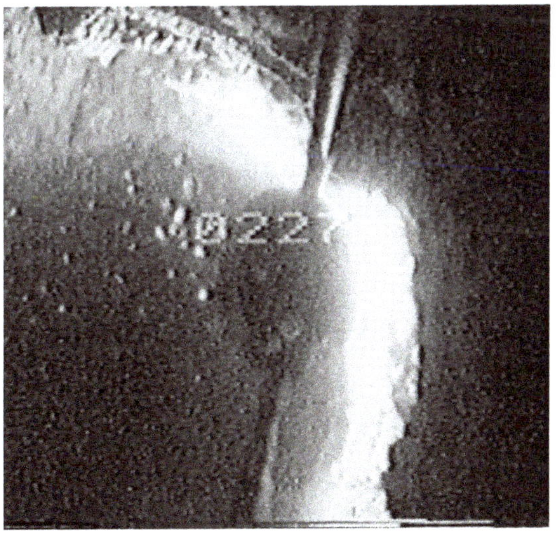

Source: Photo by John Largey

Figure 7-11 Photograph of borehole with sand entering the well

Source: Photo by Fred Bloetscher

Figure 7-12 Sand separator for removing sand from raw water lines

PLUGGING BY IRON AND MANGANESE

At a pH less than 5, iron and manganese ions remain dissolved as Fe^{+3} and Mn^{+2} in the water supply. However, in the presence of 2 or 3 mg/L of dissolved oxygen or a higher pH shift, these metals can be precipitated around the well screen in an insoluble mass. Hard nodules will form from this precipitate, which collect additional ferric or manganese precipitates. Oxygen encourages iron precipitation. Acidic groundwater (pH less than 7) may dissolve calcium carbonate from the formation materials, causing migration to the well screen or increased turbidity.

pH CONCERNS

As a side note, nitrifying bacteria are not conserved at a pH between 8.0 and 8.6. Groundwater utilities can have significant issues with nitrification in their distribution piping if they do not use sufficient chlorine and a pH above 8.6 or below 8.0.

CALCIUM CARBONATE

One of the most common well problems is encrustation of the well screen or of the gravel pack around the screen. This may be caused by the release of dissolved minerals from the native water, geochemical reactions, or microbiological activity. Calcium carbonate forms a scale on the screen and cements together particles of sand and gravel. Calcium carbonate encrustation can usually be removed by a chemical process.

CORROSION

Three general types of corrosion involved in water wells are hydraulic, chemical, and galvanic. Hydraulic corrosion is caused by turbulent flow, hard particulates, and/or wearing flow velocities, which abrade well components. Hydraulic corrosion enlarges screens and opens holes in the casing that allow larger particles into the casing. As deterioration accelerates, the casing material diminishes and potentially collapses.

1. Hydraulic corrosion is generally caused by particulate matter from incomplete well development or fine material within the formation that is not screened out. Cavitation caused by turbulent flow will aggravate corrosion by flaking off pieces of metal. Pumping at rates higher than design flow is the primary cause of hydraulic corrosion.
2. Chemical corrosion is a problem in older wells because of materials used in the past. Chemical corrosion is caused by ionization of metallic elements, typically zinc or iron, through carbonation or oxidation reduction (redox) reactions. Chloride ions that exist in raw water can form weak acids that react with metallic ions or attack metals. Sulfide ions also create acids in certain environments that may attack metal surfaces. Oxidation and reduction reactions occur in groundwater environments and can accelerate corrosion in a well. The presence of high concentrations of dissolved oxygen may accelerate desiccation of brass or other pipe material.
3. Galvanic corrosion is caused by the generation of electric currents in dissimilar metals (Figure 7-13). Galvanic corrosion is often a problem with stainless steel pumps that are connected to steel column pipes with bronze centralizers in a steel casing. Newer technologies and the use of stainless steel, bronze, and plastics over standard steel grades have reduced galvanic corrosion, although stainless steel grades may have some of the same inherent problems in high chloride waters (Bloetscher et al. 2002a, 2002b, 2002c). The higher the conductance that exists between two metals, the greater the potential for galvanic action. This corrosion is typically found where casing screen is joined, where the submersible pumps are joined to the column pipes (Figure 7-14), or where bronze spiders exist. Poor pump alignment, stressed threads as

Corroded End (Anode)

Magnesium
Magnesium Alloys
Zinc
Aluminum 25
Cadmium
Aluminum 17ST
Steel or Iron
Cast Iron
Chromium–Iron (active)
Nickel Resistant (alloys)
18–8 Chromium–Nickel–Iron (active)
18–8–3 Chromium–Nickel–Molybdenum–Iron (active)
Lead–Tin Solders
Lead
Tin
Nickel (active)
Brasses
Copper
Bronzes
Copper–Nickel Alloys
Monel
Silver Solder
Nickel (passive)
18–8 Chromium–Nickel–Iron (passive)
18–8–3 Chromium–Nickel–Iron (passive)
Silver
Gold
Platinum

Protected End (Cathode)

Figure 7-13 Galvanic corrosion series

a result of poor assembly, or poor welds may encourage this type of corrosion (Figure 7-15).

MICROBIOLOGICAL FOULING

Microbiological fouling is generally interrelated with physical and chemical processes (Bloetscher et al. 1998). Microorganisms can encrust or corrode the system, enhancing physical and chemical well deterioration problems involving some microbiological fouling (Figure 7-16). The typical symptoms of microbiological fouling problems are:

- decrease in the water quality,
- increased drawdowns,
- reduced specific capacity,
- change in the amount of iron or manganese in the water supply, and
- an apparent increase in microbiological densities, such as an observance of slimes or staining from the raw water.

Microbiological fouling encourages changes in the electrical potential and pattern of the well casing by using CO_2 on metallic surfaces to transfer ions. The bacteria attach to the steel pipe walls in the form of biofilms. A biofilm is an active ecosystem, providing an environment for survival to a variety of microorganisms

Source: Photo by Fred Bloetscher

Figure 7-14 Corrosion at a well screen/pump joint

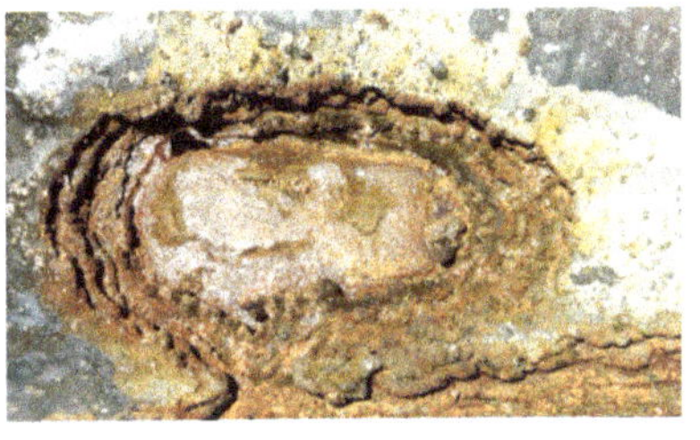

Source: Photo by Fred Bloetscher

Figure 7-15 Welds that are poorly done or use dissimilar metals may encourage corrosion.

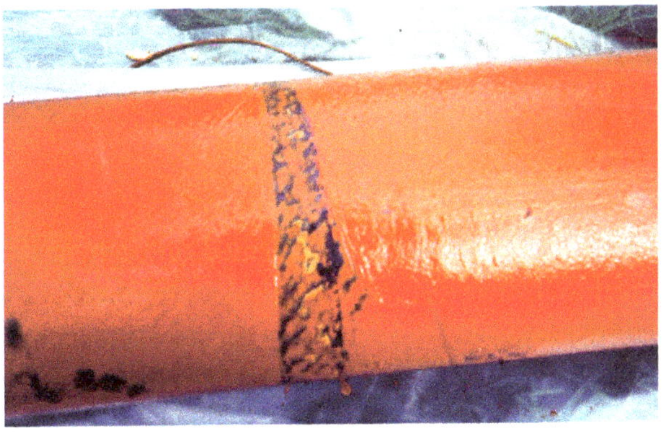

Source: Photo by Fred Bloetscher

Figure 7-16 Microbiological activity on a well column pipe

by storing and transporting nutrients. As the bacteria in the biofilm absorb nutrients, they form tubercles and biofilms that reduce the capacity of pumps and casings and may clog the well screen. Precipitates of iron, sulfur, and manganese can also exist within the biofilm. The biofilm also protects the bacteria cells from external reagents, such as chlorine, but traps iron, sulfur, manganese, and other nutrients.

It should be noted that certain microbiological activity is normal. Bacteria find aquifers to be the ideal environments as there is tremendous surface area for colonization; the temperatures are relatively constant and moderate; the flow of water provides a consistent nutrient supply; and, except for the immediate pumping zone, the water is not disturbed (Bloetscher et al. 1998). All spaces within the aquifer formation are potential areas for colonization. Vugular formations and formations with air pockets are ideal for creating large biofilms within the aquifer, but they never indicate severe plugging because of the size of the organisms in comparison to the vugs.

Several steps should be followed to look for bacteria. A downhole camera should be used to look for the colony seeds. Any equipment that is pulled out of the wells should be thoroughly cleaned so other wells are not contaminated. Operations personnel may need to obtain microbiological samples for analysis from nonwater sources (e.g., samples from the colloidal filter or from a slimy material growing on the pump or column pipe). For these analyses, swab samples should be collected. Sterile collection swabs for bacterial samples should be obtained by the operations staff and used to collect the specimen to be analyzed.

Water and swab samples should be sent to a microbiological laboratory. Conclusions based on the results of the microbiological analysis and the recommendations for control should be made by a qualified hydrogeologist with the assistance from a qualified microbiologist. The following is a list of suggested analyses to be performed on each water sample:

- total coliform count,
- fecal coliform count,
- heterotrophic plate count,
- bacterial identification,
- total fungal count,
- fungal identification,
- algal identification, and
- protozoa.

The following analyses should be performed on each swab sample: bacterial identification, fungal identification, and algal identification. *Cryptosporidium sp.* and *Giardia sp.* should be analyzed for evidence of surface water interaction or if the water supply is surface water. Biological Activity Reaction Test (BART®) tests are useful, but they are not conclusive for identifying biofouling problems. An individual BART® test only looks for limited species. Speciation of bacteria is required because most environments yield a matrix of bacterial species.

RESOLVING OPERATIONAL PROBLEMS

Proper design will reduce potentially excessive entrance velocities or improper screen placement that can allow fine-grained particles to migrate into the wells. Proper materials, such as plastics or fiberglass, instead of steel or stainless steel should be used in water where microbial activity or high chlorides are present. Dissimilar metals should not be used in close proximity. Improper construction, poor grouting, excessive screen and casing damage, or the removal of protective sealants can lead to physical deterioration of the well. The improper application of certain chemical reagents, especially chlorine, and sequestering reagents or those used during redevelopment may exacerbate deterioration. Overly aggressive pumping for redevelopment, over-pumping of the system, or the improper use of surging may cause structural damage to the well in the long term. Many older wells were installed using methods and materials no longer in use and do not meet current standards. In these cases, the problems probably cannot be fully corrected.

Physical agitation or surging is a method that often removes encrustation or reduces fine material entering the well screen or gravel pack. Tools are used that push water down into the well and pull it out, just as old hand-pumped well systems worked. Initially, the surge device is operated at less than three strokes per minute at 6–10 in. per stroke. Over time, the frequency and the stroke should be increased, which increases the surging. Care must be taken; if the casing or the formation is weak, or the screens are damaged, the well structure can collapse during surging.

The addition of chemicals to the well is the most common treatment. For many older wells, an option is acidification, dropping the pH to less than 2. Hydrochloric, sulfamic, sulfuric, and nitric acid are used, but these chemicals must be used carefully. Sulfamic acid is preferred by most hydrogeologists to improve performance. Sulfamic acid solution will remove or loosen encrustation in the screens or the column pipe, although it will not remove much biofouling.

To address biofouling, chlorine is used, although in most cases it does not kill all the bacteria; it only serves to control the biofilm. A 12 percent sodium hypochlorite solution or commercial calcium hypochlorite provides the chemical strength needed for chlorination of the bacteria. In some cases, hydrogen peroxide may be used to address biofouling problems, but certain bacteria, such as the *Pseudomonas* species, may be able to use the oxygen to their benefit—increasing rather than decreasing biological activity.

Deterioration of the well materials must be weighed against the removal of the biofilm or the encrustation. The addition of phosphates has been used, as it makes water "more slippery" and increases total well capability. However, phosphates provide a nutrient for biofilms. In all cases with chemical use, a plan for handling hazardous material and disposal must be made. None of these chemicals should be discharged to the ground; they must be hauled to an approved disposal site.

Another method, carbon dioxide injection (a form of hydraulic fracturing or fracking), uses gaseous carbon dioxide and liquid carbon dioxide under 100 psi of pressure. This technique causes the carbon dioxide to enter the formation, dropping the pH through a conversion of the CO_2 to carbonic acid. The water freezes, cracking and loosening encrustation. The formation may also crack and loosen, which can free the fractured zones or crack the bedrock formations and potentially increase yield. After the carbon dioxide is injected, the well is surged and redeveloped.

Sonar jetting is a relatively new method used to remove encrustations and may reduce biofilms. A sequence of small blasting caps is suspended and exploded, sending shock waves through the casing. Encrustations generally are blasted off the well screen, formation, and casing. After sonar jetting, surging and full redevelopment of the well must occur to remove excess debris. Acidification improves the process to remove the encrustation to some extent. The problem with this process may be the inability to get permits to do the blasting and the potential damage to the casing, cement seal, and/or the screens that may occur. Other methods that show some promise in certain specific cases are sectional flow control devices and inner sleeve installations within the casing, using entrained air to reduce fouling.

Owners and operators of wells that have increasing levels of contamination should immediately begin assessing their alternatives for correcting the problem. The major choices that may be considered in attempting to locate the source of contamination are as follows:

- determine if correction or removal of the source will allow the aquifer to return to background conditions;
- determine whether the plume of contamination flowing toward the well can be blocked or intercepted;
- determine if it is economically feasible to treat the water to remove the contamination;
- investigate whether modifying the well to draw water from a different aquifer is possible;
- investigate the feasibility of drilling a replacement well at another location where there is no contamination;
- investigate whether water from the contaminated well can be blended with water from an uncontaminated source to bring the finished water-level below the MCL;
- investigate changing to a surface water source if applicable; and
- investigate purchasing water from another water system.

Although groundwater quality does not vary much with time, certain quality features do gradually change.

WELL REDEVELOPMENT PROTOCOL

Even without plugging problems arising, wells should be redeveloped periodically to ensure efficient operation. Records of this redevelopment should be maintained. Establishing an efficient and timely protocol for redevelopment will save the utility money. A redevelopment protocol will be basically the same as the initial development protocol using the following steps:

1. Measure and record static distance to water level.
2. Measure and record flow during testing at 5-min increments.
3. Measure and record pumping distance to water level.
4. Collect water samples during development and monitor field specific conductance.
5. Turn pump off.
6. Record static water levels during recovery at 15–30 sec intervals for 15–30 min.
7. Calculate drawdown (pumping water level minus static level).
8. Calculate specific capacity (flow divided by drawdown).
9. Continue redeveloping the well by surging or air lifting using specific capacity and water quality as metrics to monitor performance.
10. Repeat steps 1–9.
11. If the well specific capacity increases by at least 20–25 percent, disinfect the well and place it back into service. Otherwise, keep repeating steps 1–9.

This protocol should be followed every time a well is taken out of service for a repair. As in the initial development process, the withdrawn water may contain chemicals, silt, sand, or other debris. The quality of this water may be such that it requires treatment. As a result, the following should be monitored in the withdrawn water during development:

- pH,
- chloride,
- toxic substances along with metals, SOCs, VOCs, and any other constituents of concern,
- silt,
- flow (i.e., estimated quantity of the water to be discharged), and
- frequency and timing when discharge will occur.

There may be a potential for environmental problems if chemicals are used in the redevelopment process or if excessive silt, sand, and other debris are in the waste. The same procedures and regulatory limitations as in the development process must be followed. Caution should be taken to minimize the potential for flooding areas near the well as the redevelopment water is discharged to the ground.

WELL ABANDONMENT

Regulatory agencies generally require utilities to abandon wells no longer in use. The well must be abandoned in accordance with the regulatory guidelines. The following is a general outline of a well abandonment plan:

- The well will be logged (i.e., geophysical logs) and television surveyed prior to abandonment.
- The nature and extent of the geophysical logging will depend on the value and cause of the well being abandoned. The following are logs that may be used:
  - Electric logs: single-point resistivity, long- and short-normal resistivity, and spontaneous potential
  - Gamma ray
  - Caliper
  - Fluid conductivity
  - Temperature
  - Flowmeter
  - Dual induction
  - 3-D velocity log (sonic log)
  - Cement bond log (if necessary)
- All geophysical logs should be performed from the bottom of the borehole to land surface (for those logs that can be effectively performed in the cased portions of the well). All logs that can be performed only in open holes should be run from the bottom of the borehole to the bottom of the casing.
- The bottom of the borehole should be filled with clean, washed, and chlorinated (disinfected) gravel. The gravel should be tapped periodically to ensure proper placement. A bentonite/sand cap should be placed on top of the gravel 5 ft below the bottom of the casing. The hole is then filled with neat cement grout from the bentonite/sand cap to land surface.
- All fluid displaced during cementing must be contained and disposed of in an environmentally safe manner.
- A final well abandonment plan should be submitted to the appropriate regulatory agencies at the time of application for a well abandonment permit, subject to field modification based on logging and conditions encountered during the abandonment.

The plan is subject to modification based on the nature and cause of the abandonment. Note the head in the well will need to be suppressed by pumping a solution of sodium chloride (salt) and/or barium sulfide (barite) into the well (if required). Blow-out prevention equipment will be on site should it be necessary for controlling the well.

REFERENCES

American Water Works Association (AWWA). 2014. M21, *Groundwater*. Denver: AWWA.

Anderson, L. 1990. "Groundwater Quality: A Regional Survey of Groundwater Quality in the Metropolitan Water District Service Area." Report No. 991. Los Angeles: Metropolitan Water District of Southern California.

Andrews, S.A. and V.Y. Taguchi. 2000. "NDMA–Canadian Issues." Proc. Water Quality Technical Conference.

Asano, T. 1991. "Planning and Implementation of Water Reuse Projects." *Water Science and Technology*. 24(9):1–10.

Block, J.C. 1989. *Viruses in Water Supplies—Detection and Identification*. New York: VCH Press.

Bloetscher, F. 2019. *Infrastructure Management*. Plantation, Fla.: J. Ross.

Bloetscher, F., A. Muniz, and G.M. Witt. 2005. *Groundwater Injection: Modeling, Risks and Regulations*. New York: McGraw-Hill.

Bloetscher, F., G.M. Witt, and A.E. Dodd. 1997. "Bacterial Issues in Raw Water Treatment." Proc. Florida Water Resource Conference, Orlando, Fla.

Bloetscher, F., G.M. Witt, A.E. Dodd, and C.P. Dodd. 1998. "Prevention of Biofouling in Membrane Water Treatment Plants." Proc. AWWA Annual Conference and Exposition. Dallas, Tex.

Bloetscher, F. and J.D. Plummer. 2011. "Evaluating the Significance of Certain Pharmaceuticals and Emerging Pathogens in Raw Water Supplies." *Environmental Practice*. 13(3):198–215.

Bloetscher, F., R.J. Bullock, R.E. Fergen, G.M. Witt, and G.D. Fries. 2002a. "Revisiting the Selection of Stainless Steel in Water and Wastewater Treatment Environments—Part 1." *Water Engineering and Management*. 149(5):36–44.

Bloetscher, F., R.J. Bullock, R.E. Fergen, G.M. Witt, and G.D. Fries. 2002b. "Revisiting the Selection of Stainless Steel in Water and Wastewater Treatment Environments—Part 2." *Water Engineering and Management*. 149(6):12–15.

Bloetscher, F., R.J. Bullock, R.E. Fergen, G.M. Witt, and G.D. Fries. 2002c. "Revisiting the Selection of Stainless Steel in Water and Wastewater Treatment Environments—Part 3." *Water Engineering and Management*. 149(7):12–26.

Bolton, J.R. and N.I. Nihaela. 2000. "UV Photodegradation as a Treatment Technology for N-Nitrosodimethylamine (NDMA)." Proc. Water Quality Technical Conference.

Bouwer, H. 1991. "The Role of Groundwater Recharge in the Treatment and Storage of Wastewater for Reuse." *Water Science and Technology*. 24(9):259–302.

Chapelle, F.H. 1993. *Groundwater Microbiology and Geochemistry*. New York: John Wiley and Sons.

Daughton, C.G. and T.A. Ternes. 1999. "Pharmaceuticals and Personal Care Products in the Environment: Agents of Subtle Change?" *Environmental Health Perspectives*. 107(6):907–938.

Durfor, C.N. and E. Becker. 1964. "Public Water Supplies of the 100 Largest Cities in the United States." 1962: US Geological Survey Water-Supply Paper 1812.

Feachem, R., H. Garelick, and J. Slade. 1981. "Enteroviruses in the Environment." *Tropical Diseases Bulletin*. 28(3):185–230.

Gerba, C.P. and G. Bitton. 1984. *Groundwater Pollution and Microbiology*. New York: John Wiley and Sons.

Hurst, C.J. 1996. *Modeling Disease Transmission and Its Prevention by Disinfection*. Cambridge, UK: Cambridge University Press.

Janson, J., L.W. Edmonds, B. Speight, and M.R. Bucens. 1989. "Movement of Viruses After Artificial Recharge." *Water Resources*. 23:293–299.

Keswick, B.H., C.P. Gerba, S.L. Secor, and I. Cech. 1982. "Survival of Enteric Viruses and Indicator Bacteria in Groundwater." *Journal of Environmental Science and Health*. A17(6):903–912.

Powelson, D.K., C.P. Gerba, and M. Yahya. 1993. "Virus Transport and Removal in Wastewater During Aquifer Recharge." *Water Resources*. 27(4):583–590.

Rose, J.B. and R.P. Carnahan. 1992. "Pathogen Removal by Full Scale Wastewater Treatment." Tampa, Fla.: University of South Florida.

Schaub, S.A. and C.A. Sorber. 1977. "Virus and Bacteria Removal from Wastewater by Rapid Infiltration Through Soil." *Applied and Environmental Microbiology*. 33(3):609–619.

Teutsch, G., K. Herbold-Paschke, D. Tougianidou, T. Hahn, and K. Botzenhart. 1991. "Transport of Microorganisms in the Underground—Processes, Experiments and Simulation Models." *Water Science & Technology*. 24(2):309–314.

Vaughn, J.M., E.F. Landry, and M.Z. Thomas. 1983. "Entrainment of Viruses from Septic Tank Leach Fields Through a Shallow, Sandy Soil Aquifer." *Applied and Environmental Microbiology*. 45(5):1474–1480.

Yates, M.V., C.P. Gerba, and L.M. Kelley. 1985. "Virus Resistance in Groundwater." *Applied and Environmental Microbiology*. 49(4):778–781.

Yates, M.V., S.R. Yates, J.M. Wagner, and C.P. Gerba. 1987. "Modeling Virus Survival and Transport in the Subsurface." *Journal of Contaminant Hydrology*. 1(3):329–345.

Yoo, L.J., S. Fitzsimmons, and Y. Shen. 2000. "Determination of N-Nitrosodimethylamine at Part-per-trillion Levels Using Positive Chemical Ionization from Aqueous Samples." Proc. Water Quality Technical Conference.

CHAPTER 8
Summary

Groundwater is water that flows downward by gravity until it contacts a layer of rock or other impenetrable material, creating an aquifer. This water may have entered the soil as a result of rainfall or snowmelt, or it may be an ancient source found well below the surface. Groundwater moves downhill, taking the path of least resistance to the flow. Therefore, if underground conduits or channels (i.e., voids and/or fractures that lead to high permeability) are present, the water will tend to flow in these pathways. These formations may yield substantial quantities of water to wells.

Groundwater can have significant advantages over more traditional surface water uses. These advantages include:

- less exposure to contamination (assuming not a surficial aquifer);
- water quality is stable;
- water temperature is stable;
- water quality changes are slow to occur;
- evapotranspiration losses are insignificant; and
- less treatment typically is required.

However, groundwater sources have the following disadvantages:

- difficult to clean up once contaminated;
- no early warning of contamination—unseen plumes of contaminants can migrate into a wellfield without warning unless sentinel wells are constructed;
- competing uses from urban industrial, commercial, agricultural, irrigation, and ecosystem users in the same area;
- determining the safe yields of material often is uncertain;
- water levels are not obvious; and

- supplies are often limited in basins, which is why most large utility systems rely on surface supplies, not groundwater.

Well siting considerations include site availability, water supply, water quality, and wellhead protection limitations. Many small water systems also include land costs as a prime consideration when selecting sites for public water supply wells; however, this may overlook significant water quality or water quantity issues. The water system must balance factors including well depth, geology of the area, characteristics of the rock formations, and dissolved minerals in the aquifer.

There are many drilling methods and many methods of well construction that have been developed for wells. A variety of wells serve many purposes:

- water production,
- oil and gas wells,
- geothermal wells,
- injection/disposal wells,
- aquifer storage and recovery wells, and
- environmental remediation wells.

Environmental remediation wells are the shallowest of this group, often less than 10 ft deep, while oil and gas wells may be thousands of feet deep. Water supply wells fall somewhere in between. The following are the primary drilling methods for water supply wells as discussed herein:

- cable tool,
- hollow stem auger,
- hydraulic rotary,
- mud rotary,
- reverse-air circulation, and
- dual tube.

All well projects go through a series of steps: predesign, regulatory criteria, design, construction and testing, and operations issues. Most testing occurs once some form of construction has begun—usually with a test well. Once the exploratory or test well is complete, geophysical logging can commence. Appropriate drilling, geologic, geophysical, and video logs and other tests—such as caliper and packer tests—must be conducted during the drilling and construction of new wells. Results of this testing permit engineers and hydrogeologists to define the appropriate zones in the formation to case and those to leave open. When construction is complete, well development may be the most important part of the well drilling process and is often underestimated with regard to the time required to properly develop the well.

Once the well is designed and constructed, two issues must be regularly monitored in a wellfield: well performance and water quality. Well-performance issues include the routine maintenance items for pumps and motors and record-keeping of withdrawals. Groundwater withdrawal wells provide good service to most utilities for many years, but all wells are subject to fouling and other performance problems. These concerns include:

- mechanical failures, including failures of electrical motors and pumps, and failures of valves;
- poor operating and maintenance procedures;
- poor well design and construction practices, including insufficient placement of grout; improper design of pumps, valves, and fittings; and excessive drawdown allowances;
- hydrogeologic constraints that cannot be evaluated at the time of design or change over time, such as:
  - sand, clay, or rock layers that are unstable and collapse into the borehole;
  - naturally occurring or induced fracturing and faulting;
  - long-term water quality changes caused by changes to the hydraulic regime, such as dams;
  - water hammer to the aquifer;
  - effects caused by mining of the water or introduction of chemicals and microorganisms; and
  - naturally occurring phenomena (such as sinkholes, karst terrain features, or faults); and
- high silt or sand content caused by failure to develop the wells fully, or intercepting sand or silt layers that have not or cannot be sealed off in the borehole or corrected in well design.

All of these problems may exist in conjunction with, or as a result of, microbiological fouling problems in wells. As treatment technologies advance, the need to review and correct well-performance problems, especially fouling concerns, has taken on greater significance.

All wells can plug or foul because of hydrogeologic, geologic, engineering, and construction factors. The problems are usually physical, mechanical, or environmental in nature. For example, performance problems are typically caused by fouling or sand and silt production in wells. These problems and their likely causes are:

- Water level decline in the well
  - reduced hydraulic efficiency in the well, most commonly plugging or encrustation of the borehole, screen, or gravel pack
  - regional water level declines
  - well interference or plugging of a gravel pack by sand, silt, or clay
- Lower specific capacity
  - drop in pumping water level
  - reduction in pumping yield caused by encrustation, formation plugging, pump corrosion, and biofouling
- Lower yield
  - dewatering or caving in of a major fracture or other water-bearing zone

    - insufficient development of the well
    - lack of connection to water-bearing fractures
    - pump wear
    - impeller detachment from shaft
    - encrustation, plugging, or corrosion and perforation of column pipe
- Sand/silt pumping
    - presence of sand or silt in fractures intercepted by well
    - leakage around casing bottom
    - inadequate screen and filter-pack selection or installation
    - screen corrosion
    - collapse of filter pack caused by excessive vertical velocity and wash-out
- Silt/clay infiltration
    - inadequate seal around the well casing or casing bottom infiltration through filter pack
    - mud seams in rock

Many of these performance problems can be traced to inadequate design and/or construction of the well. Several operational conditions that are warnings of design problems are overpumping (which results in lowering of the water table), clogging or collapse of a screen or perforation of a section, and corrosion and encrustation aggravated by excessive intake velocities. Other design and construction errors include:

- poor selection of materials that leads to significant corrosion or collapse of the well casing or screen;
- poor construction—casing cracks or leaks, leaking or missing grout, misplacement of screens and gravel pack, misalignment (enhanced corrosion or collapse can result); and
- lack of well development—poor well yield, turbidity and sand pumping, biofouling, encrustation, and excessive drawdown can result.

Long-term maintenance of wells should include periodic redevelopment and specific capacity monitoring to ensure efficient operation. Records of this redevelopment should be maintained. Establishing an efficient and timely protocol for redevelopment will save the utility money and protect the well for the long term.

As an additional benefit of this book, five appendixes have been added. Appendix A is associated with regulations for water systems. The SDWA and its associated amendments focus on protecting the public health from various contaminants in potable water supplies. This affects groundwater use—the groundwater rules, wellhead protection programs, and water quality requirements all have impacts on the selection and treatment of groundwater supplies. Treatment is a consideration for some groundwater programs and is discussed in Appendix B. The more constituents that are in the water, the more treatment is required, and the more monitoring that must be performed. Appendixes C and D relate to well casing tables, while Appendix E further discusses logging techniques. They were not included in the prior chapters, as they may affect a limited audience.

The sustainability of water supplies is a critical issue. Sustainability is discussed in M73 (AWWA 2021) and the *Sustainability Compendium* (Bloetscher 2010). Sustainability and the ability to manage groundwater supplies for the long term is the critical issue groundwater utilities face in the 21st century. Avoiding the Carolina problem is of utmost importance to protect local communities and their economic base.

The authors hope that this book provided sufficient information about the process of finding a well site, acquiring appropriate sites, drilling and developing wells, and operating them. There are additional documents available from AWWA and others if more information is desired.

REFERENCES

American Water Works Association (AWWA). 2021 (2nd ed.). M73, *Groundwater Management*. Denver: AWWA.

Bloetscher, F. (ed.). 2010. *Sustainability Compendium*. Denver: AWWA.

APPENDIX A

Regulations and Treatment

The goal of water utilities is to provide their customers with high-quality water that does not pose a public health concern and to provide it in sufficient quantities when needed (Bloetscher 2019). Public health is the major issue that utilities face each day, and it has been for as long as public water systems have been in existence. Adequate, safe water supplies are a part of the requirements for a stable society, so governments have attempted to ensure stability for society through public water works. The Roman aqueducts are an example: the government constructed facilities to transport large quantities of water from far beyond Rome to maintain public health in the city.

A functioning water system is vital to the development of modern cities and communities (Bloetscher 2019). The infrastructure is extensive and investments in the system are usually large and long term. As a result, the investment to construct many water systems, and to provide ongoing operations, is often performed by municipal governments or local authorities in the United States and Canada. However, by their very nature (buried pipes and protected facilities that are out of the public view), water and sewer utility operations are not in the forefront in the minds of elected officials or local government management and finance personnel (Bloetscher 2008, 2011, 2019). Ensuring that the utility system operates continuously is of vital importance to ensure economic stability in the community.

When these expectations are not met, problems with public trust occur. The water-contamination crisis that occurred in Flint, Mich., starting in 2014 is a prime example. Meeting public expectations involves understanding the utility operating environment, performing planning activities, and making financial decisions that will provide appropriate funding to achieve the utility's needs. To meet these goals of the utility, there is no "one way" to provide service or establish fees and, except for implementation of impact fees or operations that consistently violate regulations, there are few wrong ways to provide services or establish fees. The proof is that in the United States, for the most part, publicly owned water and sewer systems consistently comply with all regulations, but the backlog of infrastructure needs means that compliance may become an issue in the longer term.

To address the public trust issue, the regulatory perspective is to minimize risk of infection from waterborne disease. Drinking water regulations in the United States go back to 1914 and have been added to periodically as new issues arise. The current regulatory protections are accomplished in the United States and many industrialized nations by public water utilities meeting and exceeding water quality regulations and guidelines such as the Safe Drinking Water Act (SDWA) and its amendments (40 CFR Part 141 § 141.23). For much of the developed world, western Europe and North America, these efforts have been relatively successful; outbreaks of waterborne disease are relatively uncommon and tend to occur in groupings, as opposed to consistent outbreaks. Few people die each year in developed countries from waterborne illness. In contrast, the World Health Organization (WHO) reports that diseases associated with unsafe water distribution, sanitation, and hygiene cause approximately 829,000 deaths per year worldwide (WHO 2022).

"Relatively uncommon" means that incidents do occur. From 1971 to 2008, there were 733 outbreaks reported in public water systems that resulted in 579,582 cases of illness and 116 deaths (Craun 2012). Note that more than 400,000 of the cases and 62 of the deaths were associated with one incident in 1993 in the City of Milwaukee. After the Milwaukee incident, from 1997 to 2006 there were 137 waterborne disease outbreaks reported to the CDC, encompassing a total of 8,498 illnesses and 17 deaths (Barwick et al. 2000; Blackburn et al. 2004; Lee et al. 2003; Liang et al. 2000; Yoder et. al. 2008). Of the outbreaks with a known cause from 1997 to 2006 (n = 101), 17 were attributed to chemical or toxin poisoning and 84 to pathogens. Nonspecific bacteria were among the most commonly implicated pathogens. The highest number of outbreaks where the culprit was known were due to *Legionella, Giardia, Campylobacter,* norovirus, and *E. coli* O157:H7 with 78% of outbreaks attributed to groundwater systems (Bloetscher and Plummer 2011). Benedict et al. (2017) report 110 waterborne disease outbreaks associated with drinking water during 2009–2014. During the 2013–2014 survey period, 42 outbreaks associated with drinking water were reported from 19 states, resulting in 1,006 cases of illness, 124 hospitalizations (12 percent of cases) and 13 deaths (2 percent). The majority of outbreaks (57 percent) and all the deaths were due to *Legionella* (Benedict et al. 2017). During the 2009–2010 survey period, 33 outbreaks associated with drinking water were reported from 17 states, resulting in 1,040 cases of illness, 85 hospitalizations (8.2 percent of cases), and nine deaths (CDC 2013). The majority (58 percent) of outbreaks were because of *Legionella* (CDC 2013).

Over time, the initial pathogen standards were supplemented with limits on metals, synthetic and volatile organics, and more pathogens. Metals were the major concern after World War II when the first regulations were adopted in 1946. Until the 1960s, it was assumed that groundwater was not affected by surface activities. However, in the 1960s, the U.S. government found that nearly a quarter of Louisiana's wellfields were affected by surface contaminants ranging from minerals to solvents and other hydrocarbons. This discovery provided documentation that water table aquifers can be easily contaminated by spills from surface activities.

The 1974 SDWA created maximum contaminant levels (MCLs) for metals and other constituents. Water delivered to customers was not to exceed these levels. Violating the primary standards will result in enforcement actions. The SDWA also directed USEPA to established National Secondary Drinking Water Regulations (NSDWRs) that set nonmandatory water quality standards for additional contaminants. These contaminants are ***not*** considered to present a risk to human health. While USEPA does not enforce these "secondary maximum contaminant levels" (SMCLs), they were established as guidelines to assist public water systems in managing their drinking water for aesthetic considerations, such as taste, color, and odor (USEPA 2001). The pH of the water is a secondary or aesthetic standard, not a primary standard, and hence it is a guidance amount only.

In the early 1980s, USEPA's focus turned to volatile organic compounds (VOCs), commonly used as solvents and found in groundwater systems throughout the United States. The majority of VOCs were present at very low levels; pollution had occurred in water supplies previously thought to be pristine. VOCs can be acutely toxic in large concentrations, but the greater concern was the apparent carcinogenicity of these compounds with continued exposure. As part of the 1986 amendments, SDWA required the USEPA to specify procedures for utilities to meet compliance with the maximum contaminant levels (MCLs) and the use of the best available technology for those utilities that could not comply. A priority list of contaminants that may have adverse impacts on the health of people and are known to, or are anticipated to, occur in public drinking supplies was also compiled as a requirement of the 1986 amendments. USEPA was directed to look at many possible sources of these contaminants, including industrial and chemical production and used sites, landfills, septic tanks, and run-off areas. The 1986 SDWA Amendments also gave USEPA the authority to allow states to acquire primacy, the responsibility for SDWA enforcement, upon compliance with specific federal criteria.

In the late 1980s, USEPA's focus turned toward the legal use of pesticides because numerous pesticides had been found in both surficial and groundwater supplies. Efforts are still underway to survey and assess the extent of pesticide contamination in water supplies. Although the potential risk for most of the population was minimal, many of the contaminants detected were suspected carcinogens. Runoff containing atrazine was a common finding in agricultural areas.

The 1996 amendments were primarily the result of the *Cryptosporidium* outbreak in Milwaukee in 1993. As a part of the rule-making process, USEPA was required to establish National Primary Drinking Water Regulations (NPDWRs) that set mandatory water quality standards for drinking water contaminants (USEPA 1989, 2001, 2002a, 2004). These enforceable standards (MCLs) are established to protect the public against consumption of drinking water contaminants that present a risk to human health. An MCL is the maximum allowable amount of a contaminant in drinking water that is delivered to the consumer.

Today, more than 100,000 synthetic chemicals are in common commercial and industrial use in the United States—a number that grows every year. Few are tested for, yet more than 200 of these chemical substances have been found in groundwater. Others may occur in groundwater where wells are not currently being drilled or investigation has not occurred. Typically, concern does not arise until a contaminant is found in a potable supply. However, regardless of the water source or type of treatment, the SDWA has authorized the USEPA to establish health-based national drinking water regulations by setting maximum permissible levels of a significant number of pollutants in drinking water.

While the data changes with time, Table A-1 outlines the primary drinking water standards and/or treatment techniques in 2020. The definitions per USEPA's website are:

- Maximum contaminant level goal (MCLG): The level of a contaminant in drinking water below which there is no known or expected risk to health. MCLGs allow for a margin of safety and are nonenforceable public health goals.
- Maximum contaminant level (MCL): The highest level of a contaminant that is allowed in drinking water. MCLs are set as close to MCLGs as feasible using the best available treatment technology and taking cost into consideration. MCLs are enforceable standards.
- Maximum residual disinfectant level goal (MRDLG): The level of a drinking water disinfectant below which there is no known or expected risk to health. MRDLGs do not reflect the benefits of the use of disinfectants to control microbial contaminants.
- Treatment technique (TT): A required process intended to reduce the level of a contaminant in drinking water.
- Maximum residual disinfectant level (MRDL): The highest level of a disinfectant allowed in drinking water. There is convincing evidence that addition of a disinfectant is necessary for control of microbial contaminants.

With respect to treatment techniques, the definitions for these items in Table A-1 according to the USEPA are:

TT3 = USEPA's surface water treatment rules require systems using surface water or groundwater under the direct influence of surface water to:

- disinfect their water, and
- filter their water, or
- meet criteria for avoiding filtration so that the following contaminants are controlled at the following levels:
  - *Cryptosporidium*: Unfiltered systems are required to include *Cryptosporidium* in their existing watershed control provisions.
  - *Giardia lamblia*: 99.9% removal/inactivation.
  - viruses: 99.99% removal/inactivation.
  - *Legionella*: No limit, but EPA believes that if *Giardia* and viruses are removed/inactivated, according to the treatment techniques in the Surface Water Treatment Rule, *Legionella* will also be controlled.

Table A-1 Primary drinking water standards

Contaminant	MCLG (mg/L)	MCL or Treatment Technique (TT) (mg/L)*	Potential Health Effects from Long-Term Exposure Above the MCL (unless specified as short-term)
Microorganisms			
Cryptosporidium	zero	TT3	Gastrointestinal illness (such as diarrhea, vomiting, and cramps)
Giardia lamblia	zero	TT3	Gastrointestinal illness (such as diarrhea, vomiting, and cramps)
Heterotrophic plate count (HPC)	n/a	TT3	Indicator of bacterial presence in drinking water.
Legionella	zero	TT3	Legionnaire's Disease, a type of pneumonia
Total coliforms (including fecal coliform and *E. coli*)	zero	5.0%	Not a health threat in itself; it is used to indicate whether other potentially harmful bacteria may be present
Turbidity	n/a	TT3	Potential to frustrate disinfection effectiveness
Viruses (enteric)	zero	TT3	Gastrointestinal illness (such as diarrhea, vomiting, and cramps)
Disinfection Byproducts			
Bromate	zero	0.010	Increased risk of cancer
Chlorite	0.8	1	Anemia; infants and young children: nervous system effects
Haloacetic acids (HAA5)	TT6	0.060	Increased risk of cancer

(*continued*)

Table A-1 Primary drinking water standards (*continued*)

Contaminant	MCLG (mg/L)	MCL or TT (mg/L)	Potential Health Effects from Long-Term Exposure Above the MCL (unless specified as short-term)
Total trihalomethanes (TTHMs)	n/a	0.08	Liver, kidney, or central nervous system problems; increased risk of cancer
Disinfectants			
Chloramines (as Cl_2)	MRDLG = 4	MRDL = 4.0	Eye/nose irritation; stomach discomfort; anemia
Chlorine (as Cl_2)	MRDLG = 4	MRDL = 4.0	Eye/nose irritation; stomach discomfort
Chlorine dioxide (as ClO_2)	MRDLG = 0.8	MRDL = 0.8	Anemia; infants and young children: nervous system effects
Inorganic Chemicals			
Antimony	0.006	0.006	Increase in blood cholesterol; decrease in blood sugar
Arsenic	0	0.010	Skin damage or problems with circulatory systems, and may have increased risk of getting cancer
Asbestos (fiber >10 micrometers)	7 million fibers per liter (MFL)	7 MFL	Increased risk of developing benign intestinal polyps
Barium	2	2	Increase in blood pressure
Beryllium	0.004	0.004	Intestinal lesions
Cadmium	0.005	0.005	Kidney damage
Chromium (total)	0.1	0.1	Allergic dermatitis

(*continued*)

Table A-1 Primary drinking water standards (*continued*)

Contaminant	MCLG (mg/L)	MCL or TT (mg/L)	Potential Health Effects from Long-Term Exposure Above the MCL (unless specified as short-term)
Copper	1.3	TT7; Action Level = 1.3	Short-term exposure: Gastrointestinal distress Long-term exposure: liver or kidney damage
Cyanide (as free cyanide)	0.2	0.2	Nerve damage or thyroid problems
Fluoride	4.0	4.0	Bone disease (pain and tenderness of the bones); children may get mottled teeth
Lead	zero	TT7; Action Level = 0.015	Infants and children: Delays in physical or mental development; children could show slight deficits in attention span and learning abilities Adults: Kidney problems; high blood pressure
Mercury (inorganic)	0.002	0.002	Kidney damage
Nitrate (measured as Nitrogen)	10	10	Infant blue-baby syndrome
Nitrite (measured as Nitrogen)	1	1	Infant blue-baby syndrome
Selenium	0.05	0.05	Hair or fingernail loss; numbness in fingers or toes; circulatory problems
Thallium	0.0005	0.002	Hair loss; changes in blood; kidney, intestine, or liver problems

(*continued*)

Table A-1 Primary drinking water standards (*continued*)

Contaminant	MCLG (mg/L)	MCL or TT (mg/L)	Potential Health Effects from Long-Term Exposure Above the MCL (unless specified as short-term)
		Organic Chemicals	
Acrylamide	zero	TT8	Nervous system or blood problems; increased risk of cancer
Alachlor	zero	0.002	Eye, liver, kidney, or spleen problems; anemia; increased risk of cancer
Atrazine	0.003	0.003	Cardiovascular system or reproductive problems
Benzene	zero	0.005	Anemia; decrease in blood platelets; increased risk of cancer
Benzo(a)pyrene (PAHs)	zero	0.0002	Reproductive difficulties; increased risk of cancer
Carbofuran	0.04	0.04	Problems with blood, nervous system, or reproductive system
Carbon tetrachloride	zero	0.005	Liver problems; increased risk of cancer
Chlordane	zero	0.002	Liver or nervous system problems; increased risk of cancer
Chlorobenzene	0.1	0.1	Liver or kidney problems
2,4-D	0.07	0.07	Kidney, liver, or adrenal gland problems
Dalapon	0.2	0.2	Minor kidney changes

(*continued*)

Table A-1 Primary drinking water standards (*continued*)

Contaminant	MCLG (mg/L)	MCL or TT (mg/L)	Potential Health Effects from Long-Term Exposure Above the MCL (unless specified as short-term)
1,2-Dibromo-3-chloropropane (DBCP)	zero	0.0002	Reproductive difficulties; increased risk of cancer
o-Dichlorobenzene	0.6	0.6	Liver, kidney, or circulatory system problems
p-Dichlorobenzene	0.075	0.075	Anemia; liver, kidney or spleen damage; changes in blood
1,2-Dichloroethane	zero	0.005	Increased risk of cancer
1,1-Dichloroethylene	0.007	0.007	Liver problems
cis-1,2-Dichloroethylene	0.07	0.07	Liver problems
trans-1,2-Dichloroethylene	0.1	0.1	Liver problems
Dichloromethane	zero	0.005	Liver problems; increased risk of cancer
1,2-Dichloropropane	zero	0.005	Increased risk of cancer
Di(2-ethylhexyl) adipate	0.4	0.4	Weight loss, liver problems, or possible reproductive difficulties.
Di(2-ethylhexyl) phthalate	zero	0.006	Reproductive difficulties; liver problems; increased risk of cancer
Dinoseb	0.007	0.007	Reproductive difficulties
Dioxin (2,3,7,8-TCDD)	zero	0.00000003	Reproductive difficulties; increased risk of cancer
Diquat	0.02	0.02	Cataracts

(*continued*)

Table A-1 Primary drinking water standards (*continued*)

Contaminant	MCLG (mg/L)	MCL or TT (mg/L)	Potential Health Effects from Long-Term Exposure Above the MCL (unless specified as short-term)
Endothall	0.1	0.1	Stomach and intestinal problems
Endrin	0.002	0.002	Liver problems
Epichlorohydrin	zero	TT8	Increased cancer risk and, over a long period of time, stomach problems
Ethylbenzene	0.7	0.7	Liver or kidney problems
Ethylene dibromide	zero	0.00005	Problems with liver, stomach, reproductive system, or kidneys; increased risk of cancer
Glyphosate	0.7	0.7	Kidney problems; reproductive difficulties
Heptachlor	zero	0.0004	Liver damage; increased risk of cancer
Heptachlor epoxide	zero	0.0002	Liver damage; increased risk of cancer
Hexachlorobenzene	zero	0.001	Liver or kidney problems; reproductive difficulties; increased risk of cancer
Hexachlorocyclopenta-diene	0.05	0.05	Kidney or stomach problems
Lindane	0.0002	0.0002	Liver or kidney problems
Methoxychlor	0.04	0.04	Reproductive difficulties
Oxamyl (Vydate)	0.2	0.2	Slight nervous system effects

(*continued*)

Table A-1 Primary drinking water standards (*continued*)

Contaminant	MCLG (mg/L)	MCL or TT (mg/L)	Potential Health Effects from Long-Term Exposure Above the MCL (unless specified as short-term)
Polychlorinated biphenyls (PCBs)	zero	0.0005	Skin changes; thymus gland problems; immune deficiencies; reproductive or nervous system difficulties; increased risk of cancer
Pentachlorophenol	zero	0.001	Liver or kidney problems; increased cancer risk
Picloram	0.5	0.5	Liver problems
Simazine	0.004	0.004	Problems with blood
Styrene	0.1	0.1	Liver, kidney, or circulatory system problems
Tetrachloroethylene	zero	0.005	Liver problems; increased risk of cancer
Toluene	1	1	Nervous system, kidney, or liver problems
Toxaphene	zero	0.003	Kidney, liver, or thyroid problems; increased risk of cancer
2,4,5-TP (Silvex)	0.05	0.05	Liver problems
1,2,4-Trichlorobenzene	0.07	0.07	Changes in adrenal glands
1,1,1-Trichloroethane	0.2	0.2	Liver, nervous system, or circulatory problems
1,1,2-Trichloroethane	0.003	0.005	Liver, kidney, or immune system problems
Trichloroethylene	zero	0.005	Liver problems; increased risk of cancer

(*continued*)

Table A-1 Primary drinking water standards (*continued*)

Contaminant	MCLG (mg/L)	MCL or TT (mg/L)	Potential Health Effects from Long-Term Exposure Above the MCL (unless specified as short-term)
Vinyl chloride	zero	0.002	Increased risk of cancer
Xylenes (total)	10	10	Nervous system damage
Radionuclides			
Alpha particles	zero	15 pCi/L	Increased risk of cancer
Beta particles and photon emitters	zero	4 millirems per year	Increased risk of cancer
Radium 226 and Radium 228 (combined)	zero	5 pCi/L	Increased risk of cancer
Uranium	zero	30 μg/L	Increased risk of cancer, kidney toxicity

Source: USEPA 2009

*Units are in milligrams per liter (mg/L) unless otherwise noted. Milligrams per liter are equivalent to parts per million (PPM).

- Turbidity: For systems that use conventional or direct filtration, at no time can turbidity (cloudiness of water) go higher than 1 Nephelometric Turbidity Unit (NTU), and samples for turbidity must be less than or equal to 0.3 NTUs in at least 95 percent of the samples in any month. Systems that use filtration other than the conventional or direct filtration must follow state limits, which must include turbidity at no time exceeding 5 NTUs.
- Heterotrophic plate count (HPC): No more than 500 bacterial colonies per milliliter.

- Long Term 1 Enhanced Surface Water Treatment: Surface water systems or groundwater under the direct influence (GWUDI) systems serving fewer than 10,000 people must comply with the applicable Long Term 1 Enhanced Surface Water Treatment Rule provisions (such as turbidity standards, individual filter monitoring, *Cryptosporidium* removal requirements, updated watershed control requirements for unfiltered systems).
- Long Term 2 Enhanced Surface Water Treatment Rule: This rule applies to all surface water systems or groundwater systems under the

direct influence of surface water. The rule targets additional *Cryptosporidium* treatment requirements for higher risk systems and includes provisions to reduce risks from uncovered finished water storage facilities and to ensure that the systems maintain microbial protection as they take steps to reduce the formation of disinfection byproducts.

- Filter Backwash Recycling: This rule requires systems that recycle to return specific recycle flows through all processes of the system's existing conventional or direct filtration system or at an alternate location approved by the state.

TT6 = MCLGs for some of the individual contaminants:

- Trihalomethanes: bromodichloromethane (zero); bromoform (zero); dibromochloromethane (0.06 mg/L): chloroform (0.07 mg/L.
- Haloacetic acids: dichloroacetic acid (zero); trichloroacetic acid (0.02 mg/L); monochloroacetic acid (0.07mg/L). Bromoacetic acid and dibromoacetic acid are regulated with this group but have no MCLGs.

TT7 = Lead and copper are regulated by a treatment technique that requires systems to control the corrosiveness of their water. If more than 10% of tap water samples exceed the action level, water systems must take additional steps. For copper, the action level is 1.3 mg/L, and for lead is 0.015 mg/L.

TT8 = Each water system must certify, in writing, to the state (using third-party or manufacturer's certification) that when acrylamide and epichlorohydrin are used to treat water, the combination (or product) of dose and monomer level does not exceed the levels specified, as follows:

- Acrylamide = 0.05% dosed at 1 mg/L (or equivalent).
- Epichlorohydrin = 0.01% dosed at 20 mg/L (or equivalent).

Table A-2 outlines the secondary drinking water standards.

Surface Water Treatment Rule

If the source water is surface water, the Surface Water Treatment Rule (SWTR), implemented in 1989, requires utilities to filter and disinfect their water to inactivate viruses and remove *Giardia lamblia* cysts. The concept of SWTR is for the removal of turbidity and suspended solids that interfere with the disinfection process. The indication of significant amounts of suspended solids requires additional chlorine be used, which may also create conflicts with the Disinfection Byproducts Rule portion of SDWA. Waters with considerable amounts of suspended solids are generally required to be filtered to reduce chlorine demands and remove potential pathogens.

Note that certain groundwaters may be under the direct influence of surface waters and, as a result, may be required to comply with the SWTR requirements. In conjunction with wellhead protection efforts, water systems should identify any groundwater sources they are using that may be directly affected by surface water. The concern is that if there is minimal filtration occurring between the

Table A-2 Secondary drinking water standards

Contaminant	Secondary MCL	Noticeable effects above the secondary MCL
Aluminum	0.05–0.2 mg/L*	Colored water
Chloride	250 mg/L	Salty taste
Color	15 color units	Visible tint
Copper	1.0 mg/L	Metallic taste; blue-green staining
Corrosivity	Noncorrosive	Metallic taste; corroded pipes/ fixtures staining
Fluoride	2.0 mg/L	Tooth discoloration
Foaming agents	0.5 mg/L	Frothy, cloudy; bitter taste; odor
Iron	0.3 mg/L	Rusty color; sediment; metallic taste; reddish or orange staining
Manganese	0.05 mg/L	Black to brown color; black staining; bitter metallic taste
Odor	3 TON (threshold odor number)	Rotten-egg, musty, or chemical smell
pH	6.5–8.5	Low pH: bitter metallic taste; corrosion
		High pH: slippery feel; soda taste; deposits
Silver	0.1 mg/L	Skin discoloration; graying of the white part of the eye
Sulfate	250 mg/L	Salty taste
Total dissolved solids (TDS)	500 mg/L	Hardness; deposits; colored water; staining; salty taste
Zinc	5 mg/L	Metallic taste

Source: USEPA 2001

* mg/L is milligrams of substance per liter of water.

surface and the water withdrawn from wells, microbiological constituents may contaminate the water source. Water sources that meet this criterion are considered "groundwater under the direct influence of surface water" (GWUDI). If an aquifer is determined to be groundwater under the direct influence of surface water and therefore vulnerable to contamination by disease-causing organisms found in surface water, the well water must be treated under the same requirements as a surface water system, meaning mandatory disinfection and filtration.

Riverbank Filtration

Given that surface waters are required to be treated with filtration under the SWTR (except in limited circumstances), the question is whether riverbank filtration (RBF) provides the regulatory filtration needed to meet treatment goals/strategies. The issue relates directly to whether the groundwater used by RBF is designated as GWUDI, which means that the well water contains proportions of recent surface water; if so, then the riverbank filtration well must be regulated as surface water. The Long Term 2 (LT2) surface water treatment rule contemplates this question. It designates a concept called "natural filtration." Natural filtration is determined to be an effective alternative/supplement to conventional treatment (coagulation, sedimentation, and rapid sand filtration or direct filtration). Under the LT2 rule, RBF was designated as a pre-treatment alternative for systems that filter but have high *Cryptosporidium* concentrations in the raw water (USEPA 2006). As a result, under existing regulations (SWTR alternative treatment provision), any state or primacy agent can grant Bank Filtration credit for *Giardia* or *Crypto* removal, so a system may avoid construction of a filtration plant based on site-specific data. There are some limitations:

For vertical wells:

- 25-ft separation distance between river and wellhead receives 0.5 log credit (construction and operations requirements must be met).
- 50-ft separation distance between river and wellhead receives 1.0 log credit (construction and operations requirements must be met).
- Separation distance is defined as the map distance between the 100-year return period elevation or floodway boundary (as defined on FEMA flood hazard map) and the wellhead of a vertical well.

For horizontal wells:

- Horizontal well laterals must be separated from the normal flow river bottom by either 25 or 50 ft (from 0.5 or 1.0 log credit).
- Construction and operations requirements must also be met.
- Unconsolidated, predominantly sandy aquifer.

Well site drill core must be sent to an engineering laboratory for sieve analysis:

- Each recovered cored interval should be sieved to determine if at least 10% of the grains are less than 1.0 mm in diameter.
- At least 90% of the sieved, recovered, cored intervals must meet the 10% fine grained requirement.

Disinfection Byproducts Rule

The Disinfection Byproducts Rule regulates trihalomethanes and other carcinogenic organic compounds that can be produced from chlorine disinfection. These regulations apply to all drinking water. The major problem this rule poses is that to achieve the disinfection desired for raw water, the amount of chlorine byproducts is significant, creating a conflict between violating fecal coliform requirements and disinfection byproduct standards. Some discussion about what trihalomethanes are and how much of a concern they are to the public is suggested as this rule affects all utilities.

Trihalomethanes (THMs) are substances formed when chlorine is used to disinfect drinking water creating chlorinated organic compounds. Natural organic matter (NOM) includes both humic and nonhumic fractions. The humic fraction includes high molecular weight organic molecules such as humic and fulvic acids. Humic substances are terrestrially based; they are the organic portion of soil that remains after prolonged microbial decomposition formed by the decay of leaves, wood, and other vegetable matter. Therein lies the challenge—the humics are primarily a groundwater concern. Chlorine added to disinfect the water react with the organic materials to form a group of chlorinated chemical compounds (THMs). THMs (chloroform, bromodichloromethane, dibromochloromethane, and bromoform) were discovered in 1974 to be formed during the disinfection step of drinking water if free chlorine was the disinfectant (Uden and Miller 1983; Coleman et al. 1984). This, coupled with the perceived hazard to the consumer's health, led the US Environmental Protection Agency (USEPA) to amend the National Interim Primary Drinking Water Regulations to include a maximum contaminant level of 0.10 mg/L for total trihalomethanes. THMs are tasteless and odorless.

The quantity of byproducts formed is determined by several factors, such as the amount and type of organic material present in water, temperature, pH, chlorine dosage, contact time available for chlorine, and bromide concentration in the water. Most organics are not removed in water treatment plants unless the plants specifically are designed to address them (anthracite, GAC, ion exchange, ozone, membranes). THMs are just one category of disinfection byproducts (DBPs) formed in the treatment process. Because of these differences, under its Disinfectants and Disinfection Byproducts Rule (USEPA 1998), USEPA set the MCL for total trihalomethanes (TTHMs) at 0.080 mg/L. USEPA set an MCLG for the group of TTHMs although there are MCLGs for some of the individual constituents, including bromodichloromethane (zero), bromoform (zero), dibromochloromethane (0.06 mg/L), and chloroform (0.07 mg/L). USEPA considers that the MCL for TTHMs is feasible and achievable for a chlorinated drinking water supply. USEPA also believes that by meeting MCLs for TTHMs and haloacetic acids (HAA5) (another DBP), water suppliers will also control the formation of other DPBs not currently regulated that may also adversely affect human health (USEPA 2002b).

Potential health effects identified in the studies from ingestion of TTHMs in water include liver, kidney, and central nervous system problems, as well as an increased risk of cancer.

- Bromodichloromethane is likely to be carcinogenic to humans by all routes of exposure.
- Bromoform is likely to be carcinogenic to humans by all routes of exposure.

- There is suggestive evidence of carcinogenicity but not sufficient to assess human carcinogenic potential for dibromochloromethane.
- Chloroform is likely to be carcinogenic to humans by all routes of exposure under high-dose conditions that lead to cytotoxicity and regenerative hyperplasia in susceptible tissues. Chloroform is not likely to be carcinogenic to humans by all routes of exposures at a dose level that does not cause cytotoxicity and cell regeneration (USEPA 2004).

However, USEPA has not conducted a cancer assessment for the TTHMs. As a result, some researchers have raised questions about their health hazards (Jolley et al. 1990; IARC 1991). All four trihalomethanes, when administered by the administration of food or drugs through a tube to female B6C3F1 mice, were found to cause liver cancer (National Cancer Institute 1976; National Toxicology Program [NTP] 1985, 1987, 1989). The carcinogenic activity of the THMs has been proposed to be mediated through a nongenotoxic mechanism based on their weak activity in mutagenicity and genotoxicity assays (Le Curieux et al. 1995; Reitz et al. 1990; Rosenthal 1987). However, only one THM, bromodichloromethane, has demonstrated mutagenic activity apparently though a glutathione (GSH) metabolite (Pegram et al. 1997).

The route of administration appears to play a significant role in the dosimetry and metabolism of the THMs and consequently in their carcinogenic and toxic activity. Although chloroform was carcinogenic in mice when administered by via feeding tube, it was not carcinogenic and did not promote liver cancer when administered in the drinking water (Jorgenson et al. 1985; Klaunig et al. 1986; Pereira et al. 1985). Furthermore, cell proliferation was enhanced by chloroform administered by feeding tube but not in drinking water (Larson et al. 1994; Pereira 1994; Pereira and Grothaus 1997). The THMs are metabolized to reactive dihalocarbonyl metabolites that react with nucleophilic compounds and macromolecules (Ammann et al. 1998; Lilly et al. 1997; Pankow et al. 1997; Pohl et al. 1977). Thus, the dihalocarbonyl metabolite of chloroform, phosgene, binds GSH in what appears to be a detoxifying mechanism (Ilett et al. 1973; Pohl et al. 1979). The thought is that administering the THMs by feeding tube could overwhelm the ability of the liver to detoxify them, resulting in toxicity, regenerative hyperplasia, and promotion of cancer. Hence, the real risk is less clear when drinking the water from the tap.

Ground Water Rule

If the source water is groundwater, the Ground Water Rule may require that the water be disinfected on withdrawal unless the water meets the requirements for "natural disinfection" or if the system qualifies for a variance. This rule was passed in 1992, ostensibly to deal with unchlorinated well systems. However, its reach has expanded based on the desire by some groundwater utilities to reduce sampling frequency. This is the 4-log removal portion of the rule. Note that the impact of THMs and 4-log are related.

To reduce THM formation potential, ammonia is deliberately added to a chlorinated water supply to control microbial growth in finished water. This is particularly at issue in water systems that have warm water (>60°F) and organics. Chloramines include monochloramine, dichloramine, trichloramine, and organo-chloramines. For disinfection purposes, monochloramine is the preferential form. Natural sources of nitrogen generally have minimal impacts on water supply distribution systems because the concentration of nitrite nitrogen in surface and ground waters is normally low (Sawyer and McCarty 1978). The groundwater rule ONLY applies to water sources that have no ammonia in them. This rule permits reduced monitoring if 4 logs of virus removal can be demonstrated. Regulatory agencies encourage this compliance.

However, ammonia occurs naturally in some groundwater supplies, and groundwater can become contaminated with nitrogen as agriculture runoff percolates into aquifers. Where utilities have ammonia in the groundwater, the 4 log guidelines DO NOT apply. Some regulatory agencies have suggested that utilities simply chlorinate the ammonia "out of the water" and then free chlorinate the water, then add ammonia back in to create chloramines (monochloramine). However, the nitrogen does not disappear—it is just reacted with chlorine (Kirmeyer et al. 1993, 1995, 2000), which does not comply with the rule. This practice should be avoided as the total nitrogen increases, thereby increasing the potential for biofilm development because the nitrogen is still present and would potentially be available to be used as a nutrient by bacteria. As a result, the 4 log groundwater rule may be adding ammonia to the water thereby increasing the likelihood of nitrification in the water distribution system.

Nitrification can occur in water systems that contain chloramines as is typical in utilities with organic content and warm temperature, which limit the ability to maintain a chlorine residual. According to Wilczak et al. (1996), nitrification is often indirectly identified by detection via monitoring the following (USEPA 2002a):

- **Low residual disinfectant.** Low levels of residual disinfectant can allow bacteria in the biofilm to multiply. When residual disinfectant drops below your normal baseline, nitrification may soon follow. So, monitoring and mapping levels of residual disinfectant proves to be a quick and inexpensive tool you can use to pinpoint affected areas and focus your mitigation efforts.
- **Nitrite and nitrate levels in distribution.** Under the new rule, systems are required to sample nitrate and nitrite quarterly in the distribution system. One variety of nitrifying bacteria oxidizes ammonia to produce nitrite, and another will oxidize nitrite to produce nitrate. By monitoring nitrite and nitrate levels, you can know when action is needed. You may find more frequent monitoring is beneficial for your system.
- **Ammonia levels.** Ammonia is food for nitrifying bacteria. So, if ammonia levels are decreasing in at least part of your distribution system, nitrification could be the cause. Once you have determined

baseline ammonia levels for your system, look more closely at the current data from any system location for decreasing ammonia levels. If the ammonia levels in the system are less than the ammonia levels of the water leaving your treatment plant, or less than your system's baseline ammonia levels, nitrification may be occurring.
- Decrease in dissolved oxygen may indicate bacteria are consuming oxygen via the nitrification process (conversion from ammonia to nitrate).
- Drop in pH and alkalinity.
- Increase in water temperature (which facilitates bacterial growth).
- Frequently, but not always, systems that have nitrification occurring may also have increases in heterotrophic plate counts (HPC), coliform-positive test results, or both.

If these issues occur, nitrification is suggested, and a more active control strategy must be put in place.

A common control strategy is chemical control or treatment of nitrifying bacteria. Such treatment typically involves either the maintenance of high distribution system disinfectant residuals (greater than 2 mg/L) or periodic breakpoint chlorination. Analytical survey results of ten US utilities showed that greater than 90% of distribution system samples with increased nitrite and nitrate levels, indicative of nitrification, occurred in water with disinfectant residuals less than 2 mg/L (Wilczak et al. 1996). Many utilities have found that increasing disinfectant residuals by increasing chemical doses or managing water age has helped to control nitrification.

pH and Alkalinity Modification

Water pH value is an important factor in nitrification activity for two reasons. First, a reduction of total alkalinity may accompany nitrification because a significant amount of bicarbonate is consumed in the conversion of ammonia to nitrite. A model developed by Gujer and Jenkins (1974) indicates that 8.64 mg/L of bicarbonate (HCO_3^-) will be utilized for each mg/L of ammonia nitrogen oxidized. While reduction in alkalinity does not impose a direct public health impact, reductions in alkalinity can cause reductions in buffering capacity, which can impact pH stability and corrosivity of the water toward lead and copper.

Secondly, nitrifying bacteria are very sensitive to pH. *Nitrosomonas* has an optimal pH between approximately 7.0 and 8.0, and the optimum pH range for *Nitrobacter* is approximately 7.5 to 8.0. Some utilities have reported that an increase in pH (to greater than 9) can be used to reduce the occurrence of nitrification (Skadsen and Sanford 1996). According to Wilczak (2001), pH appears to be the most important factor controlling the rate of chloramine autodecomposition. Thomas (1987) stated that the rate of chloramine decay approximately doubles for a drop in pH of 0.7 units. In Florida, Cates and Lavinder (1999) noted that raising pH reduced nitrification. From Figure A-1, a pH above 8.7 limits nitrification activity.

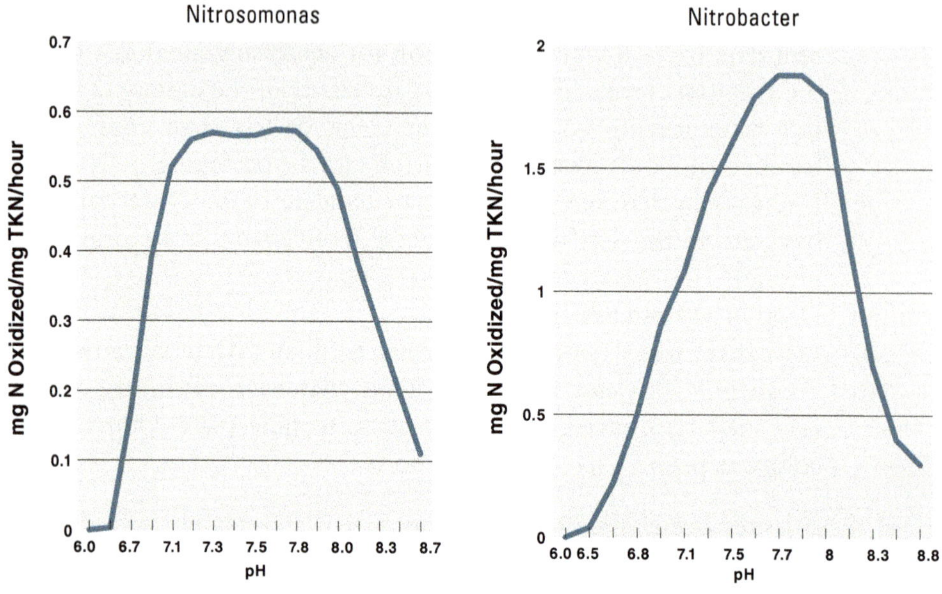

Source: Grady and Lim 1980

Figure A-1 Oxidation rate vs. pH

Utilities can use booster chlorination in the distribution system to increase disinfectant residuals. This practice is generally not employed in chloraminated distribution systems because chloramines are normally more stable than free chlorine (Woolschlager et al. 2001, Valentine et al. 1998). In addition, uncontrolled blending of chlorinated and chloraminated water could occur near a chlorine booster station; in some cases, uncontrolled blending has been shown to cause unintended breakpoint chlorination, increases in DBP levels, or decreases in disinfectant residuals (Mahmood et al. 1999, Muylwyk et al. 1999).

USE OF RISK IN REGULATORY ENVIRONMENTS

What quantities of a contaminant is permissible in drinking water? The answer is unclear, which is why there are risk factors and safety factors built into the current regulatory standards. *Risk* and *risk assessments* are buzzwords used in Washington, DC, and many state regulatory agencies. The use of these terms and their placement into laws and regulations does not mean that anyone necessarily understands how to conduct a risk assessment; however, it does create a tool for the regulatory community to use in an attempt to quantify the effects of a given activity. The most important aspect of risk is that there are no zero-risk alternatives. Only comparative risks can truly be calculated—the preference of one alternative over others and the choice by utilities to strive to minimize those risks that may exist.

Because there are no zero-risk alternatives, and because there are finite limits to the amount of resources that can be used to either define risk or maximize

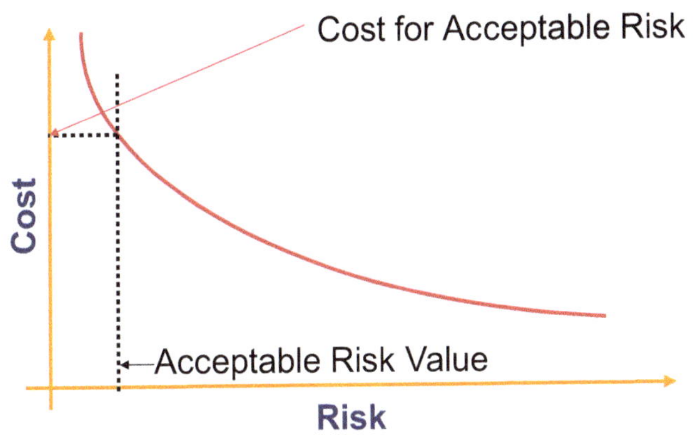

Figure A-2 Risk–cost analysis

risk reduction, the concept of *acceptable* risk has been developed. According to USEPA, an acceptable risk is the 1:1 million lifetime chance that an impact will occur to the general public as a result of an activity. Acute responses (immediate impacts), e.g., microbial water quality regulations, may be written according to the probability that less than 1 in 10,000 will contract a disease from drinking water in a given year. Carcinogens tend to use the 1:1 million lifetime (chronic) risks more frequently than acute exposures, because the effects may not manifest themselves for many years. Regulations are typically written according to these acceptable risks.

In environmental impact assessments, risk–cost assumptions may be used. Figure A-2 shows an example of such an analysis whereby the acceptable risk is defined, and the associated cost of compliance is calculated. As risk decreases, the cost rises. To some, much of the focus is whether the acceptable risk or the cost should be the limiting factor in determining acceptable risk. Obviously, if cost is used as the limit, the risk would likely be much higher. Political conflicts over resources are one of the problems encountered when determining acceptable risk. Limited data and conflicting data create confusion and uncertainty in the risk assessment process.

Risk assessments generally include two parts: the scientific investigation and the risk management portion. Scientific assessment methods include measuring the effects of exposure or the activity to the ecosystem or humans, determining the level at which the impacts are negligible, and creating methods to replicate and measure the impacts. Management of the risk includes taking the steps necessary to limit exposure. The latter is for local officials and includes the proper training of employees, maintaining appropriate records of operations, and providing those facilities and tools needed to minimize risks to the community. When making local decisions on expenditures, it should be noted that neither the preface to the Clean Water Act nor the 1996 amendments to the SDWA mention cost as a part of the laws, only protection of the public health and the ecosystem.

REFERENCES

Ammann, P., C.L. Laethem, and G.L. Kedderis. 1998. "Chloroform-Induced Cytolethality in Freshly Isolated Male B6C3F1 Mouse and F-344 Rat Hepatocytes." *Toxicology and Applied Pharmacology.* 149(2):217–225.

Barwick, R., D. Levy, G. Craun, M. Beach, and R.L. Calderon. 2000. "Surveillance for Waterborne-Disease Outbreaks—United States, 1997–1998." *Morbidity and Mortality Weekly Report.* 49(4):1–35.

Benedict, K.M., H. Reses, M. Vigar, D.M. Roth, V.A. Roberts, M. Mattioli, L.A. Cooley, et al. 2017. "Surveillance for Waterborne Disease Outbreaks Associated with Drinking Water—United States, 2013–2014." *Morbidity and Mortality Weekly Report.* 66(44):1216–1221.

Blackburn, B.G., G.F. Craun, J.S. Yoder, V. Hill, R.L. Calderon, N. Chen, S.H. Lee, et al. 2004. "Surveillance for Waterborne-Disease Outbreaks Associated with Drinking Water—United States, 2001–2002." *Morbidity and Mortality Weekly Report,* 53(08):23–45.

Bloetscher, F. 2019. *Public Infrastructure Management: Tracking Inevitable Asset Challenges and Increasing System Resiliency.* Plantation, Fla.: J. Ross.

Bloetscher, F. 2011. *Utility Management for Water and Wastewater Operators.* Denver: AWWA.

Bloetscher, F. 2008. *Water Basics for Decision Makers: What Local Officials Need to Know about Water and Wastewater Systems.* Denver: AWWA

Bloetscher, F. and J.D. Plummer. 2011. "Environmental Review and Case Study: Evaluating the Significance of Certain Pharmaceuticals and Emerging Pathogens in Raw Water Supplies." *Environmental Practice.* 13(3):198–215.

Cates, J. and S. Lavinder. 1999. "Improving Chloramine Residuals and Minimizing Nitrification." *Florida Water Resources Journal.* 2:26–28.

Centers for Disease Control and Prevention (CDC). 2013 "Surveillance for Waterborne Disease Outbreaks Associated with Drinking Water and Other

Nonrecreational Water–United States, 2009-2010." *Morbidity and Mortality Weekly Report.* 2013:714–720.

Coleman, W.E., J.W. Munch, W.H. Kaylor, R.P. Streicher, H.P. Ringhand, and J.R. Meier. 1984. "Gas Chromatography/Mass Spectroscopy Analysis of Mutagenic Extracts of Aqueous Chlorinated Humic Acid. A Comparison of the Byproducts to Drinking Water Contaminants." *Environmental Science and Technology.* 18(9):674–678.

Craun, G.F. 2012. "The Importance of Waterborne Disease Outbreak Surveillance in the United States." *Annali dell'Istituto Superiore di Sanità.* 48(4):447–459.

Grady, J.C.L. and H.C. Lim. 1980. Biological Wastewater Treatment: Theory and Applications. New York: Marcel Dekker, Inc.

Gujer, W. and D. Jenkins. 1974. "A Nitrification Model for Contact Stabilization Activated Sludge Process." *Water Research.* 9(5-6):561–566.

Ilett, K.F., W.D. Reid, I.G. Sipes, and G. Krishna. 1973. "Chloroform Toxicity in Mice: Correlation of Renal and Hepatic Necrosis with Covalent Binding of Metabolites to Tissue Macromolecules." *Experimental and Molecular Pathology.* 19(2):215–229.

International Agency for Research on Cancer (IARC). 1991. "Chlorinated Drinking-Water; Chlorination By-Products." In *IARC Monograph on the Evaluation of the Carcinogenic Risk of Chemicals to Humans.* 52:45–268. Lyon, France: IARC.

Jolley, R.L., L.W. Condie, J.D. Johnson, S. Katz, R.A. Minear, J.S. Mattice, and V.A. Jacobs (Eds.). 1990. *Water Chlorination: Chemistry, Environmental Impact, and Health Effects.* Vol. 6. Michigan: Lewis Publishing.

Jorgenson, T.A., E.F. Meierhenry, C.J. Rushbrook, R.J. Bull, and M. Robinson. 1985. "Carcinogenicity of Chloroform in Drinking Water to Male Osborne-Mendel Rats and Female B6C3F1 Mice." *Fundamental and Applied Toxicology.* 5(4):760–769.

Kirmeyer, G.J., G.W. Foust, G.L. Pierson, J.J. Simmler, and M.W. LeChevallier. 1993. "Optimizing Chloramine Treatment." Denver: AwwaRF and AWWA.

Kirmeyer, G.J., L.H. Odell, J. Jacangelo, A. Wilczak, and R. Wolfe. 1995. "Nitrification Occurrence and Control in Chloraminated Water Systems." Denver: AwwaRF and AWWA.

Kirmeyer, G.J., M. Friedman, J. Clement, A. Sandvig, P.F. Noran, K.D. Martel, D. Smith, et al. 2000. "Guidance Manual for Maintaining Distribution System Water Quality." Denver: AwwaRF and AWWA.

Klaunig, J.E., R.J. Ruch, and M.A. Pereira. 1986. "Carcinogenicity of Chlorinated Methane and Ethane Compounds Administered in Drinking Water to Mice." *Environmental Health Perspectives.* 69:89–95.

Larson, J.L., D.C. Wolf, and B.E. Butterworth. 1994. "Induced Cytotoxicity and Cell Proliferation in the Hepatocarcinogenicity of Chloroform in Female B6C3F1 Mice: Comparison of Administration by Gavage in Corn Oil vs. *ad Libitum* in Drinking Water." *Fundamental and Applied Toxicology.* 22(1):90–102.

Le Curieux, F., L. Gauthier, F. Erb, and D. Marzin. 1995. "Use of the SOS Chromotest, the Ames-Fluctuation Test, and the Newt Micronucleus Test to Study the Genotoxicity of Four Trihalomethanes." *Mutagenesis.* 10(4):333–341.

Lee, D.-G. and S.-J. Kim. 2003. "Bacterial Species in Biofilm Cultivated from the End of the Seoul Water Distribution System." *Journal of Applied Microbiology.* 95(2):317–324.

Liang, S.M., J.H. Joon, M.K. Davis, J.F. Green, and D. Remer. 2000. "Treatment of N-Nitrosodimthylamine (NDMA) by Pulsed-Ultra-Violet Irradiation and Pulsed-UV/Hydrogen Peroxide Processes." Proc. August 2000 Water Quality Technical Conference.

Lilly, P.D., T.M. Ross, and R.A. Pegram. 1997. "Trihalomethane Comparative Toxicity: Acute Renal Hepatic Toxicity of Chloroform and Bromodichloromethane Following Aqueous Gavage." *Fundamental Applied Toxicology.* 40,101–110.

Mahmood, F., J. Pimblett, N. Grace, and B. Utne. 1999. "Combining Multiple Water Sources and Disinfectants: Options for Water Quality Compatibility in Distribution Systems." Proc. 1999 AWWA Water Quality Technology Conf., Tampa, Fla.

Muylwyk, Q., A.L. Smith, and J.A. MacDonald. 1999. "Implications on Disinfection Regime when Joining Water Systems: A Case Study of Blending Chlorinated and Chloraminated Water." Proc. 1999 AWWA Water Quality Technology Conf., Tampa Fla.

National Cancer Institute. 1976. "Carcinogenesis Bioassay of Chloroform." National Tech. Inform. Service No. PB264018/AS. Bethesda, Md.: National Cancer Institute.

National Toxicology Program (NTP). 1989. "Toxicology and Carcinogenesis Studies of Bromoform in F344/N Rats and B6C3F1 Mice." National Toxicology Program TR 350, DHHS Publ. No. (NIH) 85–2850. Bethesda, Md.: National Institutes of Health.

NTP. 1987. "Toxicology and Carcinogenesis Studies of Bromodichloromethane in F344/N Rats and B6C3F1 Mice." National Toxicology Program TR 321, DHHS Publ. No. (NIH) 88–2537. Research Triangle Park, N.C.: NIH.

NTP. 1985. "Toxicology and Carcinogenesis Studies of Chlorodibromomethane in F344/N Rats and B6C3F1 Mice." National Toxicology Program TR 282, DHHS Publ. No. (NIH) 85-2538. Bethesda, Md.: NIH.

Pankow, D., B. Damme, U. Wünscher, and K. Bergmann. 1997. "Chlorodibromomethane Metabolism to Bromide and Carbon Monoxide in Rats." *Archives of Toxicology.* 71:203–210.

Pegram, R.A., M.E. Anderson, S.H. Warren, T.M. Ross, and L.D. Claxton. 1997. "Glutathione *S*-Transferase-Mediated Mutagenicity of Trihalomethanes in *Salmonella typhimurium*: Contrasting Results with Bromodichloromethane and Chloroform." *Toxicology and Applied Pharmacology.* 144(1):183–188.

Pereira, M.A. 1994. "Route of Administration Determines Whether Chloroform Enhances or Inhibits Cell Proliferation in the Liver of B6C3F1 Mice." *Fundamental and Applied Toxicology.* 23(1):87–92.

Pereira, M.A., G.L. Knutsen, and S.L. Herren-Freund. 1985. "Effect of Subsequent Treatment of Chloroform or Phenobarbital on the Incidence of Liver and Lung Tumors Initiated by Ethylnitrosourea in 15-day-old Mice." *Carcinogenesis.* 6(2):203–207.

Pereira, M.A. and M. Grothaus. 1997. "Chloroform in Drinking Water Prevents Hepatic Cell Proliferation Induced by Chloroform Administered by Gavage in Corn Oil to Mice." *Fundamental and Applied Toxicology.* 37(1):82–87.

Pohl, L.R., B. Bhooshan, N.F. Whittaker, and G. Krishna, G. 1977. "Phosgene: A Metabolite of Chloroform." *Biochemical and Biophysical Research Communications.* 79(3):684–691.

Pohl, L.R., J.W. George, J.L. Martin, and G. Krishna. 1979. "Phosgene: An In Vitro Metabolite of Chloroform." *Toxicology and Applied Pharmacology.* 48(1 II), A110.

Reckhow, D.A. and Singer, P.C. 1984. "The Removal of Organic Halide Precursors by Preozonation and Alum Coagulation." *Journal AWWA.* 76(4):151–157.

Reitz, R.H., A.L. Mendrala, R.A. Corley, J.F. Quast, M.L. Gargas, M.E. Andersen, D.A. Staats, and R.B. Conolly. 1990. "Estimating the Risk of Liver Cancer Associated with Human Exposures to Chloroform Using Physiologically Based Pharmacokinetic Modeling." *Toxicology and Applied Pharmacology.* 105(3):443–459.

Rosenthal, S.L. 1987. "A Review of Mutagenicity of Chloroform." *Environmental and Molecular Mutagenesis.* 10(2):211–226.

Sawyer, C.N. and P.L. McCarty. 1978. *Chemistry for Environmental Engineering* (3rd ed.). New York: McGraw-Hill.

Skadsen, J. and L. Sanford. 1996. "The Effectiveness of High pH for Control of Nitrification and the Impact of Ozone on Nitrification Control." Proc. 1996 AWWA Water Quality Technology Conf. Boston, Mass.

Thomas, P.M. 1987. "Formation and Decay of Monochloramine in South Australian Water Supply Systems." 12th Federal Convention, Australian Water and Wastewater Association, Adelaide, Australia.

Uden, P.C. and J.W. Miller. 1983. "Chlorinated Acids and Chloral in Drinking Water." *Journal AWWA.* 75(10):524–527.

US Environmental Protection Agency (USEPA). 2009. "National Primary Drinking Water Regulations." www.epa.gov/ground-water-and-drinking-water/national-primary-drinking-water-regulations.

USEPA. 2006. "Source Water Monitoring Guidance Manual for Public Water Systems for the Final Long Term 2 Enhanced Surface Water Treatment Rule." Washington: USEPA.

USEPA. 2004. "2004 Edition of the Drinking Water Standards and Health Advisories." EPA 822-R-04-005. Washington: USEPA.

USEPA. 2002a. "Nitrification." Cincinnati, Ohio: USEPA.

USEPA. 2002b. "Chemical Health Effects Tables." www.epa.gov/safewater/tcrdsr.html.

USEPA. 2001. "National Primary Drinking Water Regulations." 66 FR 6975. *Federal Register.* 66(14):6975–7066.

USEPA. 1998. "National Primary Drinking Water Regulations: Disinfectants and Disinfection Byproducts." 63 FR 69390. *Federal Register.* 63(241):69390–69476.

USEPA. 1989. "Drinking Water; National Primary Drinking Water Regulations; Filtration, Disinfection; Turbidity, *Giardia lamblia*, Viruses, *Legionella*, and Heterotrophic Bacteria; Final Rule." *Federal Register.* 54(124):27486–27541.

Valentine, R.L., K. Ozekin, and P.J. Vikesland. 1998. "Chloramine Decomposition in Distribution System and Model Waters." Denver: AWWA Research Foundation.

Wilczak, A., J.G. Jacangelo, J.P. Marcinko, L.H. Odell, G.J. Kirmeyer, and R.L. Wolfe. 1996. "Occurrence of Nitrification in Chloraminated Distribution Systems." *Journal AWWA.* 88(7):74–85.

Wilczak, A., 2001. "Chloramine Decay Rate: Factors and Research Needs." Proc. 2001 AWWA Annual Conf., Washington.

Woolschlager, J.E., B.E. Rittmann, P. Piriou, and B. Schwartz. 2001. "Developing an Effective Strategy to Control Nitrifier Growth Using the Comprehensive Disinfection and Water Quality Model." Proc. World Water and Environmental Resources Congress, Orlando, Fla.

World Health Organization (WHO). 2022. "Drinking Water: Key Facts." www.who.int/news-room/fact-sheets/detail/drinking-water (accessed 6/27/2022).

Yoder, J.S., M.C. Hlavsa, G.F. Craun, V. Hill, V. Roberts, P.A. Yu, L.A. Hicks, et al. 2008. "Surveillance for Waterborne Disease and Outbreaks Associated with Recreational Water Use and Other Aquatic Facility–Associated Health Events—United States, 2005-2006." *Morbidity and Mortality Weekly Report.* 57(9):1–29.

APPENDIX B
Treatment of Groundwater

Treatment is not often discussed in a book on drilling, but given the water quality issues discussed previously—and the impact that treatment may have on drilling decisions—such discussion appears warranted. Treatment of water can be costly and installations must account for treatment. In most wells systems, the treatment required is minimal: chlorination, maybe some pH or orthophosphate to reduce corrosion potential. These are simple chemical feeds that occur at the well. However, more challenging water quality issues can arise (or be discovered as a result of pilot testing) that may require the utility to invest much more money into their water system than anticipated. For example, arsenic is commonly found in water in some areas; if it is in the water withdrawn from the wells, it needs to be removed. Therefore, some discussion about the treatment options to address the water quality concerns identified in the prior chapters was deemed to be appropriate. Note that none of this discussion carries with it any significant depth, but will allow the well operator to understand the issues that may arise depending on water quality. Other references should be sought for more detailed understanding for a particular water quality consideration.

DISINFECTION

There are more than 41,000 water systems that rely on well for water supply. Many of those well exist in aquifers that are characterized as surficial, and therefore can be contaminated by surface activities. Septic tanks and runoff can carry pathogens with it that percolate into the ground. All wells have bacteria in them and the potential for viruses, protozoans, and other organisms. Hence, all wells should be assumed to have the appropriate conditions to culture pathogenic organisms. As a result, since the conditions exist for bacteria, the water must be disinfected before being discharged into the distribution system.

Disinfection is defined as the destruction of *pathogenic* microorganisms. Disinfection is not sterilization, which occurs when all living organisms are destroyed. Chlorine, in several forms, is the most used means of disinfection as it effectively kills bacteria, viruses, protozoa, amoebic cysts, algae, and helminth (worms) because of its effectiveness in killing organisms and experience over the past 100 years. Chlorine has been used since 1908, and most large systems had adopted its use by 1914 because of the reduction in waterborne disease because of chlorination. Other methods include ultraviolet radiation, maintenance of an elevated pH, iodine, or ozone, but none of these creates the lasting residual needed to continue disinfection properties in the piping system. The piping system can be contaminated so the chlorine residual is the only means to maintain water quality.

Utilities test for disinfection effectiveness by testing for the presence of coliform bacteria in the water. The maximum contaminant level (MCL) is 0 CFU/100 mL. The thought process has been that if coliforms—which are a highly prolific gut bacteria—are not present, other pathogens likely are absent as well. However, this practice has not always been effective for some enteric viruses and the protozoa *Cryptosporidium*, as occurred in Milwaukee in 1993, occurred ... in the absence of coliforms. While bacteria are the most sensitive to disinfection, some enteric viruses that lack sensitive enzyme systems are very resistant. Hence, disinfection effectiveness depends on the sensitivity of targeted microorganisms, disinfection concentration, contact time, and other water quality characteristics—the concept of "CT" (contact time) values in the groundwater rules created by USEPA in the 1980s. All groundwater systems should assume they will need chlorination and chlorine residual monitoring equipment in their system.

Chlorine

Chlorine is the most widely used disinfectant because it is effective at low concentrations, is relatively inexpensive, and can form a residual. Chlorine can be applied as a gas or as hypochlorite (Figures B-1 and B-2). Gas chlorine is liquid chlorine under pressure in the cylinder. An evaporator is used to vacuum the chlorine liquid, turn it to gas, and mix it with water immediately. The method has proven safe for many years. Gas chlorine is 100 percent chlorine and only changes to HOCl and OCl^- once in water. The pH and temperature indicate which species dominates. Compared with OCl^-, HOCl is more effective.

Hypochlorite salts are available in dry (calcium hypochlorite) or liquid (sodium hypochlorite) form, although the liquid form is preferred over solids. While stored gas chlorine does not change, sodium hypochlorite does. The normal delivery is a 12.5 percent solution. Sunlight and time will degrade the hypochlorite to a 6 percent solution. As a result, many operators will blend it immediately to a 6 percent solution. Tankage is needed to accomplish this.

Chlorine is a very reactive oxidizing agent and combines with ammonia, sulfites, metals, and organic material, thereby creating an initial chlorine *demand*. Chlorine demand is the amount of chlorine that is used up in these extraneous reactions before it becomes free available chlorine for use as a disinfectant.

Source: Photo by Fred Bloetscher

Figure B-1 Chlorine cylinder installation

Source: Photo by Fred Bloetscher

Figure B-2 Typical hypochlorite disinfection system

Chlorination reactions may produce byproducts, including trihalomethanes (THMs) and organic halides in waters that contain humics or other natural organic precursors (Reckhow and Singer 1984). If byproduct formation exceeds regulated limits, options include precursor removal or the use of alternative disinfectants.

Chloramines

Chloramines are compounds formed when chlorine reacts with ammonia, which may be naturally present or intentionally added. That means ammonia is added to water. Customers should be informed because the ammonia is toxic to fish (e.g., fish tanks). Chloramines are less effective than HOCl and OCl^- and less effective against viruses than bacteria. The benefits of chloramine disinfection are a lower generation of chlorination byproducts and a greater residual stability in the distribution system. However, ammonia contains nitrogen, which is a bacterial nutrient.

Chlorine and Chlorine Compounds

Aqueous chlorine is one of the most effective chemical oxidants for use in manganese oxidation, carbon removal, and control of taste and odors. Unfortunately, free residual chlorine may combine with organic compounds to create trihalomethanes and other byproducts.

AERATION

Some water contains metals or gases that can damage pipe or cause the water to have unattractive tastes and/or odors. Naturally occurring challenges include iron (red stains), hydrogen sulfide (rotten egg smell), swamp gas (methane), and carbon dioxide (acidity), all of which are common issues. Also of concern are the synthetic and volatile organic compounds (VOCs) that may leach into aquifers from surface activities. These are less common and harder to detect. The solution to each of these challenges is aeration.

Aeration mixes water with air to transfer gas from the water to the air using a physical principle called Henry's law. Carbon dioxide, hydrogen sulfide, and methane are easily liberated from groundwater with simple aerators (Figures B-3 and B-4) when they are in small quantities. The figures show tray aerators, which rely on the breeze and cascading water to make Henry's law work to remove the gases. Tray aeration is open to the atmosphere. Water enters the top, falls in or around trays, and is collected at the bottom. The turbulence and mixing of the cascading water provide gas transfer. There are challenges, however. The introduction of air into water containing microbiological populations can increase biological activity, potentially affecting downstream water quality. Aeration should not precede membrane filtration. The aluminum in Figure B-3 is covered in bacteria colonies, iron, and algae. Figure B-4 shows how easily iron is removed.

More-recalcitrant gases require higher volumes of air to remove the dissolved gases. These gases can be removed with packed towers (Figure B-5). Packed towers

Source: Photo by Fred Bloetscher

Figure B-3 Tray aeration (aluminum)

Source: Photo by Fred Bloetscher

Figure B-4 Cone version of a tray aerator (aluminum) on top, coated with iron

Source: Photo by Fred Bloetscher

Figure B-5 Packed tower

have water sprayed down from the top and large volumes of air blown upward to facilitate the stripping process. The concept is similar in all aerators; however, rather than relying on natural air and wind to provide the necessary air volume, packed-tower aeration is equipped with a blower. The blower forces air from the bottom of the aerator out the top. This counter-current effect provides a greater level of gas transfer than the natural draft aerator and a higher "air to water ratio." The packed tower concept is shown in Figure B-6.

To help with the air:water transfer, packed towers are "packed" with a series of small objects designed to cause the water to break apart even further than the spray droplets. Because of its design, the packed-tower aerator can achieve high levels of efficient mixing and transfer and is typically used for treating VOCs. In many cases, the off-gas containing the VOCs or hydrogen sulfide is captured for further treatment to eliminate offsite impacts.

CORROSION CONTROL

Corrosion is the deterioration of metallic or concrete structures in contact with water, usually with loss of material into solution. While many operations staff focus on external pipeline corrosion, internal corrosion creates water quality issues. Like a well, internal pipe corrosion can result from metabolic (microbial) activity, chemical dissolution, or physical abrasion by excessive fluid velocities (rare). Common things to look for include red water and tuberculation in iron, copper, and lead water pipes, valves, and appurtenances. If left unchecked, corrosion can affect both the structural capacity of the pipe and the quality of the water.

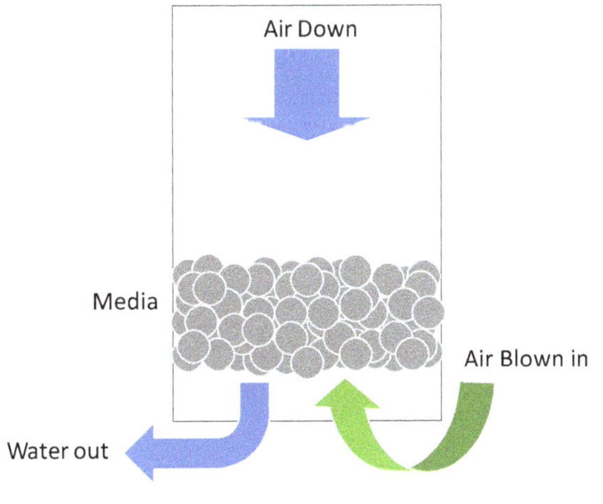

Figure B-6 Interior of how packed tower works

Water is considered chemically corrosive if:

- pH, alkalinity, hardness, silicates, and phosphates are relatively low, and/or
- dissolved oxygen, chlorine residual, total dissolved solids (or specific conductance), chlorides, sulfates, hydrogen sulfide, and carbon dioxide are relatively high.

There are several indices that are used to indicate the degree of corrosivity of water. The Langelier Saturation Index (LSI) is based on the effect of pH on the solubility of calcium carbonate. A positive LSI indicates a noncorrosive water (i.e., one that would precipitate a protective coating of calcium carbonate on pipes); a negative LSI indicates a corrosive water (i.e., one that would dissolve calcium carbonate). The calculation of the LSI is based on the alkalinity, calcium, total dissolved solids, pH, and temperature of the water. The LSI for samples from more than 20,000 groundwater sites was computed by the US Geological Survey (USGS) to identify the occurrence of potentially corrosive groundwater across the United States (Figure B-7). A second indicator, the Potential to Promote Galvanic Corrosion (PPGC), was computed for samples from about 27,000 groundwater sites. On the basis of these results, 11 states and the District of Columbia were classified as having a very high prevalence of potentially corrosive groundwater, 14 states as having a high prevalence of potentially corrosive groundwater,

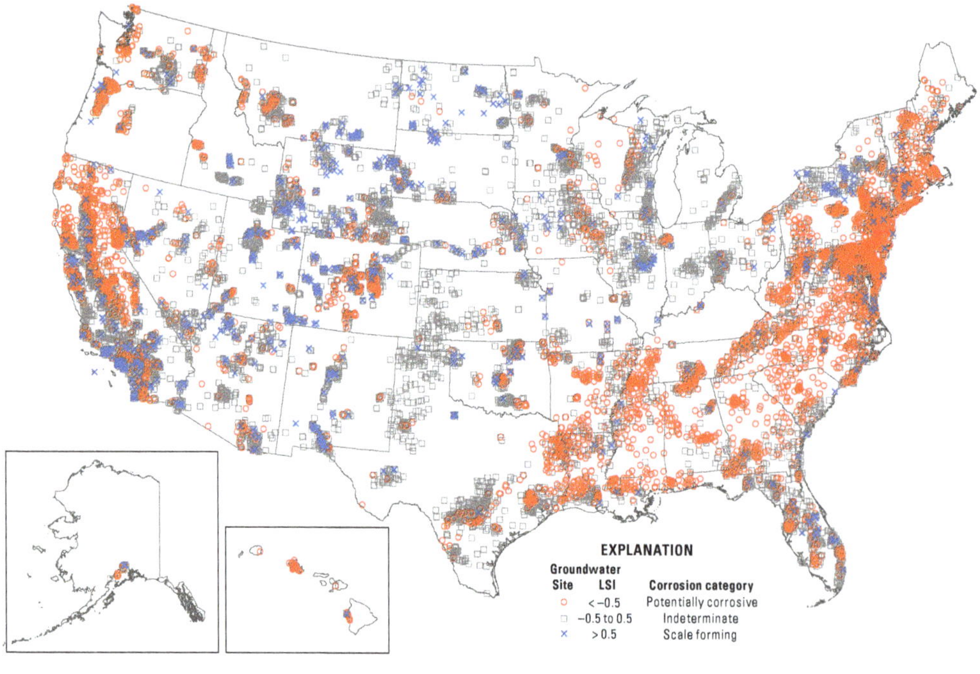

Source: Belitz et al. 2016

Figure B-7 The Langelier Saturation Index (LSI) for 20,962 groundwater sites in the United States

19 states as having a moderate prevalence of potentially corrosive groundwater, and 6 states as having a low prevalence of potentially corrosive groundwater.

If the LSI is not slightly positive, polyphosphates, zinc orthophosphates, or other corrosion inhibitors are recommended. The lead and copper testing protocols require that utilities address corrosion in the piping system—the easiest way to do it is at the plant. Figure B-8 shows the type of chemical feed systems commonly used. The amount of chemicals used is generally very small.

FLUORIDATION

A final chemical addition is fluoride. Fluoride concentrations of up to 1 mg/L in drinking water is generally considered to reduce dental decay. New recommendations suggest 0.8 mg/L is appropriate. Some groundwaters contain naturally high levels of fluoride and, in some cases, fluoride must be reduced to acceptable levels by ion exchange with activated alumina, lime softening, membranes, or coagulation.

Hydrofluosilicic acid (H_2SiF_6) is often used as an additive because it is purchased as a bulk liquid. Dry additives include sodium fluoride (NaF), calcium fluoride (CaF_2), ammonium silicofluoride (NH_4SiF_6), and sodium silicofluoride

Source: Photo by Fred Bloetscher

Figure B-8 Typical feed pumps for small volumes of chemicals

(Na_2SiF_6). The dry additives require a dry feeder (either gravimetric or volumetric) and saturation tank or must be mixed by hand. Operators should be aware that dosage is related to the available fluoride, not the dilution of the fluid, so appropriate mathematics must be employed. A small double-walled tank is typically used for hydrofluosilicic acid (Figure B-9).

METALS

Iron and Manganese Treatment Methods

Iron and manganese are common metals in groundwater (AWWA 2003). Although not considered to be health hazards, iron and manganese can impart color to water and can stain laundry and plumbing fixtures (generating "red water" complaints). Iron and manganese can precipitate in pipes and fittings and can encourage bacterial slime growths in hot water pipes, heaters, and boilers. They create red and black water concerns, respectively. There are several means to address the issue if the concentrations are small.

Sequestering. Sequestering is a strategy for controlling iron and manganese by maintaining the metals in solution. Instead of removal, the metals are made complex by reacting with a sequestering agent to prevent their oxidation and precipitation. Phosphate or silicate compounds are often used for this purpose. The success of sequestering is site specific and is generally limited to waters with a total concentration of iron and manganese below 1 mg/L. The process and equipment is similar to that described for corrosion control.

Potassium Permanganate. Potassium permanganate ($KMnO_4$) is used to oxidize organic compounds in water, primarily those that cause color. As a result, potassium permanganate is injected into the water stream before a contact or oxidation tank as a means to reduce color values through a chemical reaction. It is purple, which means small quantities must be used so as not to color the water purple. Typical doses of permanganate are 1–2 mg/L.

Oxidation and Filtration. Manganese is more difficult to oxidize than iron, and complete oxidation requires adequate oxidation contact time. A typical treatment system includes increasing the pH before filtration to reduce the solubility of the metals. Greensand continues to be a solution for the removal of iron and manganese, as the media acts like an ion-exchange system. Generally, permanganate is added as a pretreatment to greensand to oxidize the metals and regenerate the media. Chlorine may be used as a pretreatment, particularly if bacterial fouling of the filter media is a problem.

Other Metal Treatment Methods

Metals, especially arsenic, are common in groundwater. Figure B-10 shows a USGS map of arsenic concentrations in groundwater. Figure B-11 shows lead concentrations in groundwater. As lead and arsenic are potentially toxic, their removal is required before distribution of the water. A simple solution for metals

Source: Photo by Fred Bloetscher

Figure B-9 Small hydrofluosilicic acid chemical feed system (3 MGD facility)

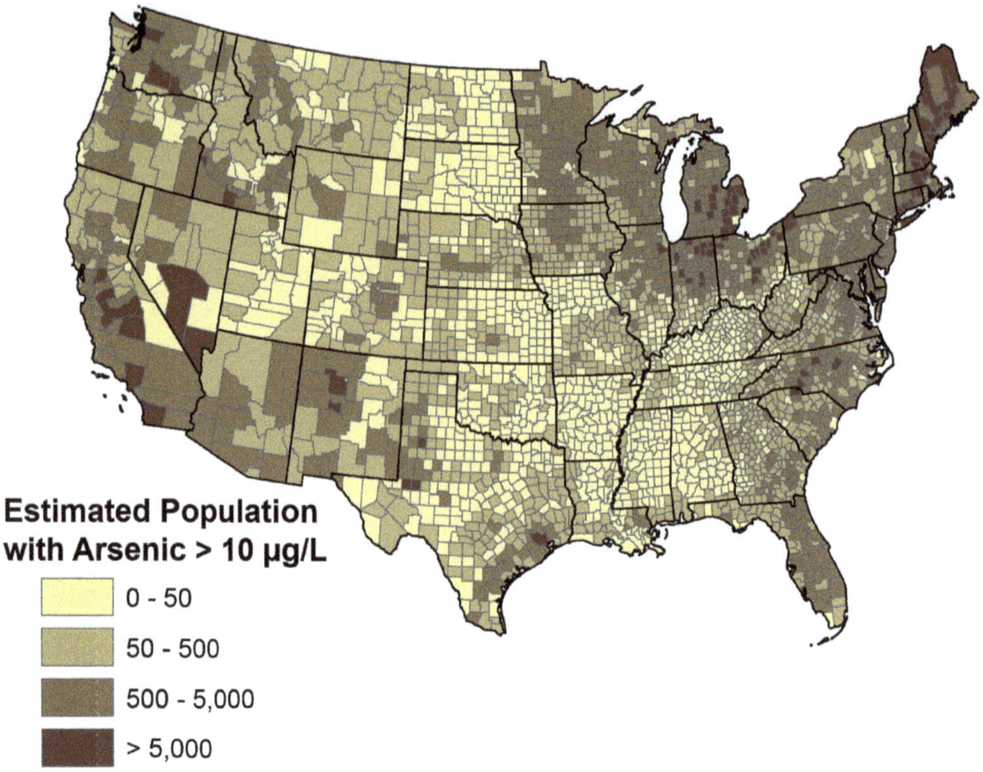

Source: USGS ND1

Figure B-10 USGS map of arsenic-laden groundwater in the United States

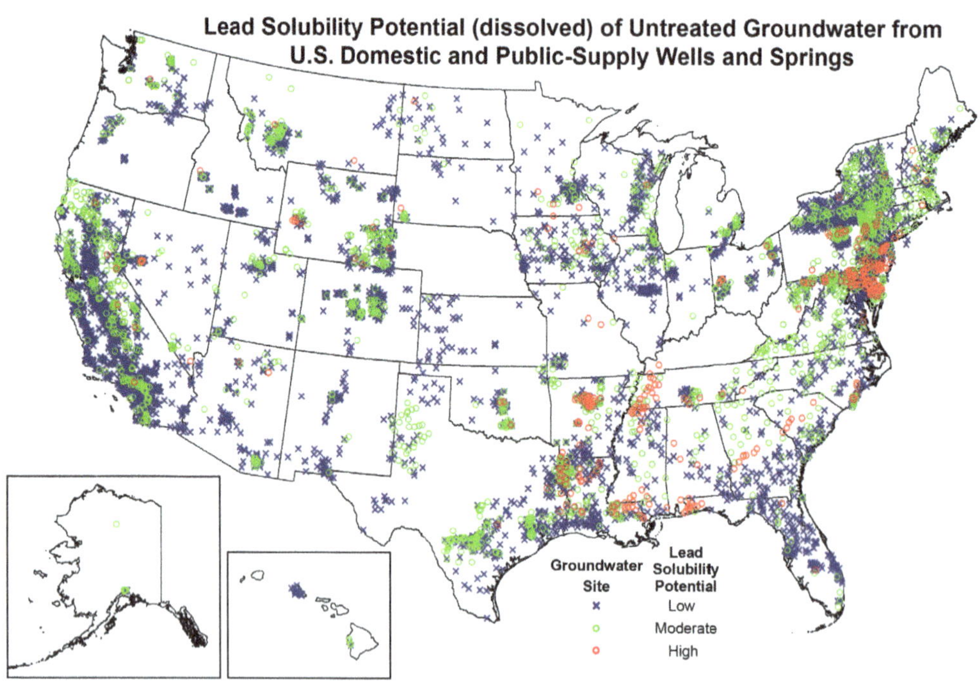

Source: USGS ND2

Figure B-11 USGS map of lead concentrations in groundwater and groundwater systems that only use chlorine for treatment in the United States.

is often ion exchange. If groundwater contains large concentrations of calcium and magnesium, lime softening or nanofiltration may be more appropriate.

ION EXCHANGE

For small systems, an option is ion exchange. Ion exchange is a chemical process that uses a catalyst in the form of resins beads to reversibly exchange undesirable ions with alternative ions—typically chlorides or sodium. The resins can be restored by washing through a strong salt solution, which results in the undesirable ions sloughing from the resin and being replaced by sodium (or chlorides). For example, ion exchange can be used to remove hardness. In the sodium-cycle ion-exchange process, calcium and magnesium are exchanged for sodium ions. The wash then removes the calcium from the resins, creates a $CaCl_2$ concentration in the concentrate, and returns the sodium ions to the resin. An example of an ion-exchange reactor tank is shown in Figure B-12.

Modern ion-exchange materials consist of a matrix or hydrocarbon network such as polystyrene, which is copolymerized with divinyl benzene (DVB). The matrix is converted to a strong-acid cation exchanger or to a strong-base anion exchanger, depending on the type of ionizable groups attached to the network. There are many variants depending on the metal that is to be removed (some

Source: Photo by Fred Bloetscher

Figure B-12 Ion-exchange tanks

Source: Photo by Fred Bloetscher

Figure B-13 Example of ion-exchange (MIEX) resins in virgin form

organics can be removed with ion exchange as well). For example, in ion-exchange softening, only strong-acid, cation-exchange resins operating in the sodium cycle are used as exchange material.

Another version of ion exchange involves magnetic ion exchange (MIEX). In MIEX processes, the water is mixed in a mixing basin to create the exchange (Figure B-13). Otherwise, the process is the same.

LIME SOFTENING TREATMENT

Hardness in drinking water increases the amount of soap needed to produce a foam or lather and causes scale in hot water pipes, heaters, and boilers. Hardness is caused by a high concentration of divalent metallic cations, mostly calcium (Ca^{2+}) and magnesium (Mg^{2+}), but also iron (Fe^{2+}), manganese (Mn^{2+}), and strontium (Sr^{2+}).

Hardness is generally high in groundwater sources because the hardness is derived largely from contact with limestone rock formations. Hard water typically originates in areas where the topsoil is thick and limestone formations are present. In other geologic formations, such as granitic-based materials, the groundwater contains less calcium and magnesium but may still contain too much iron and manganese.

The lime–soda ash process can be used at ambient or elevated temperatures. Cold lime softening is one of the oldest methods of water treatment. Lime (calcium hydroxide) is added to the water or lime in combination with soda ash (sodium carbonate). The choice of treatment depends on the composition of the water to be treated and the degree of hardness reduction desired. Examples of how the process works is described in most engineering textbooks.

However, the concept is worthy of a brief discussion since it is distinctly a groundwater treatment solution and one that is misunderstood to a degree. Lime softening relies on pH to work. The concept is to use the alkalinity diagram (Figure B-14) to move the alkalinity species from bicarbonate to carbonate, which settles. The alkalinity diagram shows that at neutral pH, bicarbonate is the predominant species. Lime is added based on the milli-equivalents of polyvalent positive ions (calcium, magnesium, iron, etc.) in the raw water. Hence, raw water quality must be known to dose properly, and pH must trend toward 10 to work effectively.

Figure B-15 shows a typical lime-softening unit (this type is called an accelator). The lime is mixed with raw water in the central mixer. It then must travel below a baffle before rising to the surface again. Because the water moves downward, and lime softening is a chemical reaction that occurs quickly, much of the calcium carbonate that settles, does so almost immediately on the down flow. Much-clearer water exits the weirs (or "launders"). While this water still needs to be filtered (Figure B-16) to remove very fine materials, the process successfully removed excessive calcium and magnesium while also removing polyvalent chemicals such as iron and manganese. The settle material is discharged from the unit as lime sludge where it is dewatered (Figure B-17). Aerators are often pre-treatment processes on lime-softening reactors as they remove hydrogen sulfide and soluble iron (Figure B-18).

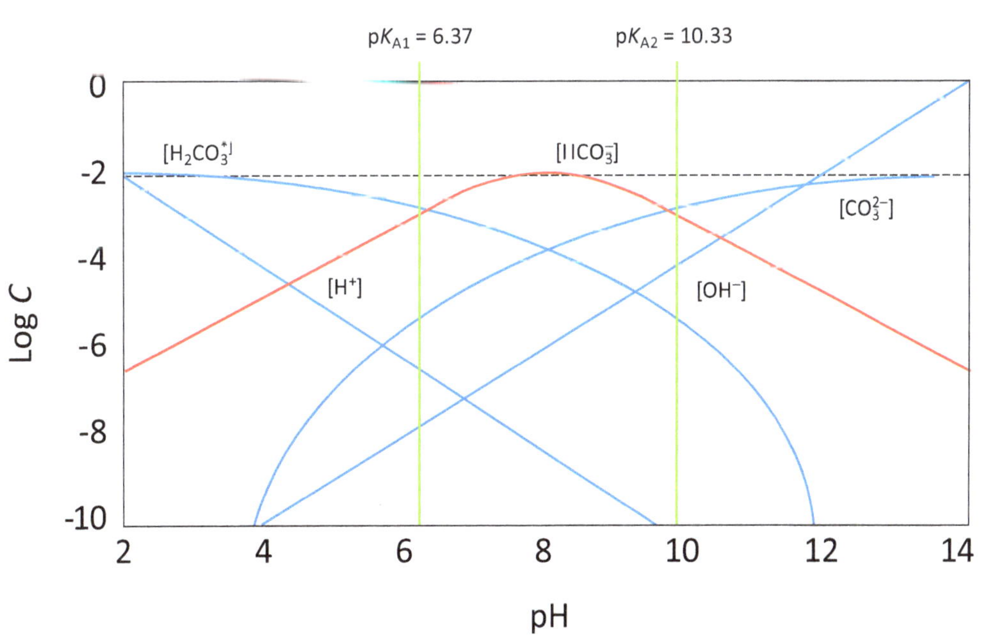

Figure B-14 Alkalinity diagram

Source: Photo by Fred Bloetscher

Figure B-15 Lime softening unit (accelator). The lime mixer is in the middle; the aerator is on the top. The clarification portion is on the outer edges.

Source: Photo by Fred Bloetscher

Figure B-16 Filter after accelator unit

Source: Photo by Fred Bloetscher

Figure B-17 Lime sludge

Source: Photo by Fred Bloetscher

Figure B-18 Aerator on lime-softening unit

MEMBRANE FILTRATION

Membranes have become increasingly important for the groundwater treatment because of their ability to remove particles as small as metal ions without coagulation. Iron, other metals, hardness, organics, salts, and other materials can be removed with membranes. Four major membrane processes are used for drinking water treatment: microfiltration, ultrafiltration, nanofiltration, and reverse osmosis. Example of a reverse osmosis or nanofiltration skid is shown in Figure B-19; they both use the same configuration of vessels—only the pump pressure and

Source: Photo by Fred Bloetscher

Figure B-19 Typical membrane plant installation

Source: Photo by Fred Bloetscher

Figure B-20 Membranes before installation in the vessels on the skid

specified membrane are different. Figure B-20 shows the membranes before they are placed in the vessels. Pressurized raw water is fed to the membrane filters to produce a filtered water stream called the "permeate" and a wastewater stream called "concentrate," which contains the concentrated contaminants rejected by the membrane.

Sand is a complicating and damaging factor for membranes. As a result, a sand separator should be installed before the membrane process (see Figure B-21). Membranes also require suspended solids to be removed prior to the membranes to avoid the fouling issue. Normally this is done with cartridge filters (Figures B-22 and B-23). In addition, membranes are subject to fouling by iron, manganese, and microbiological activity, which often requires pretreatment, filter element backwashing, and periodic cleaning of the filter elements with a mild acid solution.

One issue: They can remove virtually everything in the water. That is not desirable because water naturally wants to hold 40 mg/L of hardness; in waters that lack this concentration, water will become very corrosive. Post-treatment chemicals to add hardness and alkalinity back into the water will be needed.

Source: Photo by Fred Bloetscher

Figure B-21 Sand in the formation means a sand separator is required.

Source: Photo by Fred Bloetscher

Figure B-22 Cartridge filters are also required.

Source: Photo by Fred Bloetscher

Figure B-23 Example of a cartridge filter

Finally, the concentrated solution needs to be disposed of safely. Concentrate is the biggest barrier to membrane treatment; the concentrate is acutely toxic to aquatic organisms and often can disrupt treatment processes. This is often the barrier to membrane use because wastewater treatment plants may not be able to tolerate salt concentrations in the wastewater that results from membrane plants.

REFERENCES

AWWA (American Water Works Association). 2003. M21, *Groundwater*. Denver: AWWA.

Belitz, K., B.C. Jurgens, and T.D. Johnson. 2016. "Potential Corrosivity of Untreated Groundwater in the United States." US Geological Survey (USGS) Scientific Investigations Report 2016–5092. Reston, Va.: USGS.

Reckhow, D.A. and P.C. Singer. 1984. "The Removal of Organic Halide Precursors by Preozonation and Alum Coagulation." *Journal AWWA*. 76(4):151–157.

USGS. ND1. www.usgs.gov/media/images/estimated-population-arsenic-greater-10-micrograms-liter.

USGS. ND2. www.usgs.gov/media/images/lead-potential-solubility-untreated-groundwater.

APPENDIX C

Collapse Strength of Well Casing

This appendix is reproduced from ANSI/AWWA A100.

GENERAL

The minimum wall thickness tables for single-ply and two-ply steel well casing are provided in Table C-1. The actual wall thickness required for each well project should be based on a careful analysis of conditions, expected loads, and resultant stresses to which the casing will be subjected. Tables C-2 through C-5 may be used to assist in the determination of wall thickness requirements for several alternative well casing materials. For deep wells, these tables can also be used to assist in determination of the wall thickness required for single-ply steel well casing, and Table C-4 may be used for two-ply steel casing. Table C-5 provides similar information for polyvinylchloride (PVC) casing.

Table C-1 Minimum specifications for steel well casing (from ASTM A53)

Nominal Size (in.)	Outside Diameter (in.)	Wall Thickness (in.)	Weight Per Foot (lbs)
2	2.375	0.154	3.56
2	2.875	0.203	5.79
3	3.500	0.216	7.58
3	4.000	0.226	9.11
4	4.500	0.237	10.79
5	5.563	0.244	13.70
6	6.625	0.250	17.02
8	8.625	0.250	22.36
10	10.750	0.250	28.04

(Continued)

Table C-1 Minimum specifications for steel well casing (from ASTM A53) (*continued*)

Nominal Size (in.)	Outside Diameter (in.)	Wall Thickness (in.)	Weight Per Foot (lbs)
12	12.750	0.312	41.45
14	14.000	0.312	45.68
16	16.000	0.312	52.27
18	18.000	0.375	70.59
20	20.000	0.375	78.60

COLLAPSE STRENGTH, OR HYDRAULIC COLLAPSE PRESSURE, VALUES

The values for "collapse strength" or "hydraulic collapse pressure" set forth in Tables C-2 and C-3 were determined by the following equation:*

$$P_e^2 - P_e\left\{\frac{2S}{\left(\frac{D_o}{t} - 1\right)} + P_{cr}\left[1 + 3e\left(\frac{D_o}{t} - 1\right)\right]\right\} + \left[\frac{2SP_{cr}}{\left(\frac{D_o}{r} - 1\right)}\right] = 0 \quad \text{(Eq C-1)}$$

Where:

P_e = collapse pressure with ellipticity, psi (MPa)
S = yield strength = 35,000 psi (241.3 MPa)
D_o = casing outside diameter, in. (mm)
t = casing wall thickness, in. (mm)
P_{cr} = theoretical collapse strength of a perfectly round tube
e = casing ellipticity = 1 percent

$$= \left(\frac{2E}{1 - u^2}\right)\left[\frac{1}{\left(\frac{D_o}{t} - 1\right)}\right]^3 \quad \text{(Eq C-2)}$$

Where:

E = Young's modulus = 30×10^6 psi (206,800 MPa)
u = Poisson's ratio = 0.3

* Timoshenko, S.P. and J.M. Gere. 1961. *Theory of Elastic Stability* (p. 296, equation e). New York: McGraw Hill.

The values for "collapse strength" for two-ply steel well casing set forth in Table C-4 were determined from the following formula:†

$$P = \frac{0.65(62{,}600{,}000)}{\left(\frac{D}{T}\right)\left(\frac{D}{T} - 1\right)^2} \qquad \text{(Eq C-3)}$$

Where:

P = collapsing pressure, psi (MPa)

D = nominal diameter, in. (mm)

$$T = \sqrt{t_1^2 + t_2^2} \qquad \text{(Eq C-4)}$$

Where:

t_1 = wall thickness of inside casing (one-half casing wall thickness from Table C-4), in. (mm)

t_2 = wall thickness of outside casing (one-half casing wall thickness from Table C-4), in. (mm)

Note that Eq C-3 was empirically derived. Its validity has been established over many years by comparison with the results of test and field failure observations. Please note that this formula does not contain any safety factor. As a result, the values for collapse strength in Table C-4 are indicative of the value at which actual failure can be expected.

CASING TENSILE STRENGTH VALUES

The values for casing tensile strength set forth in Table C-3 were determined by the following equation:

$$C_{ts} = \frac{\pi t S_t (D_o - t)}{2{,}000} \qquad \text{(Eq C-5)}$$

Where:

C_{ts} = casing tensile strength, tons (kg)

t = casing wall thickness, in. (mm)

S_t = tensile strength of material, 60,000 psi (413.7 MPa)

D_o = casing outside diameter, in. (mm)

K_{uc} = coefficient for unit conversion = 2,000 for converting to tons (9.807 for converting metric units to kg)

† Cates, W.H. 1955. *Water Well Casing Manual*, Consolidated Western Steel—Division of United States Steel Corp.

CASING AXIAL COMPRESSIVE STRENGTH VALUES

The values for casing axial compressive strength for single-ply steel casing in Table C-3 were determined by the following equation:

$$C_{as} = \frac{ptS_{yp}(D_o - t)}{2{,}000} \qquad \text{(Eq C-6)}$$

Where:

C_{as} = casing axial compressive strength, tons (kg)
S_{yp} = material yield strength, 35,000 psi (241.3 MPa)
D_o = casing outside diameter, in. (mm)
K_{uc} = coefficient for unit conversion = 2,000 for converting to tons (9.807 for converting metric units to kg)

The values for axial compressive strength set forth in Table C-4 for two-ply steel casing were determined from Eq C-7 as follows:

$$P = \frac{\pi tS_{yp}(D_i - t)}{2{,}000} \qquad \text{(Eq C-7)}$$

Where:

P = compressive strength, tons (kg)
t = single-ply thickness (one-half of casing wall thickness from Table C-4), in. (mm). (assumes that load bears on inside shell only.)
S_{yp} = minimum yield strength of two-ply casing steel = 55,000 psi (379.2 MPa)
D_i = inside diameter of casing, in. (mm)
K_{uc} = coefficient for unit conversion = 2,000 for converting to tons (9.807 for converting metric units to kg)

Table C-2 Collapse strength of steel well casing

Nominal Diameter		Wall Thickness			Outside Diameter		Inside Diameter		Weight		Collapsing Strength			
in.	*(mm)*	*in.*	*in.*	*(mm)*	*in.*	*(mm)*	*in.*	*(mm)*	*lb/ft*	*(kg/m)*	*psi*	*ft water*	*(kg/cm²)*	*(m water)*
8	(203)	¼	0.250	(6.35)	8.625	(219.08)	8.125	(206.38)	22.36	(33.28)	755.54	1,745.29	(53.20)	(531.96)
8	(203)	5⁄16	0.3125	(7.94)			8.000	(203.20)	27.74	(41.29)	1,191.21	2,751.70	(83.87)	(838.72)
10	(254)	¼	0.250	(6.35)	10.750	(273.05)	10.250	(260.35)	28.04	(41.72)	461.08	1,065.10	(32.46)	(324.64)
10	(254)	5⁄16	0.3125	(7.94)			10.125	(257.18)	34.84	(51.84)	760.25	1,756.18	(53.53)	(535.28)
12	(304)	¼	0.250	(6.35)	12.750	(323.85)	12.250	(311.15)	33.38	(49.67)	306.09	707.06	(21.55)	(215.51)
12	(304)	5⁄16	0.3125	(7.94)			12.125	(307.98)	41.51	(61.78)	520.68	1,202.78	(36.66)	(366.61)
14	(355)	¼	0.250	(6.35)	14.000	(355.60)	13.500	(342.90)	36.71	(54.64)	242.43	560.02	(17.07)	(170.69)
14	(355)	5⁄16	0.3125	(7.94)			13.375	(339.73)	45.68	(67.98)	418.68	967.15	(29.48)	(294.79)
14	(355)	⅜	0.375	(9.53)			13.250	(336.55)	54.57	(81.21)	636.10	1,469.39	(44.79)	(447.87)
14	(355)	¼	0.250	(6.35)	14.500	(368.30)	14.000	(355.60)	38.05	(56.62)	221.82	512.41	(15.62)	(156.18)
14	(355)	5⁄16	0.3125	(7.94)			13.875	(352.43)	47.35	(70.47)	385.11	889.59	(27.11)	(271.15)
14	(355)	⅜	0.375	(9.53)			13.750	(349.25)	56.57	(84.19)	588.19	1,358.72	(41.41)	(414.14)
16	(406)	¼	0.250	(6.35)	16.000	(406.40)	15.500	(393.70)	42.05	(62.58)	172.25	397.90	(12.13)	(121.28)
16	(406)	5⁄16	0.3125	(7.94)			15.375	(390.53)	52.36	(77.92)	303.15	700.27	(21.34)	(213.44)
16	(406)	⅜	0.375	(9.53)			15.250	(387.35)	62.58	(93.13)	469.53	1,084.62	(33.06)	(330.59)
16	(406)	¼	0.250	(6.35)	16.625	(422.28)	16.125	(409.58)	43.72	(65.07)	155.89	360.11	(10.98)	(109.76)
16	(406)	5⁄16	0.3125	(7.94)			16.000	(406.40)	54.44	(81.02)	275.69	636.84	(19.41)	(194.11)
16	(406)	⅜	0.375	(9.53)			15.875	(403.23)	65.08	(96.85)	429.18	991.40	(30.22)	(302.18)
18	(457)	¼	0.250	(6.35)	18.000	(457.20)	17.500	(444.50)	47.39	(70.53)	126.48	292.16	(8.90)	(89.05)

(Continued)

Table C-2 Collapse strength of steel well casing (*continued*)

Nominal Diameter		Wall Thickness			Outside Diameter		Inside Diameter		Weight		Collapsing Strength			
in.	*(mm)*	*in.*	*in.*	*(mm)*	*in.*	*(mm)*	*in.*	*(mm)*	*lb/ft*	*(kg/m)*	*psi*	*ft water*	*(kg/cm²)*	*(m water)*
18	(457)	5/16	0.3125	(7.94)			17.375	(441.33)	59.03	(87.85)	225.76	521.49	(15.90)	(158.95)
18	(457)	3/8	0.375	(9.53)			17.250	(438.15)	70.59	(105.05)	354.92	819.86	(24.99)	(249.89)
18	(457)	1/4	0.250	(6.35)	18.625	(473.08)	18.125	(460.38)	49.06	(73.01)	115.51	266.84	(8.13)	(81.33)
18	(457)	5/16	0.3125	(7.94)			18.000	(457.20)	61.12	(90.96)	206.95	478.05	(14.57)	(145.71)
18	(457)	3/8	0.375	(9.53)			17.875	(454.03)	73.09	(108.77)	326.64	754.54	(23.00)	(229.98)
20	(508)	1/4	0.250	(6.35)	20.000	(508.00)	19.500	(495.30)	52.73	(78.48)	95.46	220.52	(6.72)	(67.21)
20	(508)	5/16	0.3125	(7.94)			19.375	(442.13)	65.71	(97.79)	172.25	397.90	(12.13)	(121.28)
20	(508)	3/8	0.3750	(9.53)			19.250	(488.95)	78.60	(116.97)	273.98	632.89	(19.29)	(192.90)
20	(508)	7/16	0.4375	(11.11)			19.125	(485.78)	91.41	(136.03)	399.05	921.82	(28.10)	(280.97)
20	(508)	1/4	0.2500	(6.35)	20.625	(523.88)	20.125	(511.18)	54.40	(80.96)	87.86	202.96	(6.19)	(61.86)
20	(508)	5/16	0.3125	(7.94)			20.000	(508.00)	67.79	(100.89)	159.00	367.28	(11.19)	(111.95)
20	(508)	3/8	0.3750	(9.53)			19.875	(504.83)	81.10	(120.69)	253.68	586.00	(17.86)	(178.61)
20	(508)	7/16	0.4375	(11.11)			19.750	(501.65)	94.33	(140.38)	370.69	856.31	(26.10)	(261.00)
22	(559)	1/4	0.2500	(6.35)	22.00	(558.80)	21.500	(546.10)	58.07	(86.42)	73.75	170.37	(5.19)	(51.93)
22	(559)	5/16	0.3125	(7.94)			21.375	(542.93)	72.38	(107.72)	134.22	310.05	(9.45)	(94.50)
22	(559)	3/8	0.3750	(9.53)			21.250	(539.75)	86.61	(128.89)	215.46	497.71	(15.17)	(151.70)
22	(559)	7/16	0.4375	(11.11)			21.125	(536.58)	100.75	(149.94)	316.88	732.00	(22.31)	(223.11)
22	(559)	1/4	0.2500	(6.35)	22.50	(571.50)	22.000	(558.80)	59.41	(88.41)	69.37	160.25	(4.88)	(48.84)
22	(559)	5/16	0.3125	(7.94)			21.875	(555.63)	74.05	(110.20)	126.48	292.16	(8.90)	(89.05)
22	(559)	3/8	0.3750	(9.53)			21.750	(552.45)	88.61	(131.87)	203.44	469.94	(14.32)	(143.24)

(Continued)

Table C-2 Collapse strength of steel well casing (*continued*)

Nominal Diameter		Wall Thickness			Outside Diameter		Inside Diameter		Weight		Collapsing Strength			
in.	*(mm)*	*in.*	*in.*	*(mm)*	*in.*	*(mm)*	*in.*	*(mm)*	*lb/ft*	*(kg/m)*	*psi*	*ft water*	*(kg/cm²)*	*(m water)*
22	(559)	7/16	0.4375	(11.11)			21.625	(549.28)	103.09	(153.41)	299.84	692.62	(21.11)	(211.11)
24	(610)	1/4	0.2500	(6.35)	24.00	(609.60)	23.500	(596.90)	63.41	(94.37)	58.13	134.28	(4.09)	(40.93)
24	(610)	5/16	0.3125	(7.94)			23.375	(593.73)	79.06	(117.65)	106.51	246.04	(7.50)	(74.99)
24	(610)	3/8	0.3750	(9.53)			23.250	(590.55)	94.62	(140.81)	172.25	397.90	(12.13)	(121.28)
24	(610)	7/16	0.4375	(11.11)			23.125	(587.38)	110.10	(163.84)	255.34	589.84	(17.98)	(179.78)
24	(610)	1/4	0.2500	(6.35)	24.50	(622.30)	24.000	(609.60)	64.75	(96.36)	54.92	126.88	(3.87)	(38.67)
24	(610)	5/16	0.3125	(7.94)			23.875	(606.43)	80.73	(120.14)	100.79	232.82	(7.10)	(70.96)
24	(610)	3/8	0.3750	(9.53)			23.750	(603.25)	96.62	(143.79)	163.26	377.13	(11.49)	(114.95)
24	(610)	7/16	0.4375	(11.11)			23.625	(600.08)	112.43	(167.32)	242.43	560.02	(17.07)	(170.69)
26	(660)	1/4	0.2500	(6.35)	26.00	(660.40)	25.500	(647.70)	68.75	(102.32)	46.61	107.67	(3.28)	(32.82)
26	(660)	5/16	0.3125	(7.94)			25.375	(644.53)	85.73	(127.59)	85.88	198.38	(6.05)	(60.47)
26	(660)	3/8	0.3750	(9.53)			25.250	(641.35)	102.63	(152.73)	139.73	322.78	(9.84)	(98.38)
26	(660)	7/16	0.4375	(11.11)			25.125	(638.18)	119.44	(177.75)	208.48	481.59	(14.68)	(146.79)
26	(660)	1/4	0.2500	(6.35)	26.50	(673.10)	26.000	(660.40)	70.09	(104.30)	44.21	102.13	(3.11)	(31.13)
26	(660)	5/16	0.3125	(7.94)			25.875	(657.23)	87.40	(130.07)	81.56	188.41	(5.74)	(57.43)
26	(660)	3/8	0.3750	(9.53)			25.750	(654.05)	104.63	(155.71)	132.89	306.97	(9.36)	(93.56)
26	(660)	7/16	0.4375	(11.11)			25.625	(650.88)	121.78	(181.23)	198.55	458.66	(13.98)	(139.80)
28	(711)	1/4	0.2500	(6.35)	28.00	(711.20)	27.500	(698.50)	74.09	(110.26)	37.94	87.63	(2.67)	(26.71)
28	(711)	5/16	0.3125	(7.94)			27.375	(695.33)	92.41	(137.52)	70.22	162.21	(4.94)	(49.44)
28	(711)	3/8	0.3750	(9.53)			27.250	(692.15)	110.64	(164.65)	114.83	265.25	(8.08)	(80.85)

(Continued)

Table C-2 Collapse strength of steel well casing (*continued*)

Nominal Diameter		Wall Thickness			Outside Diameter		Inside Diameter		Weight		Collapsing Strength			
in.	*(mm)*	*in.*	*in.*	*(mm)*	*in.*	*(mm)*	*in.*	*(mm)*	*lb/ft*	*(kg/m)*	*psi*	*ft water*	*(kg/cm²)*	*(m water)*
28	(711)	7/16	0.4375	(11.11)			27.125	(688.98)	128.79	(191.66)	172.25	397.90	(12.13)	(121.28)
28	(711)	1/4	0.2500	(6.35)	28.50	(723.90)	28.000	(711.20)	75.43	(112.25)	36.11	83.41	(2.54)	(25.42)
28	(711)	5/16	0.3125	(7.94)			27.875	(708.03)	94.08	(140.00)	66.91	154.55	(4.71)	(47.11)
28	(711)	3/8	0.3750	(9.53)			27.750	(704.85)	112.64	(167.63)	109.53	253.02	(7.71)	(77.12)
28	(711)	7/16	0.4375	(11.11)			27.625	(701.68)	131.12	(195.14)	164.51	380.01	(11.58)	(115.83)
30	(762)	1/4	0.2500	(6.35)	30.000	(762.00)	29.500	(749.30)	79.43	(118.21)	31.28	72.26	(2.20)	(22.02)
30	(762)	5/16	0.3125	(7.94)			29.375	(746.13)	99.08	(147.45)	58.13	134.28	(4.09)	(40.93)
30	(762)	3/8	0.3750	(9.53)			29.250	(742.95)	118.65	(176.57)	95.46	220.52	(6.72)	(67.21)
30	(762)	7/16	0.4375	(11.11)			29.125	(739.78)	138.13	(205.57)	143.85	332.20	(10.13)	(101.28)
30	(762)	1/2	0.5000	(12.70)			29.000	(736.60)	157.53	(234.44)	203.44	469.94	(14.32)	(143.24)
30	(762)	1/4	0.2500	(6.35)	30.50	(774.70)	30.000	(762.00)	80.77	(120.20)	29.86	68.99	(2.10)	(21.03)
30	(762)	5/16	0.3125	(7.94)			29.875	(758.83)	100.75	(149.94)	55.55	128.31	(3.91)	(39.11)
30	(762)	3/8	0.3750	(9.53)			29.750	(755.65)	120.65	(179.55)	91.31	210.91	(6.43)	(64.29)
30	(762)	7/16	0.4375	(11.11)			29.625	(752.48)	140.47	(209.04)	137.73	318.15	(9.70)	(96.97)
30	(762)	1/2	0.5000	(12.70)			29.500	(749.30)	160.20	(238.41)	194.99	450.43	(13.73)	(137.29)

Table C-3 Axial-compressive strength and tensile strength of steel well casing

Nominal Diameter		Wall Thickness			Outside Diameter		Inside Diameter		Axial-Compressive Strength		Tensile Strength	
in.	*(mm)*	*in.*	*in.*	*(mm)*	*in.*	*(mm)*	*in.*	*(mm)*	*tons*	*(kg)**	*tons*	*(kg)*
08	(203)	¼	0.2500	(6.35)	8.625	(219.08)	08.125	(206.38)	115.11	(104,426.03)	197.33	(179,014.76)
08	(203)	5/16	0.3125	(7.94)			08.000	(203.20)	142.81	(129,555.05)	244.82	(222,096.96)
10	(254)	¼	0.2500	(6.35)	10.750	(273.05)	10.250	(260.35)	144.32	(130,924.90)	247.40	(224,437.49)
10	(254)	5/16	0.3125	(7.94)			10.125	(257.18)	179.32	(162,676.36)	307.41	(278,877.65)
12	(304)	¼	0.2500	(6.35)	12.750	(323.85)	12.250	(311.15)	171.81	(155,863.40)	294.52	(267,184.04)
12	(304)	5/16	0.3125	(7.94)			12.125	(307.98)	213.68	(193,847.23)	366.31	(332,310.83)
14	(355)	¼	0.2500	(6.35)	14.00	(355.60)	13.500	(342.90)	188.99	(171,448.84)	323.98	(293,909.70)
14	(355)	5/16	0.3125	(7.94)			13.375	(339.73)	235.16	(213,333.55)	403.13	(365,713.37)
14	(355)	⅜	0.3750	(9.53)			13.250	(336.55)	280.90	(254,828.18)	481.55	(436,854.79)
14	(355)	¼	0.2500	(6.35)	14.50	(368.30)	14.000	(355.60)	195.86	(177,681.20)	335.76	(304,596.33)
14	(355)	5/16	0.3125	(7.94)			13.875	(352.43)	243.75	(221,126.27)	417.86	(379,076.20)
14	(355)	⅜	0.3750	(9.53)			13.750	(349.25)	291.21	(264,181.26)	499.22	(452,884.75)
16	(406)	¼	0.2500	(6.35)	16.00	(406.40)	15.500	(393.70)	216.48	(196,387.34)	371.10	(336,656.24)
16	(406)	5/16	0.3125	(7.94)			15.375	(390.53)	269.52	(244,504.42)	462.04	(419,155.62)
16	(406)	⅜	0.3750	(9.53)			15.250	(387.35)	322.14	(292,240.48)	552.23	(500,974.61)
16	(406)	¼	0.2500	(6.35)	16.625	(422.28)	16.125	(409.58)	225.07	(204,180.06)	385.83	(350,019.07)
16	(406)	5/16	0.3125	(7.94)			16.000	(406.40)	280.26	(254,247.58)	480.44	(435,847.82)
16	(406)	⅜	0.3750	(9.53)			15.875	(403.23)	335.02	(303,925.02)	574.32	(521,014.32)
18	(457)	¼	0.2500	(6.35)	18.00	(457.20)	17.500	(444.50)	243.96	(221,316.78)	418.22	(379,402.79)

(Continued)

Table C-3 Axial-compressive strength and tensile strength of steel well casing (*continued*)

Nominal Diameter		Wall Thickness			Outside Diameter		Inside Diameter		Axial-Compressive Strength		Tensile Strength	
in.	*(mm)*	*in.*	*in.*	*(mm)*	*in.*	*(mm)*	*in.*	*(mm)*	*tons*	*(kg)**	*tons*	*(kg)*
18	(457)	5/16	0.3125	(7.94)			17.375	(441.33)	303.88	(275,675.29)	520.94	(472,588.80)
18	(457)	3/8	0.3750	(9.53)			17.250	(438.15)	363.37	(329,643.70)	622.92	(565,103.49)
18	(457)	1/4	0.2500	(6.35)	18.625	(473.08)	18.125	(460.38)	252.55	(229,109.50)	432.95	(392,765.62)
18	(457)	5/16	0.3125	(7.94)			18.000	(457.20)	314.62	(285,418.45)	539.35	(489,290.07)
18	(457)	3/8	0.3750	(9.53)			17.875	(454.03)	376.25	(341,328.24)	645.01	(585,143.20)
20	(508)	1/4	0.2500	(6.35)	20.00	(508.00)	19.500	(495.30)	271.45	(246,255.29)	465.35	(422,158.40)
20	(508)	5/16	0.3125	(7.94)			19.375	(492.13)	338.24	(306,846.15)	579.84	(526,021.98)
20	(508)	3/8	0.3750	(9.53)			19.250	(488.95)	404.60	(367,046.93)	693.60	(629,223.31)
20	(508)	7/16	0.4375	(11.11)			19.125	(485.78)	470.53	(426,857.62)	806.63	(731,762.39)
20	(508)	1/4	0.2500	(6.35)	20.625	(523.88)	20.125	(511.18)	280.04	(254,048.00)	480.07	(435,512.16)
20	(508)	5/16	0.3125	(7.94)			20.000	(508.00)	348.98	(316,589.32)	598.25	(542,723.25)
20	(508)	3/8	0.3750	(9.53)			19.875	(504.83)	417.49	(378,740.54)	715.69	(649,263.02)
20	(508)	7/16	0.4375	(11.11)			19.750	(501.65)	485.57	(440,501.67)	832.40	(755,140.54)
22	(559)	1/4	0.2500	(6.35)	22.00	(558.80)	21.500	(546.10)	298.94	(271,193.79)	512.47	(464,904.94)
22	(559)	5/16	0.3125	(7.94)			21.375	(542.93)	372.60	(338,017.02)	638.75	(579,464.23)
22	(559)	3/8	0.3750	(9.53)			21.250	(539.75)	445.84	(404,459.23)	764.29	(693,352.19)
22	(559)	7/16	0.4375	(11.11)			21.125	(536.58)	518.64	(470,502.27)	889.10	(806,577.92)
22	(559)	1/4	0.2500	(6.35)	22.50	(571.50)	22.000	(558.80)	305.81	(277,426.15)	524.25	(475,591.58)
22	(559)	5/16	0.3125	(7.94)			21.875	(555.63)	381.19	(345,809.74)	653.48	(592,827.06)
22	(559)	3/8	0.3750	(9.53)			21.750	(552.45)	456.14	(413,803.23)	781.96	(709,382.15)

(Continued)

Table C-3 Axial-compressive strength and tensile strength of steel well casing (*continued*)

Nominal Diameter		Wall Thickness			Outside Diameter		Inside Diameter		Axial-Compressive Strength		Tensile Strength	
in.	*(mm)*	*in.*	*in.*	*(mm)*	*n.*	*(mm)*	*in.*	*(mm)*	*tons*	*(kg)**	*tons*	*(kg)*
22	(559)	7/16	0.4375	(11.11)			21.625	(549.28)	530.67	(481,415.70)	909.71	(825,274.99)
24	(610)	1/4	0.2500	(6.35)	24.00	(609.60)	23.500	(596.90)	326.43	(296,132.30)	559.60	(507,660.56)
24	(610)	5/16	0.3125	(7.94)			23.375	(593.73)	406.97	(369,196.96)	697.65	(632,897.41)
24	(610)	3/8	0.3750	(9.53)			23.250	(590.55)	487.07	(441,862.45)	834.98	(757,481.08)
24	(610)	7/16	0.4375	(11.11)			23.125	(587.38)	566.74	(514,137.86)	971.56	(881,384.37)
24	(610)	1/4	0.2500	(6.35)	24.50	(622.30)	24.000	(609.60)	333.30	(302,364.66)	571.38	(518,347.19)
24	(610)	5/16	0.3125	(7.94)			23.875	(606.43)	415.56	(376,989.67)	712.38	(646,260.24)
24	(610)	3/8	0.3750	(9.53)			23.750	(603.25)	497.38	(451,215.53)	852.65	(773,511.03)
24	(610)	7/16	0.4375	(11.11)			23.625	(600.08)	578.77	(525,051.29)	992.18	(900,090.52)
26	(660)	1/4	0.2500	(6.35)	26.00	(660.40)	25.500	(647.70)	353.92	(321,070.81)	606.72	(550,407.10)
26	(660)	5/16	0.3125	(7.94)			25.375	(644.53)	441.33	(400,367.82)	756.56	(686,339.66)
26	(660)	3/8	0.3750	(9.53)			25.250	(641.35)	528.30	(479,265.68)	905.66	(821,600.90)
26	(660)	7/16	0.4375	(11.11)			25.125	(638.18)	614.85	557,782.51)	1,054.03	(956,199.89)
26	(660)	1/4	0.2500	(6.35)	26.50	(673.10)	26.000	(660.40)	360.79	(327,303.17)	618.50	(561,093.74)
26	(660)	5/16	0.3125	(7.94)			25.875	(657.23)	449.92	(408,160.54)	771.29	(699,702.49)
26	(660)	3/8	0.3750	(9.53)			25.750	(654.05)	538.61	(488,618.75)	923.33	(837,630.85)
26	(660)	7/16	0.4375	(11.11)			25.625	(650.88)	626.88	(568,695.94)	1,074.65	(974,906.04)
28	(711)	1/4	0.2500	(6.35)	28.00	(711.20)	27.500	(698.50)	381.41	(346,009.32)	653.84	(593,153.64)
28	(711)	5/16	0.3125	(7.94)			27.375	(695.33)	475.69	(431,538.69)	815.46	(739,772.84)
28	(711)	3/8	0.3750	(9.53)			27.250	(692.15)	569.54	(516,677.97)	976.35	(885,729.78)

(*Continued*)

Table C-3 Axial-compressive strength and tensile strength of steel well casing (*continued*)

Nominal Diameter		Wall Thickness			Outside Diameter		Inside Diameter		Axial-Compressive Strength		Tensile Strength	
in.	*(mm)*	*in.*	*in.*	*(mm)*	*in.*	*(mm)*	*in.*	*(mm)*	*tons*	*(kg)**	*tons*	*(kg)*
28	(711)	7/16	0.4375	(11.11)			27.125	(688.98)	662.96	(601,427.17)	1,136.50	(1,031,015.41)
28	(711)	1/4	0.2500	(6.35)	28.50	(723.90)	28.000	(711.20)	388.28	(352,241.68)	665.62	(603,840.28)
28	(711)	5/16	0.3125	(7.94)			27.875	(708.03)	484.28	(439,331.41)	830.19	(753,135.67)
28	(711)	3/8	0.3750	(9.53)			27.750	(704.85)	579.84	(526,021.98)	994.02	(901,759.74)
28	(711)	7/16	0.4375	(11.11)			27.625	(701.68)	674.98	(612,331.53)	1,157.11	(1,049,712.49)
30	(762)	1/4	0.2500	(6.35)	30.000	(762.00)	29.500	(749.30)	408.90	(370,947.82)	700.97	(635,909.26)
30	(762)	5/16	0.3125	(7.94)			29.375	(746.13)	510.05	(462,709.56)	874.37	(793,215.09)
30	(762)	3/8	0.3750	(9.53)			29.250	(742.95)	610.77	(554,081.20)	1,047.03	(949,849.60)
30	(762)	7/16	0.4375	(11.11)			29.125	(739.78)	711.06	(645,062.75)	1,218.96	(1,105,821.86)
30	(762)	1/2	0.5000	(12.70)			29.000	(736.60)	810.92	(735,654.22)	1,390.15	(1,261,122.81)
30	(762)	1/4	0.2500	(6.35)	30.50	(774.70)	30.000	(762.00)	415.77	(377,180.18)	712.75	(646,595.89)
30	(762)	5/16	0.3125	(7.94)			29.875	(758.83)	518.64	(470,502.27)	889.10	(806,577.92)
30	(762)	3/8	0.3750	(9.53)			29.750	(755.65)	621.08	(563,434.27)	1,064.71	(965,888.62)
30	(762)	7/16	0.4375	(11.11)			29.625	(752.48)	723.09	(655,976.18)	1,239.58	(1,124,528.01)
30	(762)	1/2	0.5000	(12.70)			29.500	(749.30)	824.67	(748,128.01)	1,413.72	(1,282,505.15)

* Conversion factor used: 1 ton (2,000 lb) = 907.1847 kg

Table C-4 Collapse strength and axial-compression strength of two-ply steel well casing

Nom. Diam.	US Std.	Casing Wall Thickness		Outside Diameter		Inside Diameter		Weight		Collapsing Strength				Axial-Compression Strength	
in.	Gauge	*in.*	*(mm)*	*in.*	*(mm)*	*in.*	*(mm)*	*lb/ft*	*(kg/m)*	*psi*	*ft water*	*(kg/cm²)*	*m water*	*tons*	*0(kg)**
8	12	0.218	(5.54)	8.436	(214.27)	8	(203.2)	19.13	(28.47)	302.65	698.97	(21.28)	(213.05)	76.36	(69,272.62)
8	10	0.282	(7.16)	8.504	(216)	7.94	(201.68)	24.76	(36.85)	662.74	1,530.58	(46.6)	(466.53)	98.44	(89,303.26)
10	12	0.218	(5.54)	10.436	(265.07)	10	(254)	23.79	(35.4)	153.75	355.07	(10.81)	(108.23)	95.20	(86,363.98)
10	10	0.282	(7.16)	10.504	(266.8)	9.94	(252.48)	30.79	(45.82)	335.88	775.71	(23.61)	(236.44)	122.80	(111,402.28)
12	12	0.218	(5.54)	12.436	(315.87)	12	(304.8)	28.45	(42.33)	88.51	204.41	(6.22)	(62.31)	114.03	(103,446.27)
12	10	0.282	(7.16)	12.504	(317.6)	11.94	(303.28)	36.81	(54.78)	193.06	445.88	(13.57)	(135.9)	147.16	(133,501.3)
12	8	0.344	(8.74)	12.568	(319.23)	11.88	(301.75)	44.91	(66.83)	353.07	815.41	(24.82)	(248.54)	179.09	(162,467.71)
14	12	0.218	(5.54)	14.496	(368.2)	14.06	(357.12)	33.24	(49.47)	55.53	128.25	(3.9)	(39.09)	133.43	(121,045.65)
14	10	0.282	(7.16)	14.564	(369.93)	14	(355.6)	43.01	(64.01)	120.99	279.43	(8.51)	(85.17)	172.26	(156,271.64)
14	8	0.344	(8.74)	14.628	(371.55)	13.94	(354.08)	52.48	(78.1)	221.03	510.47	(15.54)	(155.59)	209.70	(190,236.63)
16	12	0.218	(5.54)	16.496	(419)	16.06	(407.92)	37.9	(56.4)	37.1	85.68	(2.61)	(26.12)	152.26	(138,127.94)
16	10	0.282	(7.16)	16.564	(420.73)	16	(406.4)	49.04	(72.98)	80.77	186.52	(5.68)	(56.85)	196.62	(178,370.66)
16	8	0.344	(8.74)	16.628	(422.35)	15.94	(404.88)	59.83	(89.03)	147.42	340.47	(10.36)	(103.78)	239.42	(217,198.16)
16	6	0.406	(10.31)	16.692	(423.98)	15.88	(403.35)	70.62	(105.09)	243.72	562.86	(17.14)	(171.56)	282.06	(255,880.52)
18	12	0.218	(5.54)	18.496	(469.8)	18.06	(458.72)	42.56	(63.33)	26	60.05	(1.83)	(18.3)	171.10	(155,219.3)
18	10	0.282	(7.16)	18.564	(471.53)	18	(457.2)	55.06	(81.94)	56.57	130.64	(3.98)	(39.82)	220.98	(200,469.68)
18	8	0.344	(8.74)	18.628	(473.15)	17.94	(455.68)	67.17	(99.97)	103.1	238.3	(7.25)	(72.64)	269.14	(244,159.69)
18	6	0.406	(10.31)	18.692	(474.78)	17.88	(454.15)	79.29	(118)	170.48	393.71	(11.99)	(120.01)	317.14	(287,704.56)
20	12	0.218	(5.54)	20.556	(522.12)	20.12	(511.05)	47.35	(70.47)	18.92	43.7	(1.33)	(13.32)	190.50	(172,818.69)

(Continued)

Table C-4 Collapse strength and axial-compression strength of two-ply steel well casing (*continued*)

Nom. Diam.	US Std.	Casing Wall Thickness		Outside Diameter		Inside Diameter		Weight		Collapsing Strength				Axial-Compression Strength	
in.	Gauge	*in.*	*(mm)*	*in.*	*(mm)*	*in.*	*(mm)*	*lb/ft*	*(kg/m)*	*psi*	*ft water*	*(kg/ cm^2)*	*m water*	*tons*	*0(kg)**
20	10	0.282	(7.16)	20.624	(523.85)	20.06	(509.52)	61.27	(91.17)	41.14	95.02	(2.89)	(28.96)	246.08	(223,240.01)
20	8	0.344	(8.74)	20.688	(525.48)	20	(508)	74.74	(111.23)	75.02	173.25	(5.27)	(52.81)	299.75	(271,883.25)
20	6	0.406	(10.31)	20.752	(527.1)	19.94	(506.48)	88.22	(131.29)	123.88	286.09	(8.71)	(87.2)	353.27	(320,481.14)
22	10	0.282	(7.16)	22.624	(574.65)	22.06	(560.32)	67.29	(100.14)	30.86	71.26	(2.17)	(21.72)	270.44	(245,339.03)
22	8	0.344	(8.74)	22.688	(576.28)	22	(558.8)	82.09	(122.16)	56.24	129.87	(3.95)	(29.59)	329.47	(298,890.14)
22	6	0.406	(10.31)	22.752	(577.9)	21.94	(557.28)	96.9	(144.19)	92.82	214.38	(6.53)	(65.34)	388.34	(352,296.11)
24	10	0.282	(7.16)	24.624	(625.45)	24.06	(611.12)	73.31	(109.1)	23.73	54.8	(1.67)	(16.7)	294.80	(267,438.05)
24	8	0.344	(8.74)	24.688	(627.08)	24	(609.6)	89.44	(133.1)	43.23	99.85	(3.07)	(30.43)	359.19	(325,851.67)
24	6	0.406	(10.31)	24.752	(628.7)	23.94	(608.08)	105.57	(157.1)	71.34	164.76	(5.02)	(50.22)	423.42	(384,120.15)
26	10	0.282	(7.16)	26.624	(676.25)	26.06	(661.92)	79.34	(118.06)	18.64	43.05	(1.31)	(13.12)	319.17	(289,546.14)
26	8	0.344	(8.74)	26.688	(677.88)	26	(660.4)	96.79	(144.03)	33.95	78.41	(2.39)	(23.9)	388.91	(352,813.2)
26	6	0.406	(10.31)	26.752	(679.5)	25.94	(658.88)	114.24	(170.1)	56.01	129.35	(3.94)	(39.43)	458.49	(415,935.11)

* Conversion factor used: 1 ton (2,000 lb) = 907.1847 kg

Table C-5 Hydraulic collapse pressure and unit weight of PVC well casing

Nom. Size, *in.*	Actual Outside Diameter, *in.*	SDR/ SCH*	Min. Wall Thickness, *in.*	Inside Diameter, *in.*	Dimension Ratio DR†	Weight in Air, *lb/100 ft*	Weight in Water, *lb/100 ft*	RHCP,‡ *psi*
2	2.375	SCH40	0.154	2.049	15.4	69	20	291
3	3.500	SCH40	0.216	3.042	16.2	143	41	250
4	4.500	SDR26	0.173	4.133	26.0	151	43	58
		SDR21	0.214	4.046	21.0	185	53	111
		SCH40	0.237	3.998	19.0	203	58	152
4.5	4.950	SDR26	0.190	4.547	26.0	182	52	58
		SCH40	0.248	4.424	20.0	235	67	130
		SDR17	0.291	4.333	17.0	273	78	215
5	5.563	SCH80	0.375	4.768	14.8	391	112	329
		SDR26	0.214	5.109	26.0	231	66	58
		SCH40	0.258	5.016	21.6	276	79	103
		SDR21	0.265	5.001	21.0	283	81	111
		SDR17	0.327	4.869	17.0	345	98	215
6	6.625	SDR32.5	0.204	6.193	32.5	264	75	29
		SDR26	0.255	6.084	26.0	327	94	58
		SCH40	0.280	6.031	23.7	358	102	77
		SDR21	0.316	5.955	21.0	402	115	111
		SDR17	0.390	5.798	17.0	489	140	215
6.9 OD	6.900	SDR27.6	0.250	6.370	27.6	335	96	48
		SDR21	0.329	6.203	21.0	435	124	111
		SDR17	0.406	6.039	17.0	531	152	215
8	8.625	SDR26	0.332	7.921	26.0	555	159	58
		SDR21	0.410	7.756	21.0	678	194	111
		SDR17	0.508	7.548	17.0	830	237	215
10	10.750	SDR26	0.413	9.874	26.0	860	246	58
		SDR21	0.511	9.667	21.0	1,054	301	111
		SDR17	0.632	9.410	17.0	1,287	368	215
12	12.750	SDR26	0.490	11.711	26.0	1,211	346	58
		SDR21	0.606	11.465	21.0	1,482	424	111
		SDR17	0.750	11.160	17.0	1,811	518	215
14	14.000	SCH40	0.437	13.074	32.0	1,195	341	30
		SDR17	0.823	12.255	17.0	2,183	624	215
16	16.000	SCH40	0.500	14.940	32.0	1,563	446	30

(Continued)

Table C-5 Hydraulic collapse pressure and unit weight of PVC well casing (*continued*)

Nom. Size, *in.*	Actual Outside Diameter, *in.*	SDR/ SCH*	Min. Wall Thick-ness, *in.*	Inside Diame-ter, *in.*	Dimen-sion Ratio DR†	Weight in Air, *lb/100 ft*	Weight in Water, *lb/100 ft*	RHCP,‡ *psi*
		SDR26	0.616	14.694	26.0	1,910	546	58
		SDR21	0.762	14.385	21.0	2,339	668	111
		SDR17	0.941	14.005	17.0	2,852	815	215
17.4 OD	17.400	SDR17	1.024	15.229	17.0	3,375	964	215

* SDR: standard dimension ratio, SCH: schedule

† DR: dimension ratio (actual OD^3 minimum wall thickness)

‡ RHCP: resistance to hydraulic collapse pressure (per ASTM F480, Appendix X.2) based on PVC cell class 12454 per ASTM D1784

APPENDIX D

Screen Length and Entrance Velocities

This appendix is reproduced from ANSI/AWWA A100.

GENERAL

The common practice historically used for sizing well-screen length and diameter was based on screen open area and entrance velocity through the screen slots. Accordingly, previous editions of ANSI/AWWA's A100 standard used this approach for determining the recommended minimum screen length. However, the recommended upper limit for the entrance velocity has varied considerably among designers. One commonly reported design recommendation limited well-screen entrance velocities to not exceed 0.1 ft/sec (0.03 m/sec). Others have advocated velocities on an order of magnitude larger or more, and the most recent edition of the A100 standard recommended an upper limit of 1.5 ft/s (0.46 m/s). As demonstrated in the Design Examples section, the broad ranges of these entrance velocities and screen open areas, where used as the sole design approach, can result in large variations in recommended well screen length. These examples are presented merely as guidelines and not as rigorous design standards.

THEORETICAL RELATIONSHIPS

The flow through well screens is quantified by the continuity equation:

$$Q = A \times V_e \qquad \text{(Eq D-1)}$$

Where:

Q = discharge of the well, L^3/T

A = open area of the screen through which flow occurs, L^2

V_e = entrance velocity through the screen open area, L/T

Note: Units are shown in general dimensions of length (L) and time (T).

The open area of the well screen through which flow occurs can be further defined:

$$A = \frac{A_p \times A_b \times \pi \times D \times L}{12} \qquad \text{(Eq D-2)}$$

Where:

- A = open area of the well screen through which flow occurs, ft^2
- A_p = percentage open area of screen as per manufacturer, decimal fraction
- A_b = percent blockage of well-screen open area, decimal fraction
- D = outside diameter of well screen, in.
- L = design length of well screen, ft

The design length of the well screen can be further defined:

$$L = \frac{(12 \times Q)}{(\pi \times D \times q_{uc})} \qquad \text{(Eq D-3)}$$

Where:

- Q = discharge of the well [gpm]
- q_{uc} = unit capacity of the face of the well screen $60 \times 7.48 \times A_p \times A_b \times V_e$, gpm/ft^2
- V_e = screen entrance velocity, ft/s

Combining the above two equations:

$$L = Q / \left(117.5 \times D \times A_p \times A_b \times V_e\right) \qquad \text{(Eq D-4)}$$

DESIGN EXAMPLES

Consider the variation of minimum screen length for an 18-in. (0.45-m) diameter well producing 2,000 gpm (126 L/s), using a common range of screen open area and entrance velocity.

Example 1: 40 percent open area and entrance velocity = 0.1 ft/s

Design discharge rate of well (Q), gpm:	2,000
OD of well screen (D), in.:	18
Percentage open area of well screen (A_p):	0.40
Percentage blockage of well screen (A_b):	0.50
Unit capacity of screen face (q_{uc}), gpm/ft^2:	9.0
Screen entrance velocity (V_e), ft/s:	0.1
Design length of well screen (L), ft:	47

Example 2: 40 percent open area and entrance velocity = 1.5 ft/s

Design discharge rate of well (*Q*), gpm:	2,000
OD of well screen (*D*), in.:	18
Percentage open area of well screen (A_p):	0.40
Percentage blockage of well screen (A_b):	0.50
Unit capacity of screen face (q_{uc}), gpm/ft²:	135
Screen entrance velocity (V_e), ft/s:	1.5
Design length of well screen (*L*), ft:	3

Example 3: 3 percent open area and entrance velocity = 0.1 ft/s

Design discharge rate of well (*Q*), gpm:	2,000
OD of well screen (*D*), in.:	18
Percentage open area of well screen (A_p):	0.03
Percentage blockage of well screen (A_b):	0.50
Unit capacity of screen face (q_{uc}), gpm/ft²:	0.7
Screen entrance velocity (V_e), ft/s:	0.1
Design length of well screen (*L*), ft:	630

Example 4: 3 percent open area and entrance velocity = 1.5 ft/s

Design discharge rate of well (*Q*), gpm:	2,000
OD of well screen (*D*), in.:	18
Percentage open area of well screen (A_p):	0.03
Percentage blockage of well screen (A_b):	0.50
Unit capacity of screen face (q_{uc}), gpm/ft²:	10.0
Screen entrance velocity (V_e), ft/s:	1.5
Design length of well screen (*L*), ft:	42

As can be seen in the previous examples, the design length of well screen may vary widely, depending on the percentage of open area of the screen slots and the selected design screen entrance velocity. Selection of the appropriate screen length and position would undoubtedly need to consider aquifer thickness and stratigraphic layering, in conjunction with the calculations. However, two of the calculated screen lengths, 3 ft and 630 ft (in Examples 2 and 3), are relatively extreme and would likely not be used in common practice.

The other two lengths, 47 ft and 42 ft, agree reasonably well despite substantial differences in screen open area and entrance velocity. In both of these latter cases, the unit capacity of the screen surface area is relatively similar, 9 and 10 gpm/ft², respectively, regardless of screen open area or entrance velocity. This relative agreement corresponds to a similar flow velocity (specific discharge) through the filter and aquifer surrounding the well, as flow approaches the screen face. To the extent that this specific discharge (q_{uc}) can be derived with respect to aquifer permeability, hydraulic gradient, Reynolds number (tendency for turbulence), or other design considerations, supplemental criteria for well design can be established as part of the design process.

REFERENCES

American National Standards Institute (ANSI)/American Water Works Association (AWWA). 2006. A100-06. Water Wells. Denver: AWWA

APPENDIX E
Well Logging Methods

(reproduced and edited from Chapter 2 of M21 [3rd ed.])

FIELD METHODS FOR LOCATING SUITABLE GROUNDWATER SUPPLIES

The goal of a groundwater exploration exercise is to locate productive aquifers that yield sustainable, high-quality water. The amount of production available in an aquifer and the water quality (and therefore the treatment required) will guide exploration efforts, but effort will generally focus on locating large deposits of the following:

- unconsolidated sands and gravels of alluvial or glacial origin,
- sandstones and conglomerates,
- limestones and dolomites, and
- porous or fractured volcanic rocks.

Many other types of rock can yield small quantities of water to meet the domestic needs of a single household on its own well but will not be sufficient to provide public water supply needs for a community.

In many regions, general cross-sections and hydrogeologic interpretations already exist, along with well logs and production information. In these cases, favorable aquifers will have already been identified, and exploration efforts can fill in the details of the local hydrogeologic environment. Surface geophysical methods, borehole geophysical methods, aquifer testing, exploratory drilling, and lithologic logging are the most common methods.

Geophysical Methods without Drilling

Surface geophysical methods, principally electrical resistivity and seismic reflection and refraction, can be used to provide a more complete picture of subsurface structure—given some prior knowledge obtained from surface geology and borehole logs. A general model of the subsurface geology should be developed to provide the basis for a proper interpretation of the surface geophysical data. The information generated can be used to locate sites for further investigation with test drilling. In circumstances of well-defined subsurface geology and well-documented groundwater movement, contaminant plumes may be projected using geophysical methods of measurement.

Successful application of any surface geophysical method depends on the presence of sharp changes in the physical properties. The detectable physical properties provide indirect estimates of the hydrogeologic properties, with the accuracy of such estimates depending on how closely these physical properties relate to each other. In addition, the structural configurations amenable to investigation must be relatively simple to separate from one another. A comprehensive reference on the use of surface geophysical methods for groundwater investigations is given by Zohdy, Eaton, and Mabey (1974).

Note that there are land parcels that may not lend themselves to surface geophysical methods. These include areas consisting of large cobbles in alluvium, areas of severely distressed geology, or areas in which high hydraulic energy was dissipated (in geologic time scale).

Electrical resistivity. Electrical resistivity is probably the most commonly used surface geophysical method for groundwater investigations. It is economical to apply, and it produces useful information.

In the direct-current resistivity method, electrodes placed into the ground transmit current through the earth, and voltage potential is measured between two points near the center of the generated field. Figure E-1 schematically illustrates the most common electrode arrangements: the Schlumberger array and the Wenner array. Sets of resistivity readings can be gathered along the transect with constant electrode spacing (horizontal profiling) or at one location with expanding electrode spacing (electrical sounding). The method will show apparent resistivities of materials at roughly the same depth along the transect, while the second method produces a depth profile of resistivity.

Electrical resistivity is strongly affected by water content, thus interpretations involving the unsaturated zone are quite difficult because of the undefined distribution of moisture. In the saturated zone, resistivity is largely determined by the rock-matrix density and porosity or by the saturating-fluid salinity (electrical conductivity).

Holding other factors constant, coarse sediments with low clay content will have higher resistivity than fine-grained sediments. This difference makes possible the mapping of buried stream channels or the depth profiling of shale-sandstone sequences. Such changes in mineral quality are detectable because of the relation between dissolved solids and electrical conductivity. Resistivity values for earth materials range over more than 16 orders of magnitude, and resistivity surveys can provide useful results in most environments with simple geologic structure and distinct resistivity contrasts.

Electrical resistivity may be used in existing or new well fields, which require evaluation of suspect chemical migration from a polluting source. Plumes of discharged chemicals commonly have relatively high dissolved-solids content.

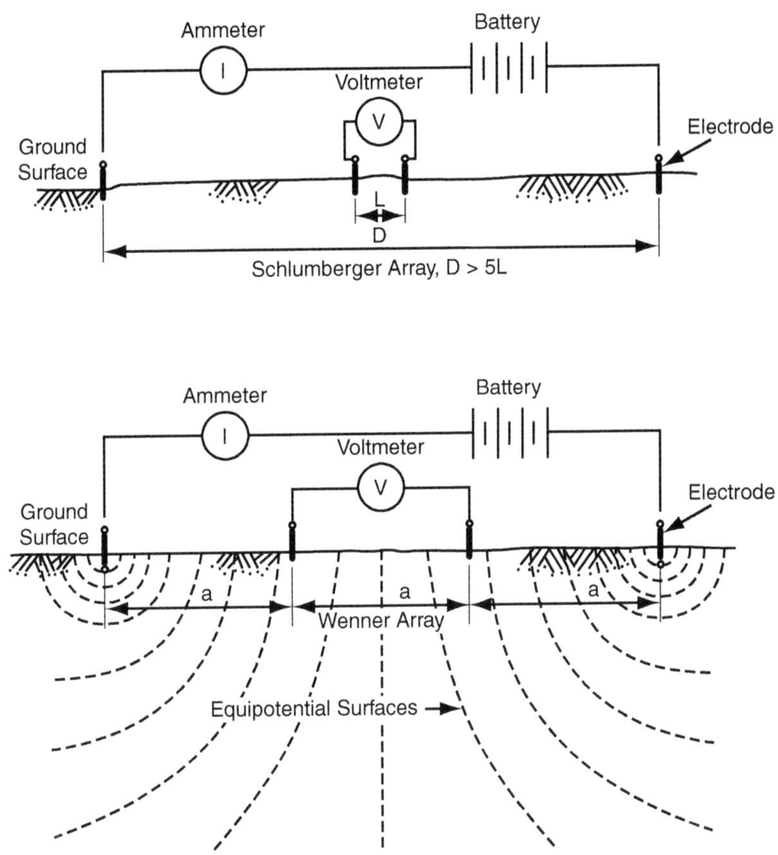

Source: AWWA 2003

Figure E-1 Schlumberger and Wenner electrode arrays for measuring earth resistivity

The boundaries of such plumes may be mapped with resistivity surveys, under the proper conditions.

The Wenner array (see Figure E-1) provides for the simplest interpretation of this method. The apparent resistivity R_a is given by:

$$R_a = 2\pi a\, V/I \qquad \text{(Eq E-1)}$$

Where:

a = the electrode spacing
V = voltage
I = direct current

The apparent resistivity characterizes a volume of earth extending below the electrode array to some effective depth of penetration. The depth of penetration is related to the electrode spacing. For the Wenner array, the effective depth of penetration is commonly assumed to be equal to a. However, this assumption can lead to serious errors in calculated depths. The depths to horizontal boundaries are best determined by matching the theoretical apparent resistivity and electrode spacing curves of various model conditions to the curves obtained from field measurements. The Schlumberger array provides better resolution of subsurface features, but it is slightly more difficult to analyze. Interpretations should be performed by a trained analyst familiar with local conditions.

The resistivity method is generally limited to use in simple geologic environments, with two or three distinct layers, and where depth of penetration is limited to about 1,500 ft (460 m). Best results are obtained when the depth to groundwater is small because of the complications of unsaturated materials. Also, the method is less effective in urban areas because of the presence of buried metal pipes, wires, and similar obstructions, which dominate measurements with unwanted noise.

Other surface geophysical methods fall within the electrical methods category, including telluric, magneto-telluric, electromagnetic, and induced polarization methods. However, their use is not generally applicable to groundwater supply investigations. Electromagnetic methods have recently become popular for shallow groundwater investigations, especially those involving groundwater contamination. The results are similar to those derived from the direct-current resistivity method, although resolution is poorer, and exploration is generally limited to the upper 180 ft (55 m). The method has received attention because it involves no direct contact with electrodes, making it quick and easy to apply in the field. The other methods mentioned have some application in specialized research and are discussed by Zohdy, Eaton, and Mabey (1974).

Seismic refraction and reflection. Seismic methods are perhaps the most useful geophysical tools for hydrogeologic investigations, although costs are relatively high. Seismic methods use contrasts in the velocities of elastic wave propagation between different earth materials. For example, unconsolidated sands and gravels exhibit low propagation velocities, whereas crystalline rocks exhibit the highest propagation velocities. Propagation velocities are higher in saturated materials, providing for detection of the water table.

Elastic waves are commonly initiated with the use of explosive "shots" in shallow borings, although truck-mounted hydraulic earth vibrators (thumpers) are becoming widespread. Lines of geophones are laid out on the ground surface to detect waves refracted or reflected from various subsurface discontinuities to measure travel time. Travel time records can then be analyzed to produce a picture of the subsurface. As with all surface geophysical methods, the interpretation of seismic data requires an assumed model of subsurface structure; the more preexisting information from surface geologic data and borehole logs that is available, the more reliable the results from seismic surveying will be.

Reflection and refraction are two methods of seismic exploration. Seismic reflection is preferred for petroleum exploration and is the most widely used method. The cost and complexity of equipment and analyses required to apply the reflection method is greater than that required for the refraction method. However, for deep exploration in multilayered environments, the reflection method is generally superior.

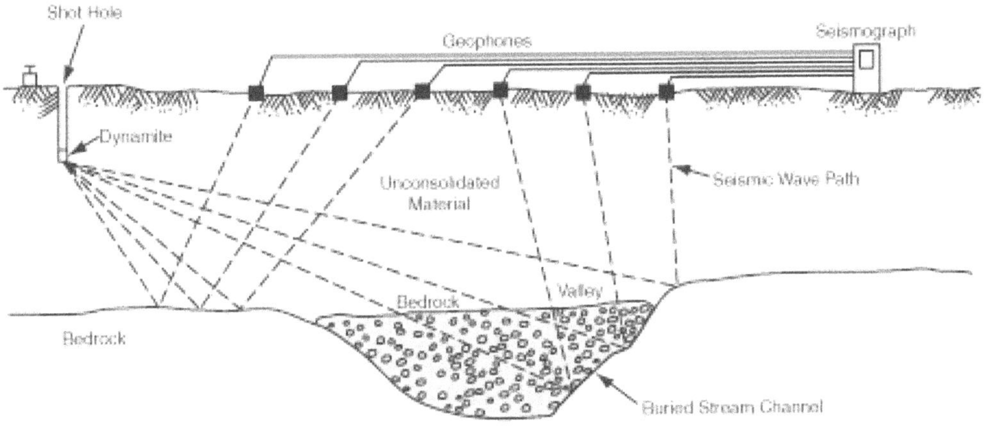

Source: AWWA 2003

Figure E-2 Application of seismic refraction method for reconnaissance mapping

Seismic refraction is the only method that has been commonly applied in groundwater investigations. Besides cost considerations, this method can have advantages in environments where deep alluvial or glacial fill exists. Less knowledge is required to apply refraction seismology, and good results can usually be achieved in most groundwater investigations. The most severe limitation of seismic refraction is that return signals can only be obtained as long as each successively deep layer has a higher propagation velocity than the overlying layer. In areas of scarce water, where water wells are drilled to great depths (e.g., parts of the arid southwestern United States), this limitation may prove too restrictive, and the higher cost of seismic reflection may be warranted.

Figure E-2 shows the use of the seismic refraction method for reconnaissance mapping of the depth to bedrock and the location of a buried stream channel. Seismic refraction can be used to determine the thickness of surficial fracture zones in crystalline rock and to map the depth and thickness of subsurface layers, up to at least the equivalent of three layers. As in the resistivity method, the analyses become very difficult and the results less reliable for more than three layers.

If the geologic structure is simple, depths of a few thousand feet (600–700 m) can be explored with seismic refraction, given a sufficient explosive and deep shot. Seismic reflection can be used to gather information from more than 10,000 ft (3,000 m) deep. For small area applications, where only very shallow materials need to be explored, a sledgehammer struck on metal plate at ground surface might be sufficient to explore the first 100 ft (30 m).

Other methods. Other surface geophysical methods can be useful in groundwater investigations, although none have been widely used because of the low benefit-to-cost ratio associated with their use. Some relatively new methods, such as ground-penetrating radar, have proven useful in groundwater contamination studies; however, the depths of penetration may be too shallow for general use in water supply applications. Gravity methods and magnetic methods have been used for general hydrogeologic investigations, although they must be viewed as supplementary methods to be applied in situations where maximum information is desired and cost is not a limiting factor. These methods are discussed by Zohdy, Eaton, and Mabey (1974).

Borehole Geophysical Logging

Borehole geophysical logging has become a standard tool in groundwater exploration. The technology of borehole logging is quite involved, and experts using special equipment are needed to perform the logging and interpretation. The discussion here introduces the capabilities and limitations of the logging tools found to be most useful in groundwater exploration. There are many texts on borehole logging, but the reader should be cautioned that most of these texts present information tailored for use in petroleum exploration. The presence of low or nonsaline water in a formation mandates the use of special analyses for interpretation. For a detailed discussion of this subject, the reader is referred to Keys and MacCary (1971) or to Keys (1989).

Many borehole logging techniques are available, although electrical, naturally occurring gamma radiation, and caliper measurements are the most widely used. All geophysical log measurements are obtained by lowering a probe down the borehole and recording continuous measurements with depth. Logging is frequently performed during drilling operations, and quick analyses of the logs by qualified personnel provide the basis for decisions regarding well completion, including depth of casing and screened intervals. Some types of logs must be performed in an uncased well, while other logging can be done in cased wells, allowing data collection from existing wells.

Multiple logs in a single well will provide confidence in interpretations. Each type of log measures different physical properties, and combined analysis may resolve ambiguities that might exist from a single log. The greater the number of wells logged in an area, the greater the statistical confidence in the data and interpretations as being representative of the subsurface environment.

Determining the number of wells and the types of logs to be used in an investigation is often difficult. Most groundwater investigations obtain adequate information using caliper, resistivity, spontaneous potential, natural gamma, and lithologic logging. The cost of these techniques should be evaluated in the context of the time available, accuracy needed, and the basic purpose of the survey.

Caliper logging. Caliper logging is used to measure the diameter of the borehole. A probe that usually has either three or four levered arms is brought up through the hole, while a record of the depth and degree of extension of the arms is made at the surface. Four armed probes can be used to measure the diameter of the hole in two directions and thus permit the evaluation of the asymmetry of the borehole. Caliper logs are of importance in groundwater exploration and well construction since the interpretation of other borehole geophysical logs requires data on the borehole diameter. Caliper logs can provide indications of the presence of high-permeability fractured or cavernous zones, as well as the occurrence of swelling clays and well-lithified layers in friable or unconsolidated rock or sediment. Caliper logs are commonly run before the cementing of well casings to determine the volume of the annular space and to locate uniform diameter areas of boreholes in which to set cement baskets and packers.

Electrical resistivity logging. Electrical resistivity logging measures the apparent resistivity of the formations in a borehole. According to Ohm's law, resistance (ohms) is equal to potential (volts) divided by current (amperes). Electrical resistivity logs use Ohm's law to determine the apparent resistance of the formations in the borehole. A current is applied between a downhole electrode and an electrode at the surface. The potential (voltage) is then measured between a downhole electrode and a surface electrode or between two or more downhole electrodes. Variations in the measured resistance are related to the composition of the rocks and sediments and the salinity of pore waters. Silt, clay, and shale tend to have the lowest resistivities; sands, sandstones, and limestones with nonsaline pore waters have the highest resistivities.

The simplest and least-expensive electric resistivity log is the single point resistance log, illustrated in Figure E-3. The single point resistivity log measures the potential drop between a surface and downhole electrode, which are also the current electrodes. The single point resistivity log is used primarily for geological

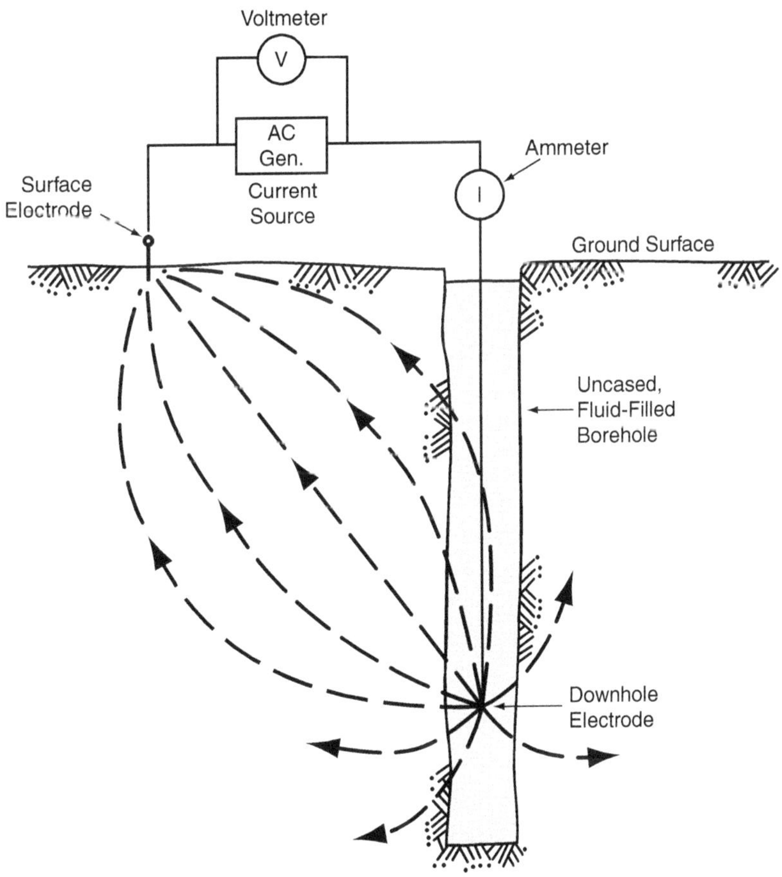

Source: AWWA 2003

Figure E-3 Single-point electrode arrangement for resistance and spontaneous potential logging

correlation and the location of bed boundaries, changes in lithology (rock characteristics), and the fracture zones. Single point resistivity logs have a very good vertical resolution of lithologic changes, but they do not provide quantitative data on formation porosity or pore water salinity.

Separate current electrode and potential electrode(s) are located on the downhole probe in normal resistivity logs. Normal resistivity logs measure the apparent resistivity of a volume of the formation perpendicular to the borehole electrodes. The size of the volume of investigation is proportional to the electrode spacing. The probes are commonly configured so that short normal (16-in. electrode spacing) and long normal (64-in. spacing) resistivities are measured simultaneously. Normal resistivity logs are commonly used in groundwater investigations as a source of qualitative information on water quality. True formation resistivity and salinity can be calculated from the measured apparent resistivities, but the calculations require the application of several correction factors, whose values are estimates. Other resistivity log types that are less-commonly used in groundwater investigations are discussed by Keys (1989).

Spontaneous potential logging. Spontaneous potential (SP) logging was the first type of downhole geophysical log. Spontaneous potentials are the naturally occurring electrical potentials that develop at the contacts between beds of different types of geological materials, such as between shale and sandstone beds. The SP logging apparatus consists of a surface and downhole electrode connected to a voltmeter. The SP logging equipment is usually incorporated into the electrical resistivity log apparatus. SP logs provide information regarding geologic correlation, bed-thickness determination, and changes in lithology. Log definition depends on the contrast in fluid conductivity between the borehole and the geologic formation penetrated.

Gamma logs. Gamma logs record naturally occurring gamma radiation emitted by the earth materials. The most significant natural resource of gamma radiation is the decay of potassium-40 isotope and the daughter products of the uranium and thorium decay series. Rocks and sediments with relatively high concentrations of potassium, uranium, and thorium give high gamma ray counts. Clay-rich rocks, such as shales and phosphatic rocks in particular, give high gamma ray counts, whereas nonphosphatic limestones and dolostones and clean quartz sandstones tend to give low gamma ray counts. The gamma log is by far the most commonly used nuclear log and is very valuable for geological correlation and lithologic determination. (It contains no nuclear material but measures background gamma radiation in the formation.) The gamma log is useful because it can be run in cased wells.

Gamma-gamma logs. Gamma-gamma logs (also called radioactive tracer logs or surveys) are obtained by introducing a gamma radiation-emitting material into the borehole (usually cesium-137 or cobalt-60) and measuring the intensity of the backscattered radiation. The backscattered radiation detected by the probe is proportional to the density of electrons in the formation, which is in turn correlated with the bulk density of the rock. Gamma-gamma logs provide information on lithology and porosity. This log can also be used to locate cavities in the cement outside of a well casing. Gamma-gamma logs are widely used in the oil industry, but they are not often used in groundwater investigations because of their high costs and liabilities associated with the potential loss or rupturing of the radioactive source within the aquifer.

Neutron logging. Neutron logs are obtained by introducing a radioactive source that emits neutrons into the borehole. The emitted neutrons interact with hydrogen atoms and release gamma radiation, which is measured by a detector on the logging probe. The intensity of the measured gamma radiation is proportional to the hydrogen atom concentration and thus water content and porosity of saturated rocks. Neutron logs are widely used in the oil industry to determine the porosity and water content of formations and lithology. As is the case with gamma-gamma logs, neutron logs are not often used in groundwater investigations because of their high costs and liabilities associated with the potential loss or rupturing of the radioactive source within the aquifer.

Acoustic logging. Acoustic or sonic logging involves the recording of transit time of acoustic pulses radiated from a probe in a borehole to one or more receivers also located on the probe. Transit times are related to matrix mineralogy and the porosity of the rock. Transit times decrease and acoustic velocities increase in sedimentary rocks with increasing hardness and cementation. Most rock types have a limited range of travel times, which allows for acoustic logs to be used to determine lithology. Porosity and fracturing can also be approximately determined from acoustic logs. The cement bond log is a type of acoustic log that is used to determine how well a casing has been cemented to the formation.

Fluid logs. Fluid logs include temperature, fluid resistance, and flow meter logs. The probe used for temperature logs usually contains a glass bead thermistor. Temperature logs can be used to identify the boundaries of aquifer zones in boreholes because the normal geothermal gradient will vary as water flows through permeable zones. Temperature logs can also be used to detect inter-aquifer flow either up the borehole or down the borehole, depending on the differential head pressure that is present. The presence of cement grout in the annular space of a well can be determined by running a temperature log within 24 hours of grouting because the heat of hydration of the cement raises the fluid temperature in the casing in cemented areas.

The fluid resistivity (or conductivity) probe contains two internal electrodes that measure the capacity of the borehole fluid to conduct electricity. The conductivity of a fluid increases with increasing salinity and temperature. A salinity-versus-depth profile can be constructed for an open borehole using the results of a temperature and fluid resistivity log. The calculated salinity-versus-depth profile is subject to considerable error because of the flow within the well.

Flow-meter logs are used to measure flow velocity within the wells. The most common construction is the impeller type; the rate of rotation of the impeller is proportional to the relative flow velocity of the probe. The relative flow velocity includes the actual flow velocity of water in the well and the rate at which the probe is being raised or lowered into the well. Caliper logs must be run with the flow-meter log because the flow velocity is a function of the cross-sectional area of the borehole. The relative contribution of individual aquifer zones to the total flow from a well can be calculated using data from flow-meter and caliper logs.

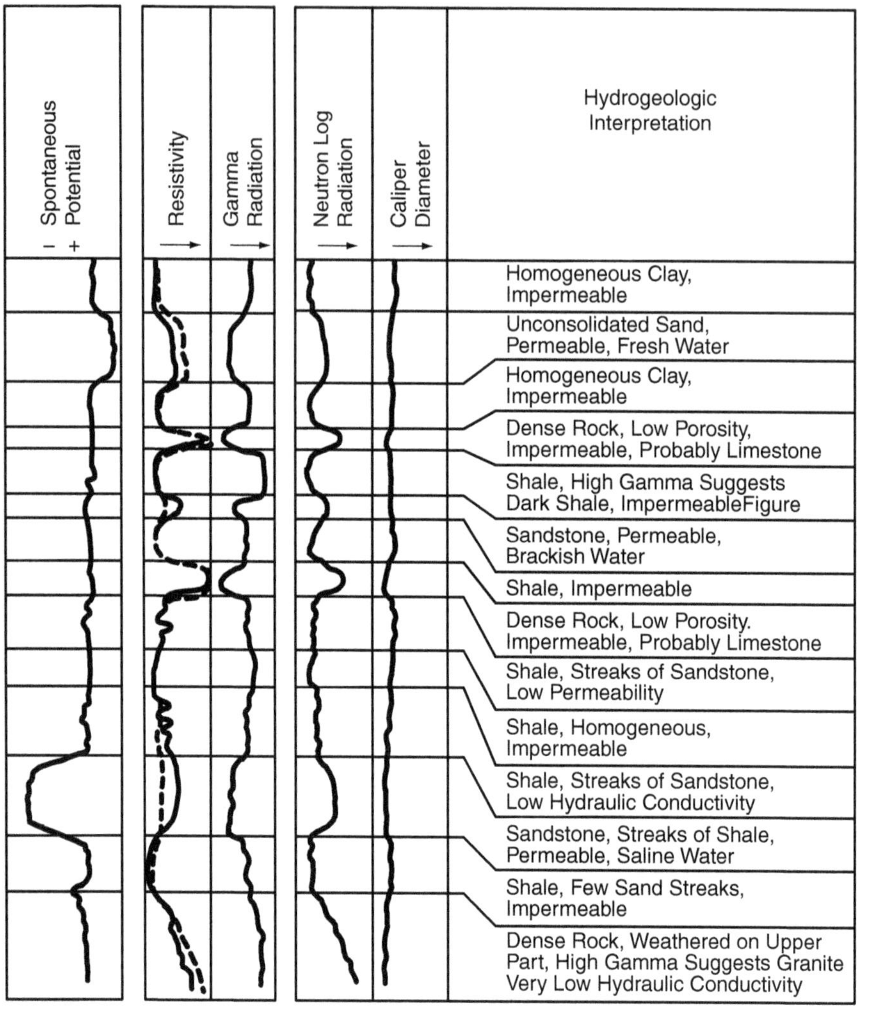

Source: AWWA 2003

Figure E-4 Qualitative interpretation of a suite of geophysical logs

Log suites. Individual borehole geophysical logs by themselves do not (in most instances) provide unequivocal lithological information. A high gamma ray count, for example, could be produced by a shale bed or phosphatic limestone layer. By running a suite of logs, more accurate qualitative and quantitative information about formation porosity, hydraulic conductivity, bulk grain density, and fluid conductivity could be extracted. Figure E-4 depicts a qualitative interpretation of a suite of geophysical logs. Shale layers, for example, are identifiable by a low resistivity and high gamma log response.

Hydrophysical logging. Hydrophysical logging is a new form of borehole logging that introduces demineralized water into the borehole and continuously logs the dispersion of the demineralized water with respect to time. This allows a more accurate measurement of the variations in the formation's hydraulic conductivity.

Aquifer Testing

Aquifer testing is very useful during exploration of test holes, and many effective methods are available for performing such testing. In the simplest form, a record of flow rates of the water produced at different depths while drilling with air is a form of aquifer testing, yielding valuable information.

Exploratory Drilling

In some instances, existing wells may not be located in a potential water supply aquifer and drilling one or more exploratory wells into the aquifer is necessary. Exploratory drilling is performed to determine an aquifer's characteristics, including hydraulic conductivity, water quality, thickness, and areal extent. The newly drilled wells can also be used for borehole geophysical logging and aquifer testing as described previously. The number and spacing of the wells drilled will depend on the size of the aquifer and the amount of water supply development planned.

Lithologic Logging

Accurate lithologic logging during drilling is crucial, and a hydrogeologist experienced in well logging should be employed. Lithologic logs and drillers' logs, including the drilling rate, are used with geophysical logs to obtain the most information about the groundwater. Geophysical logs can more accurately place the depth of discontinuities than a lithologic log.

Application

There is no established order in which exploration methods should be applied; a balanced program of appropriate combinations will produce the most information. However, the knowledge gained through a geophysical investigation only extends the information gained from test drilling and adequate sampling.

METHODS FOR MONITORING GROUNDWATER QUALITY

Before developing a groundwater supply, the water quality must be currently acceptable and expected to remain so in the foreseeable future. This initial assessment can provide a basis for legal action against anyone who contaminates the water supply after development. After the initial water quality assessment is performed and groundwater development is assured, a system for monitoring water quality should be maintained, and a reassessment of up gradient contamination risks should be performed periodically.

Monitoring Wells

Where contamination risks are high, sentinel monitoring wells should be installed. These wells, located at various depths, will provide definition for the initial groundwater assessment. Sentinel wells also serve as an early-warning system to detect changes in water quality and water elevations before they affect the water supply wells. In recent years, equipment for sampling from monitoring wells has become widely available. Small submersible or portable pumps can be installed into well casings as small as 1 in. (25 mm) in diameter, although it is common for monitoring wells to be at least 4 in. in diameter and made with polyvinyl chloride (PVC) casings.

The number of wells needed and their locations, depths of completion, and construction details must be specified as part of an integrated plan. The plan must account for likely sources of contamination, local hydrogeology, and the hydraulic effects of the proposed groundwater development. For example, what was previously considered down gradient from the well can become up gradient either after pumping begins or as influenced by nearby surface water. These changes should be simulated with computer modeling to aid in designing a monitoring-well network.

The initial assessment may indicate that one monitoring well is sufficient to begin with, but an increased contamination threat in future years (e.g., because of local growth and development) could indicate a need for additional monitoring wells. This potential for development demonstrates the importance of continual evaluation of changes that might affect the groundwater supply. One suggestion is to reverse the potential pollution-site-monitoring requirement of one well up gradient and three down gradient for a permitted hazardous waste site. Therefore, the water supply agency would be responsible for three wells up gradient and one down gradient.

Sampling

Multilevel sampling capability that ensures against the possibility of a monitoring well acting as a conduit for vertical migration of contaminants should be used. Several techniques are available, including locating several wells of differing depths near one another (cluster wells) and using multiple-completion monitoring wells, which consist of a nest of piezometers installed in a single borehole, as shown in Figure E-5.

Proper construction of multiple-completion wells is not easy, but it can offer cost savings. Materials must be properly placed into the well bore. For example, the perforated portion of each piezometer must be isolated from the others in the nest so that fluid pressures and water quality can be monitored correctly at that isolated level. Three-dimensional data collected using a network of multilevel monitoring wells should provide a useful definition of contaminant distribution, although the cost of such information may be high.

Analysis. Samples taken from monitoring wells should be analyzed for suspected contaminants, including severe mineralization. The mineral quality of water will limit the range of possible water uses. For example, hard water (high concentrations of calcium and magnesium) will be unsuitable for boiler feed water. Water containing high concentrations of sodium or boron will be unsuitable for irrigation. Although the biological quality of deeper groundwater is usually good, testing for fecal bacteria and other microbiological indicators should be performed periodically.

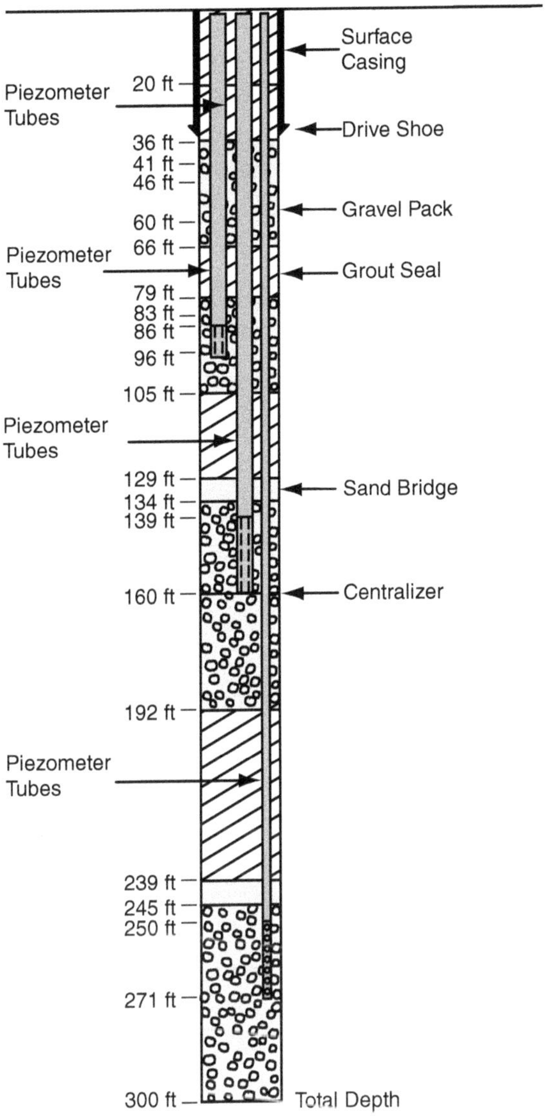

Source: M21 2003

Figure E-5 Schematic of a multiple-completion monitoring well

In recent years, a wide variety of constituents, which can be harmful even in extremely low concentrations, have become a concern. Unfortunately, analyses to detect all these chemicals can be expensive, so analytical methods used should be directed toward detection of suspected compounds. Knowledge of probable sources of contaminant chemicals used in the area and selection of any key indicator constituents should be used in the design of the sampling and analysis program to reduce cost without loss of study credibility. Guidance for selecting chemicals to be tested may be obtained from state and federal regulatory officials responsible for facility permits. Indicator parameters, referred to as "priority pollutants," often can be used to determine the likely presence or absence of chemicals that are a concern to groundwater development.

Fortunately, groundwater quality in many locations does not change because of shallow gradients (i.e., movement of groundwater is very slow, especially compared with surface water quality). Therefore, once water quality has been established, the frequency of groundwater sampling normally need not exceed quarterly or even semiannual checks, except for potable water sources or areas of suspected contamination.

FIELD LOGISTICS AND DOCUMENTATION

Land costs, permits, professional services, and documentation are very important to the evaluation of potential water supply sites.

Land Costs

Legal access to property for preliminary groundwater investigation will commonly be granted by the landowners on request. However, as soon as a property has been identified as attractive for detailed exploration, purchase or lease options should be obtained. The cost of land may be more reasonable before good water supplies are confirmed. Drilling-site preparation and restoration can be added costs for the groundwater development that should not be overlooked.

Permits

Exploratory drilling permits must be secured and fees paid, often at the state or county level. Obtaining permits is usually the responsibility of the well driller or the engineer in charge. In addition to the driller, a qualified hydrogeologist often supervises the drilling and well-construction activities. The hydrogeologist's responsibilities will usually include procuring well-construction materials, well logging, conducting or overseeing geophysical logging, interpreting logs, well designing, and certifying as-built drawings.

Part of the drilling permit will include proper abandonment once the well has served its purpose. Public agencies are concerned about well construction and post use of a well because of possibility of cross-contamination between shallow zones and deeper high-quality aquifers. The documentation of all field work can be very valuable in later phases of groundwater development or protection. Proper land survey location, global positioning system (GPS) locations, description of the wells, and complete as-built drawings of construction are desirable.

Professional Services

Firms specializing in hydrogeology are valuable to the groundwater developer for their knowledge and experience in carrying out field work and preparing the necessary reports. Some firms offer total services in groundwater development, as well as highly specialized equipment or services. Budget constraints, the complexity of the project, and the adequacy of the groundwater developer's staff can determine the most appropriate team. Some states require registered engineers, certified geologists, and other professionals to verify the accuracy and completeness of field work.

Documentation

Documentation of initial investigations and water supply development must be detailed and complete. Because of the complex nature of the information, the use of graphics is helpful. Cross sections (Figure E-6) showing the hydrogeologic interpretation, with lithologic logs, geophysical logs, and as-builts for wells, present a good summary of information. Maps showing pre-development groundwater contours versus the contours as affected by the new pumping, including surrounding land uses and any potential sources of groundwater contamination, should be available. Figures E-7 and E-8 illustrate such maps.

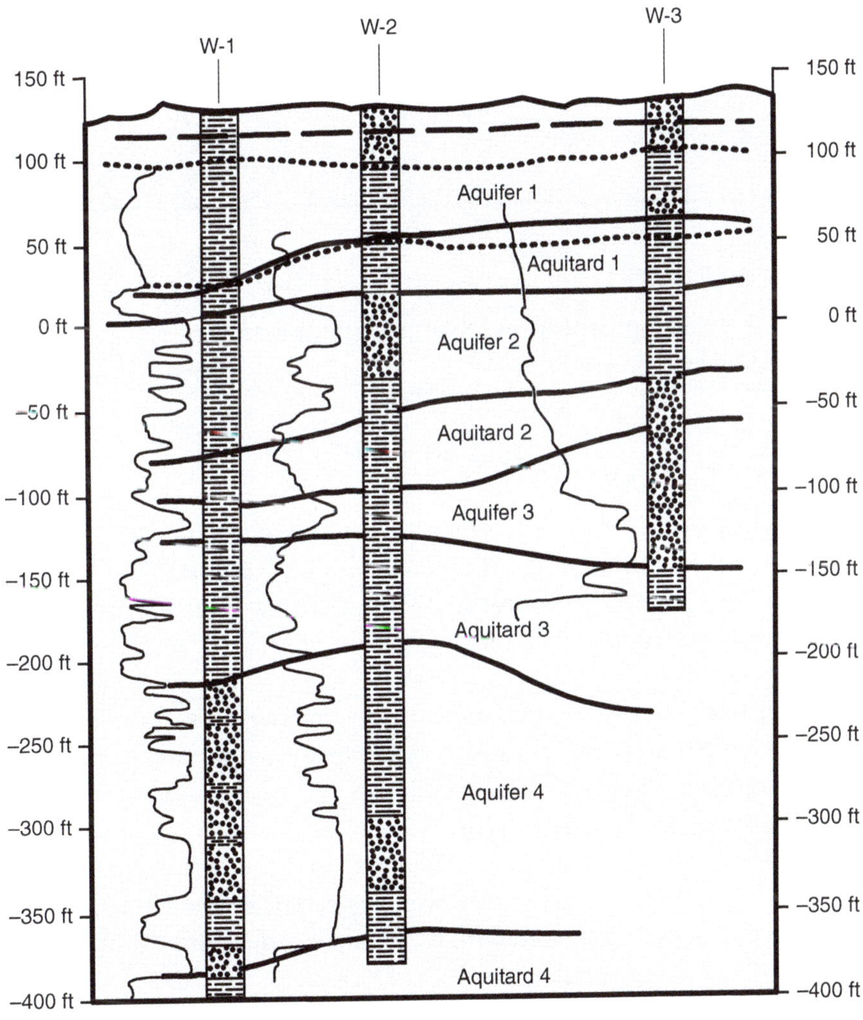

Source: AWWA 2003

Figure E-6 Graphic detail of a hydrogeologic cross section

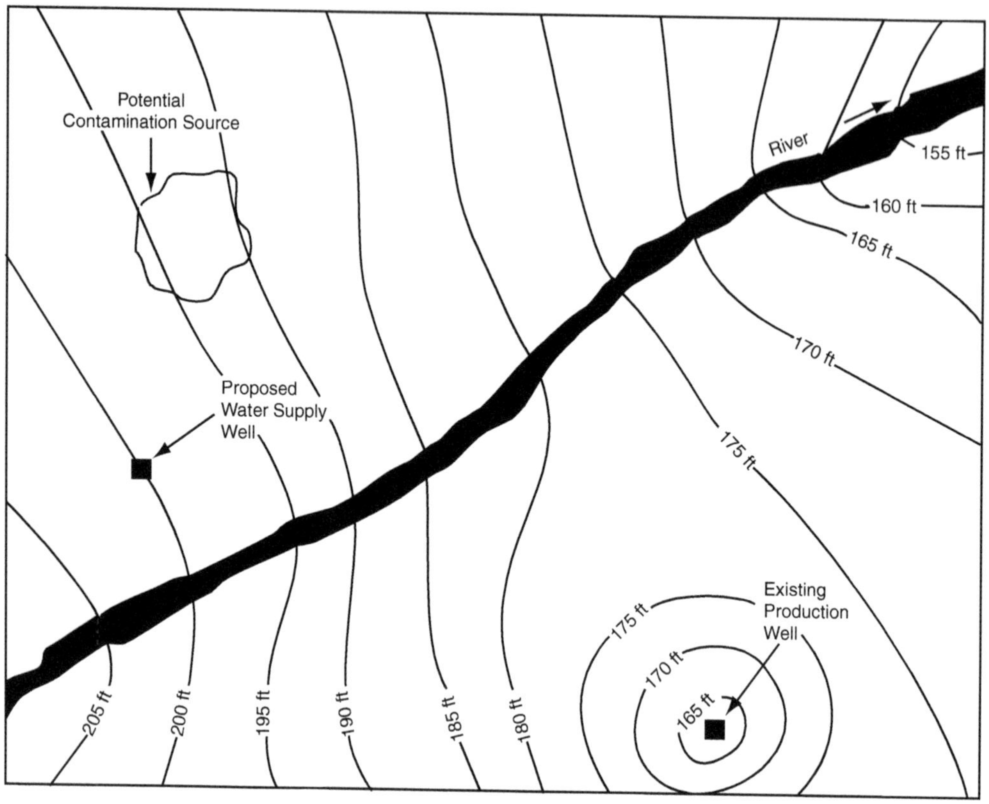

Source: AWWA 2003

Figure E-7 Predevelopment groundwater contours showing potential contamination source downgradient

Legal documents. Multiple copies of reports pertaining to groundwater development may be required by different levels of government for differing purposes. If the required reports are not to be submitted on issued forms, a reporting system that meets the needs of all federal, state, and local organizations should be designed.

Federal, state, and local laws increasingly require more information to be filed on a periodic basis. Permits must be filed, fees must be paid, and monitoring must be conducted. These activities should be part of the planned groundwater development program. It may be desirable to consult an attorney knowledgeable in all aspects of the law, including groundwater, land use, and permit procedures, before finalizing reporting procedures. Reports can be of great value in litigation and, therefore, should be prepared with care and be subjected to appropriate legal, technical, and managerial review.

Application. In addition to the traditional groundwater quantity reporting (along with the few inorganic analytical tests), more extensive testing for organic contaminants is being required. Water conservation interest groups are using the

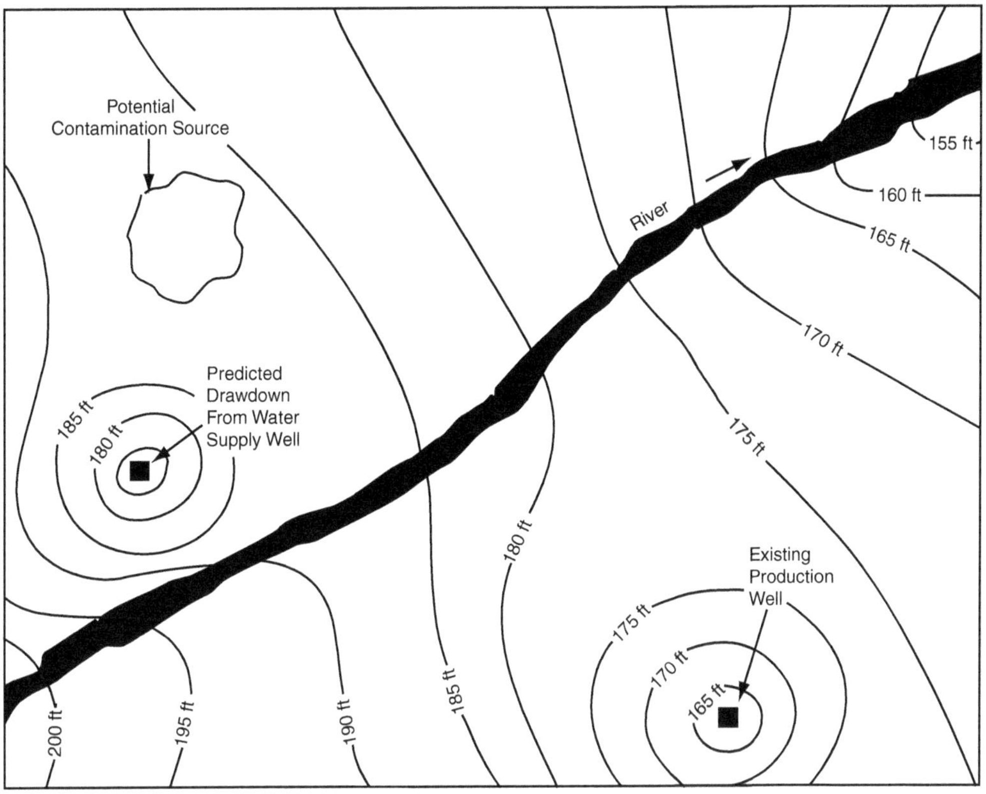

Source: AWWA 2003

Figure E-8 Predevelopment groundwater contours showing predicted effects of ill-advised development

pumping data to predict available groundwater supply and look for indications of groundwater mining. Information is also being filed with various government agencies involved in groundwater monitoring of facilities that produce, handle, store, treat, or dispose of chemicals determined to be hazardous to health or the environment. This information, when combined with information provided by the groundwater developer, can increase understanding of the regional groundwater system under study and its potential and reliability as a water supply.

REFERENCES

American Water Works Association (AWWA). 2003 (3rd ed.). M21, *Groundwater*. Denver: AWWA.

Keys, W.S. 1989. *Borehole Geophysics Applied to Groundwater Investigations*. Westerville, Ohio: National Water Well Association.

Keys, W.S. and L.M. MacCary. 1971. "Application of Borehole Geophysics to Water-Resources Investigations." In *Techniques of Water-Resources Investigations*, Book 2, Chap. E1. Arlington, Va.: US Geological Survey.

Zohdy, A.R., G.P. Eaton, and D.R. Mabey. 1974. "Application of Surface Geophysics to Ground-Water Investigations." In *Techniques of Water-Resources Investigations*, Book 2, Chap. D1.

ABOUT THE AUTHORS

Dr. Frederick Bloetscher is a professor of civil, environmental, and geomatics engineering and associate dean for undergraduate studies and community outreach at Florida Atlantic University (FAU) in Boca Raton, Fla. He came to academia after 13 years as a city manager for two communities in North Carolina that were undergoing significant infrastructure improvements and after utility management positions in Florida. In Florida, he was the deputy director for the Collier County Water–Sewer District and the City of Hollywood, and the director of engineering, operation, and planning for the Florida Governmental Utility Authority—systems that included groundwater, aquifer storage, reclaimed water systems, wastewater plants, and many miles of piping that needed to be maintained. He continues to consult with local governments.

His research focus has been urban infrastructure systems—particularly public water, stormwater, and sewer systems and their sustainability—in part because of his prior experience with city and county infrastructure organizations. Dr. Bloetscher teaches the capstone senior design sequence at FAU, plus classes in water/wastewater, construction, environmental engineering and modeling, hydraulics, and infrastructure management.

Dr. Bloetscher received his bachelor's degree in civil engineering from the University of Cincinnati and earned his master of public administration from the University of North Carolina at Chapel Hill. His Ph.D. is in civil, environmental, and architectural engineering from the University of Miami in Coral Gables, Fla.

He is a longtime member of the American Water Works Association (AWWA) and has served as a chair for the Water Resource Division Trustees, a member and three-time chair of the Groundwater Resource Committee, a member of the Aquifer Storage and Recovery Sub-Committee, and a member of AWWA's Technical and Education Council. He was the 2021 chair for the Florida Section of AWWA and has been the Technical Program chair since 2004. He was AWWA's inaugural Volunteer of the Year awardee among numerous other professional recognitions. Dr. Bloetscher can be contacted at P.O. Box 221890, Hollywood, FL 33022-1890 or h2o_man@bellsouth.net. His website is www.h2o-pe.com, and his blog is at publicutilitymanagement.com.

Mr. Albert Muniz is a vice president with Hazen and Sawyer, P.C., managing the firm's Boca Raton office. Hazen and Sawyer is one of the top environmental engineering firms in the Western Hemisphere, involved in water resources, water treatment, wastewater treatment, and injection programs. Mr. Muniz has decades of experience with injection well programs, including water supply and disposal injection programs. Mr. Muniz is a former chair of AWWA's Groundwater Committee and serves on its Aquifer Storage and Recovery subcommittee. He has also served on the accreditation board for both the University of Florida and Florida Atlantic University schools of engineering. Mr. Muniz can be reached at 2101 Corporate Blvd., Suite 301, Boca Raton, FL 33431-7343.

Mr. John Largey is the owner/operator of Drilling Geo Water Well Technical Services. He is a hydrogeologist with extensive field experience related to water resources, water supply, and injection programs. Mr. Largey has more than 40 years of experience involving all aspects of drilling wells for a variety of applications. He specializes in large-diameter deep water wells including water supply; Class I, Class III, and Class V injection wells; and aquifer storage and recovery wells. Previous drilling experience includes oil and gas; geotechnical investigation; and environmental investigation and remediation for Superfund sites (CERCLA), the Department of Energy, and the United States Agency for International Development.

INDEX

Note: *f* indicates figure; *n* indicates (foot)note; *t* indicates table.